Change Point Analysis

Change point analysis is a crucial statistical technique for detecting structural breaks within datasets, applicable in diverse fields such as finance and weather forecasting. The authors of this book aim to consolidate recent advancements and broaden the scope beyond traditional time series applications to include biostatistics, longitudinal data analysis, high-dimensional data, and network analysis.

The book introduces foundational concepts with practical data examples from literature, alongside discussions of related machine learning topics. Subsequent chapters focus on mathematical tools for single- and multiple-change point detection along with statistical inference issues, which provide rigorous proofs to enhance understanding but assume readers have foundational knowledge in graduate-level probability and statistics. The book also expands the discussion into threshold regression frameworks linked to subgroup identification in modern statistical learning and apply change point analysis to functional data and dynamic networks—areas not comprehensively covered elsewhere.

Key Features:

- Comprehensive Coverage of Diverse Applications: This book expands the scope of change point analysis to include biostatistics, longitudinal data, high-dimensional data, and network analysis. This broad applicability makes it a valuable resource for researchers and students across various disciplines.
- Integration of Theory and Practice: The book balances rigorous mathematical theory with practical applications by providing extensive computational examples using R. Each chapter features real-world data illustrations and discussions of relevant machine learning topics, ensuring that readers can see the relevance of theoretical concepts in applied settings.
- Accessibility for Students: The content is designed with graduate-level students in mind, providing clear explanations and structured guidance through complex mathematical tools. Rigorous proofs are included to facilitate understanding without overwhelming readers with overly advanced theories early on.

The book incorporates computational results using R, showcasing various packages tailored for specific methods or problem domains while providing references for further exploration. By offering a selection of widely adopted methodologies relevant in scientific research as well as business contexts, this text aims to equip junior researchers with essential tools needed for their work in change point analysis.

Change Point Analysis

Theory and Application

Baisuo Jin and Jialiang Li

CRC Press
Taylor & Francis Group
Boca Raton London New York

CRC Press is an imprint of the
Taylor & Francis Group, an **informa** business
A CHAPMAN & HALL BOOK

First edition published 2026
by CRC Press
2385 NW Executive Center Drive, Suite 320, Boca Raton FL 33431

and by CRC Press
4 Park Square, Milton Park, Abingdon, Oxon, OX14 4RN

CRC Press is an imprint of Taylor & Francis Group, LLC

ISBN: 978-1-032-64904-7 (hbk)
ISBN: 978-1-032-65704-2 (pbk)
ISBN: 978-1-032-65707-3 (ebk)

DOI: 10.1201/9781032657073

Typeset in Nimbus Roman font
by KnowledgeWorks Global Ltd.

Publisher's note: This book has been prepared from camera-ready copy provided by the authors.

Contents

Preface

Change point analysis is a vital statistical method for detecting structural breaks or shifts in datasets, with applications spanning diverse fields such as finance, where it is used to identify market regime changes, and weather forecasting, where it helps detect shifts in climatic patterns or extreme events. There is a rich history for statistical research on change point analysis, with its roots tracing back to early discussions in the 1950s. Over the decades, numerous theories and methods have been developed to address the problem of detecting change points, and our understanding of the topic has deepened significantly in the 21st century. As the field continues to expand, it has increasingly intersected with other disciplines, particularly personalized medicine, where identifying structural changes plays a critical role in tailoring treatments to individual patients.

The two authors of this book initiated this project during Li's visit to the University of Science and Technology of China in 2023. Recognizing the lack of a comprehensive textbook on change point detection suitable for undergraduate and graduate students, we set out to create a monograph to consolidate the latest advancements in this field and their application to subgroup identification. Unlike traditional treatment of this subject mainly in the time series domain, this book extends the scope to encompass broader areas in modern statistical research, including biostatistics, longitudinal and functional data analysis, high-dimensional data, and network analysis. The selection of topics also reflect our research interests in the recent decade.

While we have endeavored to cover a wide range of methods, it is possible that our coverage may not be exhaustive, owing in part to our own limited knowledge and research experiences. Furthermore, there is no universal consensus on the optimal method for change point analysis, as different approaches may prove useful in different applications and under different conditions. Our intention is to introduce a selection of methodologies that are widely adopted in both scientific and business contexts. We believe that these materials will provide junior researchers with a solid foundation to begin their work in this field.

The concept of subgroup analysis has emerged as a new and dynamic research area in recent decades. In many clinical settings, the assumption of sampling from a homogeneous population is often unrealistic. Identifying subpopulations or dividing a heterogeneous population into relatively more homogeneous subgroups can be highly beneficial, enabling focused studies within each subgroup and potentially yielding more accurate and targeted insights. A key methodology for identifying subgroups within a population is through the identification of change points or change planes across the covariate space. The authors of this book are actively engaged in research projects in this field and aim to share their findings and insights within its pages.

The first chapter introduces some basic notions and concepts, with a wide selection of data examples from the literature as illustrations. Some related research topics on machine learning are also discussed. The following chapters may be divided into several groups. Chapters 2 through 4 delve into the technical construction of various mathematical tools for addressing single- and multiple-change point detection problems, along with statistical inference issues. The objective is to provide readers, especially students, with a sufficient background to understand the mathematical concepts and formulations used in this field. Rigorous proofs of all the theoretical results are provided immediately to facilitate understanding. Probability theories are extensively used in the technical development of these chapters. Students should generally have had at least half a year of graduate-level probability as well as a year of graduate level mathematical statistics before working through these materials. Chapter 5 generalizes the earlier results to the threshold regression framework, highlighting its close connection to the subgroup identification issues in modern statistical learning. Finally, Chapter 6 extends the applications of change point analysis to functional data and dynamic network data, demonstrating its broader applicability. These two chapters introduce relatively recent research that has not yet been presented elsewhere in a systematic textbook exposition. We focus more on methodology and applications and leave out advanced mathematical development. All the proofs are thus omitted and referred to in the references.

Some computational results from `R` are included in this book and we make the assumption that readers all have a basic working knowledge of `R`. Currently, there exists a wide array of packages for change point analysis, each tailored to specific computational methods or problem domains. In this book, we will highlight quite a few representative packages, providing necessary references for all the tools used in our illustrations. This approach ensures that readers can identify the appropriate resources for their applications and explore the methodological nuances of each package. Please feel free to inform us if you see any errors or inconsistencies.

We wish to extend our sincere thanks to our collaborators, students, colleagues, family and friends who supported our research work. It would be impossible to complete this book without their help. Both of us have greatly benefited from the many teachers, advisors, and mentors throughout our lives and professional careers. Their wisdom and guidance have played an enormously important role in shaping our statistical research careers. We particularly thank Zhidong Bai, Hock Peng Chan, Louis Chen, Xiru Chen, Ming-Yen Cheng, Kjell Doksum, Jason Fine, Peter Hall, Tailen Hsing, Michael Kosorok, Anthony Kuk, Tze Leung Lai, Mei-Ling Ting Lee, Wei Liem Loh, Wei-Yin Loh, Baiqi Miao, David Nott, Rick Nordheim, Mari Palta, C.R. Rao, John Rush, Jun Shao, Jon Wellner, Weng Kee Wong, Yuehua Wu, Yingcun Xia, Qiwei Yao, Chunming Zhang, Jin-Ting Zhang, Wenyang Zhang, Xiao-Hua Zhou, Hongtu Zhu and Ji Zhu. We apologize for not being able to list all our great teachers, advisors and mentors, and we take the opportunity to thank them all. We thank our parents for their love and sacrifices in our upbringings and educational opportunities.

Authors

Baisuo Jin is a professor at University of Science and Technology of China. His research fields include spatial statistics, random matrix and change point. His research works have been accepted for publication in premium journals including *The Proceedings of the National Academy of Sciences* (PNAS), *The Annals of Statistics*, and *Biometrika.*

Jialiang Li is a professor in the Department of Statistics and Data Science, National University of Singapore. He was an Elected Member of the International Statistical Institute (ISI) in 2019, Fellow of American Statistical Association (ASA) in 2020 and Fellow of the Institute of Mathematical Statistics (IMS) in 2022. He has served on the editorial board for *Annals of Applied Statistics, Annual Review of Statistics and Its Application, Biometrics, Biostatistics & Epidemiology, Lifetime Data Analysis* and *Statistical Methods in Medical Research.*

1

Overview

This book is for biostatisticians, econometricians, consulting statisticians, students taking a first course on change point, and anyone with an interest in the field of change point research. It deals with statistical methods for analyzing data with discontinuous change in distribution parameters such as mean, regression function or the distribution itself. There are plenty of examples and applications where we would question the constancy of the parameters and distributions. The methods introduced in this book can provide a unified solution to these problems.

Change points may also be referred to as thresholds, splitting points, or breakpoints, depending on the context and application. These terms are considered interchangeable within this book. Common examples of change point include the moment in time when a financial crisis occurs, the demarcation lines between different regions or groups, and other noteworthy breakpoints associated with special events.

We offer an introduction in this chapter by describing the basic change point problem. While we refrain from providing detailed mathematical derivations in this chapter, our primary focus remains on frequentist change point detection methodologies throughout this book. Hence, change points are regarded as unknown parameters within the distribution of a random variable. It is worth noting that Bayesian statisticians approach change points differently, treating them as random variables. Although Bayesian statisticians have contributed numerous optimal change point detection methods, they fall outside the scope of this book. Interested readers can refer to the excellent review in [243] for more information on Bayesian change point detection methodologies.

1.1 Earlier developments

The scientific study of change point has a long history by now. Most authors agree that the subject first began to emerge in the 1920–1930s motivated by considerations of quality control. The early development of change point methodology is related to the monitoring and evaluation of dynamical systems which give rise to sequences of correlated observations. The behavior of some suitably chosen quality characteristic of the subject, eg. a product or a human patient, at time t follows a fixed probability distribution that depends on some parameter θ_t, whose value remains at a specified level or within certain specified limits. As an illustration, in Figure 1.1, we have serial observations of a random variable Y collected at different time points $t \in [0, 1]$, depicted by the hollow circles in the plot. In this simple example, this random process shifts its mean parameter θ from 1 to 2 at a change point $t = 0.5$. As in all statistical problems, hidden parameter and its change are not directly observable, and we have to employ the sample data to recover their values. To this end we may adopt the so-called control chart method. Such a process inspection scheme takes samples of fixed size at regular intervals of time and computes from the sample available up to time t a suitably chosen statistic X_t. These sequential statistics X_t are presented graphically in the form of a control chart. As soon as a sample point falls outside the control limits drawn on the chart, the monitoring scheme signals that the process has departed from the normal state and a change point is declared. Choosing

DOI: 10.1201/9781032657073-1

X_t to be the sample mean gives Shewhart's $\bar{X}$-chart for monitoring changes in the process mean. Many improved versions of control charts have been developed since the original Shewhart charts. Using `R` package `qcc`, we can obtain Figure 1.2 for the data shown in Figure 1.1. Now each point in the graph is a cumulative sum (CUSUM) statistic calculated from the past data. The so-called CUSUM method easily detects the correct change point position in this example where the chart exceeds the control limit (referenced by two dashed lines as upper and lower limits) exactly at the underlying change point.

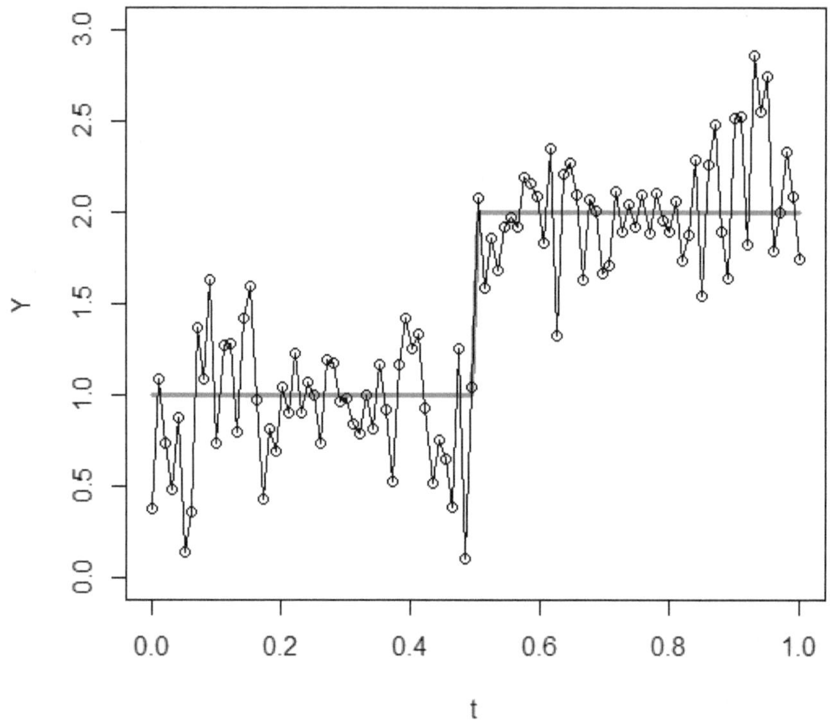

Figure 1.1: Change in mean of a process. Hollow circles are the observed data, while the red solid line is the underlying mean process.

Let us slightly diverge on the topic of control chart, which is not the primary focus of this book. More generally, a measurement control chart is a chart for which a measured process characteristic is being monitored, such as the blood pressure of a patient in a clinical study. The basic idea of the standard Shewhart charts was first sketched out by Dr Walter A. Shewhart in [228]. In addition to mean chart, Shewhart also proposed to choose X_t to be the sample range to produce a range chart for monitoring within-sample variations. Furthermore, using weighted mean from a moving window may be more desirable than using the equal weight average since it may be more informative to put more weight on the current and immediate past observations than on those remote data. See [153, 154].

We are also not going to examine the related testing problems with more details in this book. The decision of detection of a change point is similar to the test comparing two parameters and

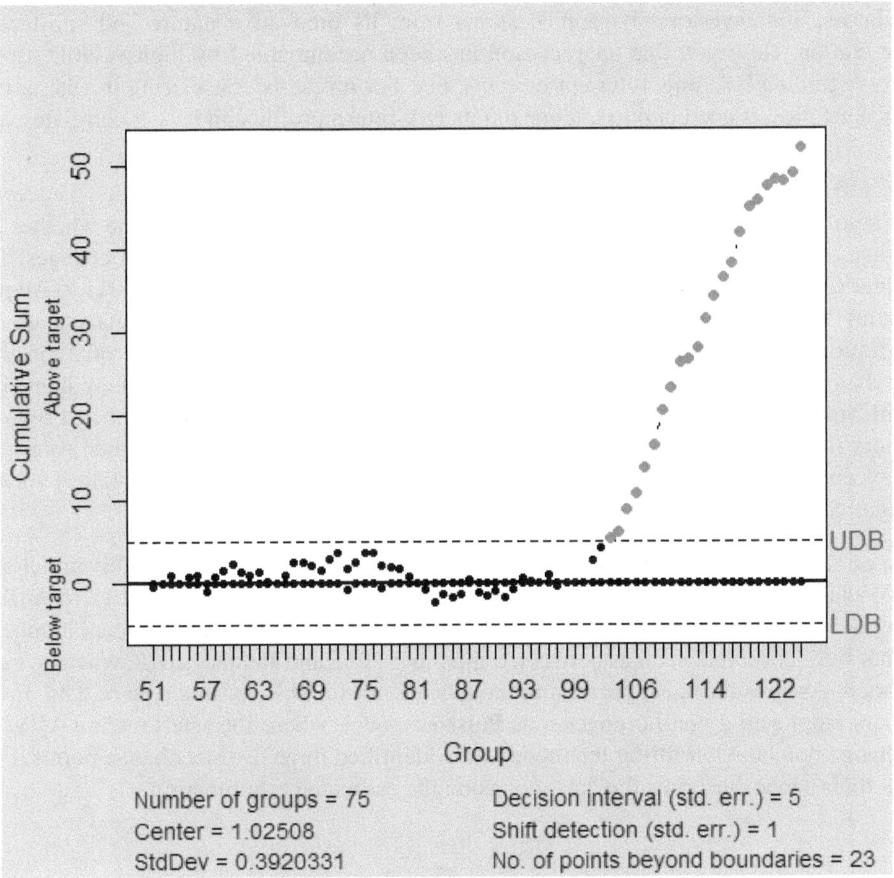

Figure 1.2: CUSUM control chart.

we do such a test repeatedly over the sequence of data. Sequential hypothesis tests appear in the construction of some change point detection plans and in determining properties of others. The question of efficiency of sequential and group sequential hypothesis tests is now largely resolved, and these results can shed light on change point detection procedures. Although Wald's sequential probability ratio test ([254, 255]) provides the most efficient sequential test between two simple hypotheses $\theta = \theta_0$ versus $\theta = \theta_1$, it does not have particularly low power at parameter values other than the two points θ_0 and θ_1. Tests with high efficiency over a range of θ values have been, sought and Schwartz (1962) later derived the shapes of sequential testing boundaries, which minimize, asymptotically, the expected sample size at all values of θ as error rates tend to zero. We refer to [243] for more comprehensive review on this topic.

Traditionally change point detection is closely associated with time series analysis. Time series usually refers to a set of observation collected sequentially in time, often at equally spaced time points. These data are naturally indexed by time and are present in a variety of areas of application. For example, they are used in epidemiology, economy, environment, finance, among others. When studying time series we may consider the readings as results of some random experiment recorded according to some time scale (e.g., hours, days, weeks, years). We now provide the following examples to illustrate various change point applications:

- Bitcoin, the first digital currency to operate via a peer-to-peer network without a central authority, has garnered significant attention from various sectors, including the media, governments, regulators, and investors. Its appeal stems from its innovative nature and substantial price appreciation. However, this appreciation has been accompanied by high volatility and uncertainty regarding Bitcoin's future price trajectory. For investors considering including Bitcoin in their portfolios, it is crucial to understand its risk-return profile and how it correlates with other assets.

Typically, average returns and volatility are estimated using historical data. However, given the rapidly changing nature of the Bitcoin market, it is essential to examine whether Bitcoin's average return and volatility remain stable over time or experience frequent changes. Thies and Molnar (2018) [244] analyzed Bitcoin's price time series from September 2011 to August 2017, utilizing 2170 daily observations and applying a Bayesian change point analysis on Bitcoin's log returns. They identified 48 structural breaks over the entire period, equating to roughly one change point every 1.5 months on average. These segments varied in length and exhibited significant differences in both mean and volatility. After clustering the identified segments, the authors found that, in most cases, higher volatility was associated with higher average returns. An exception occurred in 2013, where a brief 10-day regime exhibited the highest volatility and a negative return.

- Disease monitoring often results in discrete count data. For example, de Oliveira et al. (2022) [192] analyzed the monthly incidence of measles in Brazil from January 2012 to August 2018, using data sourced from the World Health Organization (WHO). A significant number of zero counts were observed, such as between September 2015 and January 2018, when no cases were reported. As a result, linear regression models are not suitable for this type of data. Instead, the authors employed a non-homogeneous Poisson model, where the rate function $\lambda(t)$ is subject to change points. After fitting the model, they identified three distinct change points. Figure 1.3 from their paper illustrates the data alongside the estimated rate function.

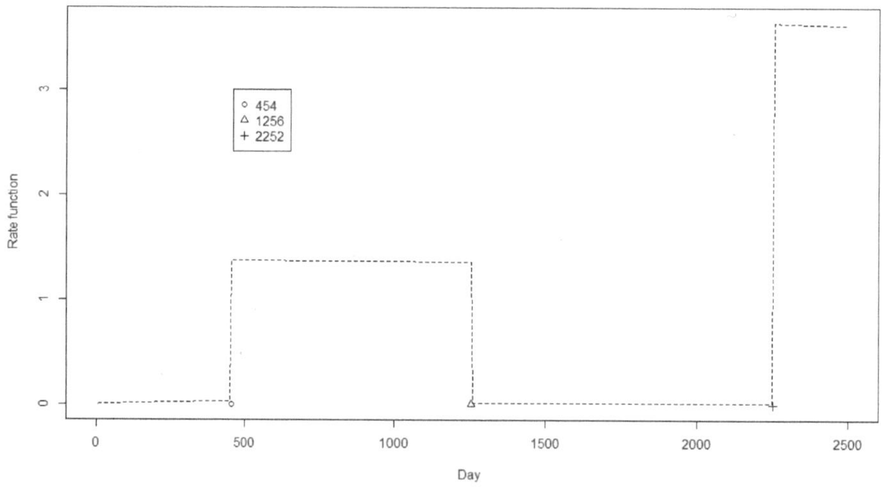

Figure 1.3: Estimated rate function of the non-homogeneous Poisson model applied to Brazil measles data in [192]. Different symbols depict three change points.

- During the outbreak of the novel coronavirus disease 2019 (COVID-19), swift and decisive actions were essential from both individuals and society to curb the virus's spread. Given the high transmissibility of the virus and the lack of immunity in the population, the epidemic would grow exponentially without intervention, overwhelming public health systems and causing significant disruption to the economy and daily life. In response, many countries implemented emergency measures, such as regional lockdowns and widespread testing. Dehning et al. (2020) [78] analyzed data from Germany using the susceptible-infected-recovered (SIR) model to study the rate of spread, $\lambda(t)$. Their analysis identified three key change points. The first occurred on March 7, 2020, when $\lambda(t)$ dropped from an initial value of $\lambda_0 = 0.43$ to $\lambda_1 = 0.25$, following the first government intervention that included the cancellation of large events and heightened public awareness. A second change point was observed around March 16, 2020, where the rate further decreased to $\lambda_2 = 0.15$, coinciding with measures such as school closures and the shutdown of some businesses. Finally, on March 24, 2020, a third change point brought the rate down to $\lambda_3 = 0.09$, following additional government actions, leading to a sustained decline in new infections.

- Reliable estimates of volatility and correlation are crucial in economics and finance for understanding the effects of macroeconomic events on markets and for guiding future investment and policy decisions. The dependence across financial returns is often subject to sudden structural breaks, particularly during major global events like the COVID-19 pandemic. Franzolini et al. (2024) [99] developed a graphical model change point detection method and applied it to the weekly returns of nine industry portfolios from January 2019 to December 2021. The data, sourced from Kenneth R. French's Data Library at $https://mba.tuck.dartmouth.edu/pages/faculty/ken.french/data_library.html$, reveal key changes in the conditional dependence structure over time, with two prominent change points.

  The first change point, detected on February 24, 2020, marked the market's reaction to the initial significant rise in global COVID-19 cases and deaths, particularly outside of China. During the weekend of February 21–23, 2020, Italy, the first European country severely impacted by the virus, reported its initial local cases. Following this change point, the Dow Jones and S&P 500 plummeted by 11% and 12%, respectively, representing their steepest weekly declines since the 2008 financial crisis. Notably, a strong correlation emerged between the energy and utilities sectors after this change point (as seen in the middle panel of Figure 1.4), a relationship that was absent prior to the event (top panel of Figure 1.4).

  The second change point occurred in the final week of June 2020, signaling a partial stabilization of financial markets following the initial, most volatile period of the pandemic. After this point, new dependencies appeared, such as the connection between the energy sector and retail (shops) (bottom panel of Figure 1.4), indicating evolving market dynamics as the pandemic continued to unfold.

- DNA copy number refers to the quantity of DNA in a specific genomic region, measured relative to a control sample or a pool of population reference samples. Copy Number Variants (CNVs) are regions of the genome where the copy number differs among individuals, representing a common form of genetic polymorphism. In contrast, tumor cells often exhibit somatic structural mutations, such as deletions or duplications, that lead to copy number differences between tumor cells and normal cells from the same individual. These changes are referred to as Copy Number Aberrations or Copy Number Alterations (CNA). Both CNVs in healthy individuals and CNAs in tumors are of significant scientific interest. The goal in both cases is to identify the boundaries of the genomic regions that have experienced a copy number change (i.e., breakpoints) and estimate the copy numbers within these regions. Shen and Zhang (2012) [227] addressed this problem using change point modeling for DNA copy numbers from next-generation sequencing data. In their approach, the sequenced reads for tumor cases and healthy controls are modeled as two non-homogeneous Poisson processes with distinct mean functions, μ_t and λ_t. The event

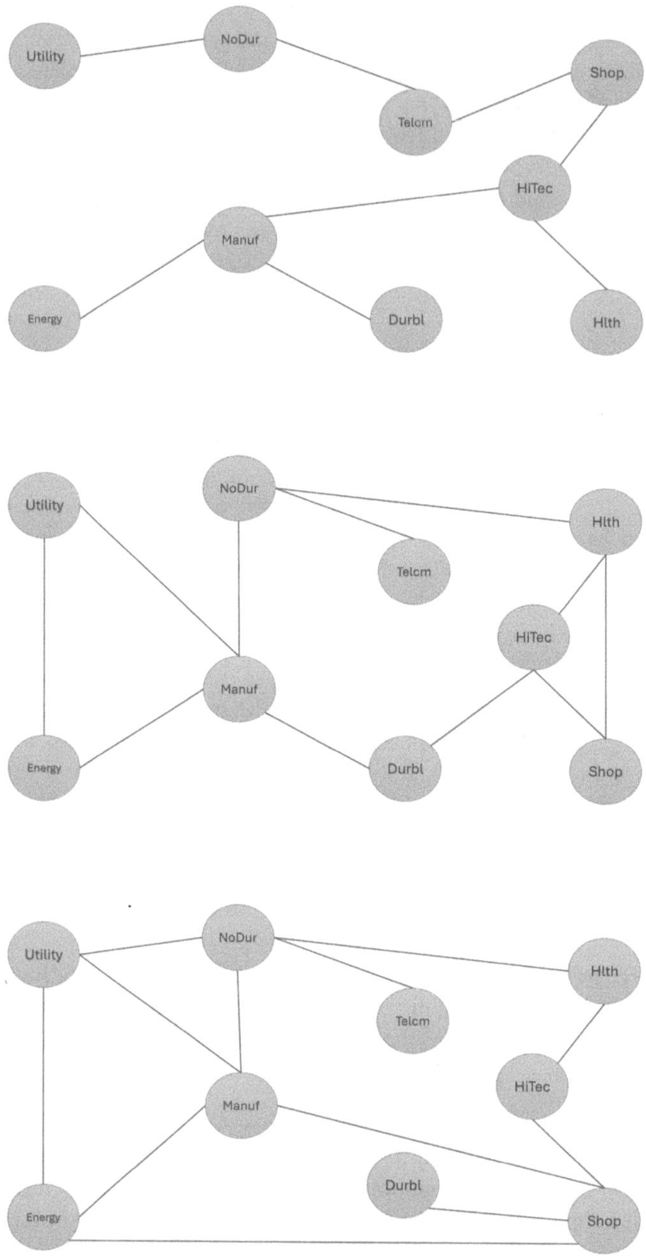

Figure 1.4: Estimated graphs during the three periods in [99]: December 31, 2019 to February 21, 2020 (top); February 24, 2020 to June 26, 2020 (middle); June 29, 2020 to December 31, 2021 (bottom). The nine industry portfolios are consumer nondurables (NoDur), consumer durables (Durbl), manufacturing (Manuf), energy, business equipment (HiTec), telecommunications (Telcm), shop, health (Hlth), and utilities (Utils). Each edge indicates a conditional dependence.

probability at position t for being a case is given by $p(t) = \mu_t/(\mu_t + \lambda_t)$. Figure 1.5 illustrates the application of this change point model to genomic data, where the probability curve remains constant between change points but jumps at each detected change point. In this context, the index t represents spatial locations on the genome rather than time, as in previous examples.

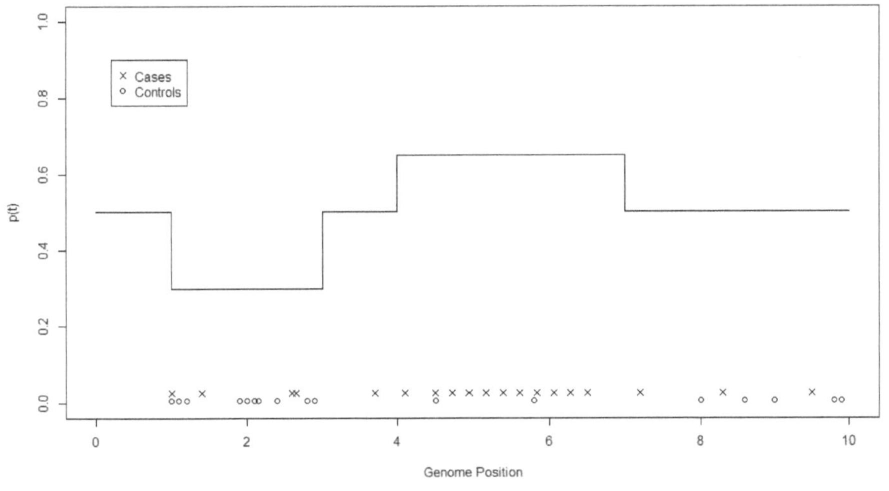

Figure 1.5: Illustration of paired Poisson processes for DNA copy numbers in [227]. The piecewise function $p(t)$ is the probability that a read at genome position t is from the tumor case sample, conditional on the existence of a read at position t.

1.2 More up-to-date problems

Many classical works concentrate on changes in mean or variance parameter over a discrete index set as in time series studies, as in Figure 1.1. In this book we will adopt a more general setting with change over a continuum. This allows a more useful framework for a large number of current scientific inquiries in fields such as personalized medicine, online learning, network data analysis and causal inference. Consider a regression setting with data $\{X_i, Y_i, i = 1, \cdots, n\}$ where we want to examine the relationship between the response variable Y_i and the predictor X_i for a sample of n points. These data are not necessarily independent and identically distributed (i.i.d.). For example, in Figure 1.6, the dependence is clearly different in the three segments over the range of the predictor variable. In particular, when the predictor value x lies in the interval $[0, 0.2]$, the regression slope is $\beta = 1$; when x moves into the interval $[0.2, 0.8]$, the slope jumps to be $\beta = 4$; after x further increases above 0.8, the slope becomes negative with $\beta = -1$. In practice, we do not observe the true regression curve and only have the empirical data. When the change points are known to us, we can easily fit a piecewise linear regression model to consistently estimate the three different regression slopes. However, in most scientific studies, the number of change points is unspecified, and their locations are unknown. Therefore, it is imperative to estimate these quantities with a statistical approach. The control chart method introduced to detect the change in the mean is no longer appropriate for such a complicated change problem. We have to choose methods covered in this book to carefully address this question. Using R package cpop, we can produce the detected results in Figure 1.7. We will provide more details in Chapter 5.

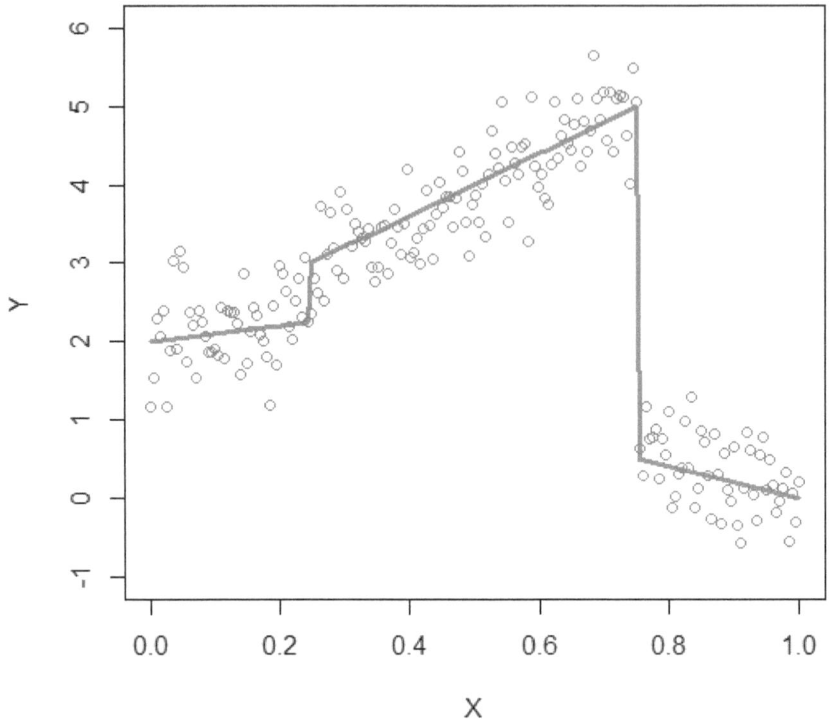

Figure 1.6: Change in slope of regression lines. Hollow circles are the observed data, while the red solid curve is the underlying regression function.

The results from change point detection lead to segmentation or separation of the data, where subjects in the same segment share similar mean or regression coefficients. Finding such a local homogeneity in a general heterogeneous population is very important. In fact, many modern clinical trials focus on identifying the subgroup of patient population which may benefit from a particular drug with enhanced treatment effects than average population. Such an identification process invokes the change point detection method since the resulting subgroups are usually constructed by a splitting of the covariate space at certain covariate values. Let us consider two examples from pharmaceutical studies.

- Phase I clinical trials for monotherapy using cytotoxic agents aim to determine the maximum tolerated dose (MTD), which is defined as the highest dose that can be administered with acceptable levels of toxicity. To find the MTD, model-based dose-finding methods such as the continual reassessment method (CRM) are often employed. These methods estimate the probability of toxicity based on accumulated toxicity data from trial participants and select the MTD from among the candidate dose levels. A key assumption behind these methods is that both the toxicity and efficacy of the drug increase monotonically with the dose escalation, which leads to dose adjustments based primarily on observed toxicity outcomes.

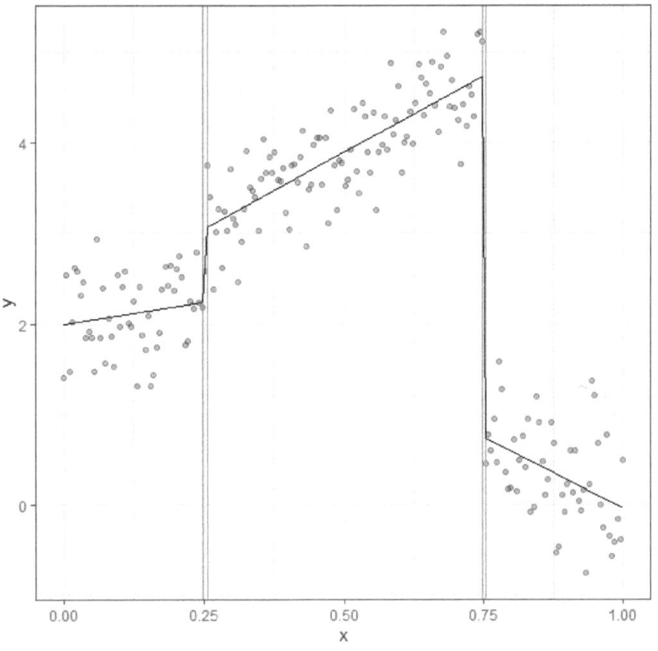

Figure 1.7: Detected changes in slope of regression lines.

In recent years, however, the clinical development of molecularly targeted agents (MTAs), which aim to inhibit specific molecular pathways, has emerged as part of personalized medicine efforts. Unlike traditional cytotoxic agents, MTAs do not necessarily exhibit a monotonic dose-response relationship. This non-monotonicity presents a challenge for conventional dose-finding methods, as the assumption that higher doses always result in greater efficacy and toxicity does not hold for MTAs. Sato et al. (2016) [216] considered a change point logistic regression model where two separate logistic models are constructed depending on whether the actual dose of the agent administered to the patient is greater than a threshold or not. In this example, Y_i is a binary outcome variable and the model may be written as

$$P(Y_i = 1) = 1/(1 + \exp(-\beta_0 - \beta_1 X_i - \beta_2 X_i I(X_i > x^*))), \quad (1.1)$$

where β_0, β_1, and β_2 are the regression coefficients of the logistic model and x^* is the change point. The authors estimated the model parameters, including the change point location, by using a Bayesian approach. The dose-finding algorithm was also designed accordingly. Their simulation results demonstrated the superior performance of this model, comparing to the naive model without a change point.

- Traditionally, the main focus was on determining the maximum tolerated dose (MTD). However, with the advent of these new therapies, the objectives have broadened to include the simultaneous evaluation of both toxicity and efficacy, often accompanied by expansion cohorts. This shift is driven by the development of novel therapies and the increasing pressure to expedite drug approval processes. Furthermore, the assumption that higher doses correlate with greater efficacy is no longer universally valid in the context of MTAs and immunotherapies. Evidence shows that the dose-activity relationship can reach a plateau after a certain dose level, as depicted in Figure 1.8. Altzerinakou and Paoletti (2021) [4] examined a French multicenter phase IB trial in patients with platinum-resistant epithelia ovarian carcinoma. A maximum of eight cycles

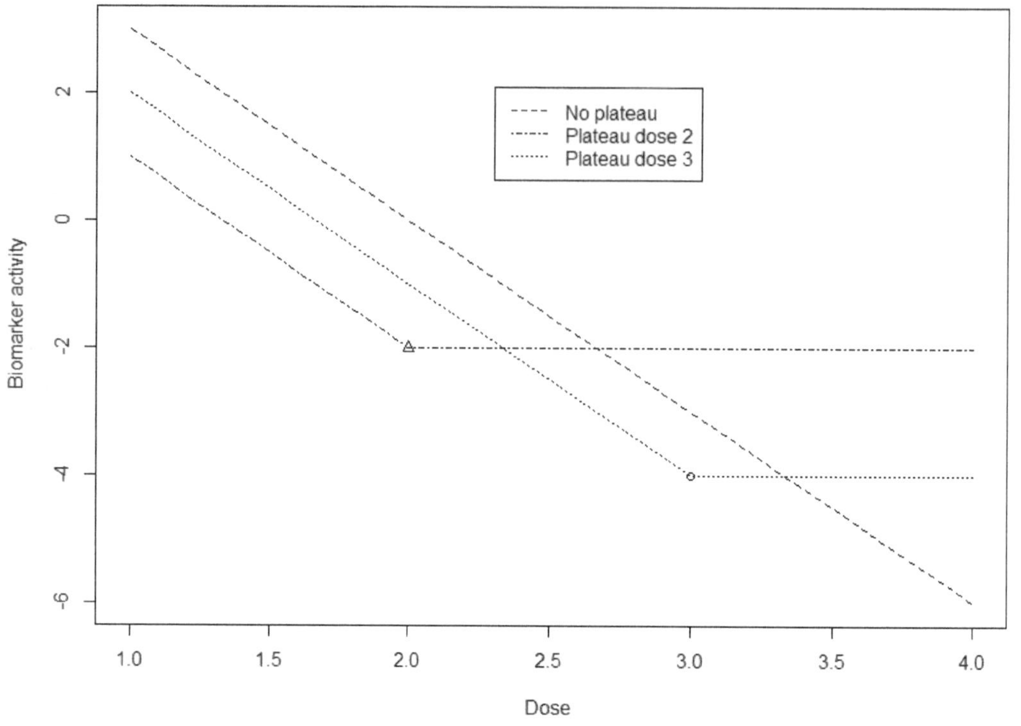

Figure 1.8: Dose-activity relationships for certain biomarkers. The effects of drug could be monotone decreasing or flatting out after a plateau value.

with a duration of three weeks each and a maximum of 42 patients were planned. Activity was evaluated radiologically and with CA 125 measurements. CA 125 is a common biomarker used in ovarian cancer to assess the activity of a treatment regimen. Authors thus considered a linear mixed effects model for the repeated measures of the biomarker, and the time effects were thresholded by a dose level.

Another modern application is to find change in a function. The function could be density function, distribution function, regression function, and covariance function, among others. The field of function estimation usually assumes that the unknown functions are smooth everywhere, mainly for the purpose of mathematical convenience. There are possible applications where the function may be better treated as piecewise smooth. Thus it is meaningful to conduct change point detection for such settings. In Chapter 6, we will explore the estimation of mean function for functional data where discontinuities exist.

We further point out the meaningful question of change in network evolving. Network data is becoming more and more common in statistical research and machine learning. A network is defined by edges and nodes. If we observe the network over time, we may obtain a dynamic network series. At each time point, the network may follow a community structure where groups of nodes share similar characteristics. However, community memberships may change over time, leading to structural shifts in the network. Detecting these change points is analogous to identifying changes in traditional time series data. This process helps us understand how network structures evolve over time. Further details on this topic can be found in Chapter 6.

Applications of change point detection span numerous disciplines. In addition to the aforementioned examples, we will provide other useful examples throughout this book.

1.3 Methodology and structure of the book

There are many useful and well developed methods available in the literature. We will introduce the standard approaches in Chapter 2, 3 and 4 of this book.

In chapter 2 we consider the detection of a single change point. In a simplest setting investigators assume there are at most one change point in the time interval considered, usually of a short length. The traditional CUSUM method will be introduced. The null distribution for the related test statistic is actually quite complicated when the change location is unknown. The correct asymptotic quantification of the max of CUSUM statistics is often unwieldy as the scenario is not readily scalable to an extreme value distribution, even though the statistic is a maximum. We will see that the limit distribution often converges to the supremum of some Gaussian process.

In chapter 3 we extend our discussion to the problem with multiple changes. It is more realistic to assume more than one change points, especially for a long follow up period. In fact, there are two main stream approaches to the multiple change point detection. The first class is based on the binary segmentation (BS) method while the second will be referred to as penalized (Pen) estimation method in this book. The BS method directly generalizes the CUSUM method for univariate change detection. Initially BS examines the entire series for a single change point. If a change point is identified, the series is then split into two subsegments about the identified change point and the two subsegments are further scrutinized for a single change point. The recursive partitioning procedure continues iteratively until all subsegments are declared change point free. At every splitting step, there are many possible criterion functions to consider. In addition to the popular CUSUM statistics, one can also consider likelihood-based statistic or other scan statistics. This method could be rather time-consuming due to its iterative feature. On the other hand, Pen method will be much faster. For Pen method the choice of penalty function is also quite abundant in the literature, including L1, L0, SCAD and other non-convex penalty functions. The literature of comparing these methods are abundant ([230]) but we do not attempt to give an answer on which method is superior over others.

In chapter 4 we consider the inference issues for change point detection. Dealing with uncertainty is very important for statistical practice. In addition to the bootstrap resampling approach, we offer some theoretical discussions to contribute analytic solutions.

In Chapter 5, we will discuss regression models involving change points. This essentially introduces the threshold regression model formally. Such models have been studied in econometrics for a long time and recently attracted attentions from biostatisticians as well. The model developed under this approach is also similar to tree-based method and thus opens a door to machine learning. We provide theoretical discussion and some empirical examples for this model. A further extension on change plane model is also included in this chapter.

Chapter 6 offers some further applications of change point methods in recent statistical research and machine learning. This chapter include change points analysis for functional data and network data. Functional data may be viewed as an extension of longitudinal data, observed over a range of time. Alternatively it could be viewed as a multi-dimensional time series. When there is sudden change, it is better to treat the curve with change points. We will focus on a half kernel approach to deal with the change point detection. Network is another type of interesting data in recent literature. We especially are concerned with dynamic networks and discuss a new and simple approach based on binary segmentation.

Because the change point problem usually is driven by the analysis of serially dependent data such as time series. Ignoring positive autocorrelation in a series will often produce spurious change point conclusions [230]. In fact, series that are heavily positively correlated tend to make long sojourns above and below the long-term mean of the series, inducing the appearance of a change point. Ignoring correlation may induce the spurious conclusion that a change point exists when in truth it does not. Therefore, pre-whitening autocorrelated series is necessary before implementing any method to be introduced in this book. For a similar reason, series with trends or seasonality should also be treated with caution. Directly applying change point detection on those series are simply declaring that the series' mean is time varying. We thus should remove the trends and seasonality from the series by using standard approaches [199].

1.4 Connection to other fields

The change point problem discussed in this book has strong connections to several other research areas in modern statistics and machine learning. In this section, we will briefly outline three relevant areas. First, identifying group membership for a subject can be viewed as a classification problem when a gold standard exists, and a variety of modern classifiers can be applied to accomplish this task. Second, even in the absence of ground truth, subjects can still be grouped into distinct clusters using statistical techniques such as cluster analysis and factor analysis. Finally, the development of subgroup identification methods has progressed rapidly within the statistical community, with a particular focus on tree-based methods, which will be reviewed in greater detail in Chapter 5.

Some authors use the terms "classification" and "clustering" interchangeably in their works. However, in this book, we make a clear distinction between the two. Classification refers to a supervised learning task, where information about the response or outcome is available, allowing for subjects to be assigned to predefined groups. Clustering, on the other hand, is an unsupervised learning task, where no response or outcome is observed, and the goal is to group subjects based on inherent similarities without prior knowledge of group membership.

1.4.1 Classification

Classification is a supervised learning problem. Statistical classification is needed particularly in clinical settings where the accurate diagnosis of a subject's status is crucial for proper treatment. An assessment of patient disease conditions and evaluation of the prognosis of patients can be achieved by analyzing clinical and laboratory data using appropriate learning tools. Medical decision making sometimes may involve more than two categories. We note that in a medical classification problem it is often necessary to involve more than one biomarkers. A statistical model is learned from a training sample and using the learned model one can produce predictive probability for each individual. For categorical outcome, the traditional learning programs include logistic regression, linear discrimination analysis, and support vector machines among others.

We first introduce some mathematical notation. Consider a set of predictors $\Omega = \{X_1, \cdots, X_p\}$, where $X_j \in \mathbb{R}$ $(j = 1, \cdots, p)$ is the jth predictor which can be discrete or continuous. Suppose we have a training sample of n subjects with measurements $\{x_{ij}, i = 1, \cdots, n; j = 1, \cdots, p\}$ and their class labels $\{y_i, i = 1, \cdots, n\}$. Denote $x_i = (x_{i1}, \cdots, x_{ip})$ as the ith subject. Researchers want to make use of the markers to accurately classify or predict the categorical outcome y. Suppose the multi-category outcome y takes values from $\mathcal{Y} = \{1, 2, \cdots, M\}$. These categories are known and in general not ordered. Of course it is also easy to imagine classifying patients into low-, medium-, and high-risk groups, for instance. We define the binary random variable $\delta_m = I(y = m)$ and let the prevalence for the mth category be $\rho_m = E(\delta_m) = P(y = m)$.

In practice in order to incorporate multiple markers we have to construct statistical models and make classification decision based on the model-based risk prediction. Suppose a model $\mathcal{M}_1$ is learned based on a set of predictors $\Omega_1 \subset \Omega$. Such a model $\mathcal{M}_1$ can generate an M−dimensional probability assessment vector $p_i(\mathcal{M}_1) = (p_{i1}(\mathcal{M}_1), \cdots, p_{iM}(\mathcal{M}_1))^T$ for the ith subject such that $\sum_{m=1}^{M} p_{im}(\mathcal{M}_1) = 1$. Each component $p_{im}(\mathcal{M}_1)$ in the vector is also commonly referred to as the *risk score* and indicates the predicted probability of the mth class membership. A greater value of $p_{im}(\mathcal{M}_1)$ suggests by the model $\mathcal{M}_1$ that the ith subject is more likely to be in the mth category. A classifier thus may want to assign the ith subject into the category corresponding to the greatest value of $p_{im}(\mathcal{M}_1)$. This is a very common take-the-winner classification rule. Different learning methods in general may generate quite different probability assessment vector for the same set of predictors Ω_1 and thus lead to different classification results with varied accuracy.

For the sample of n subjects, we can stack the vectors $p_i(\mathcal{M}_1)$ to form an $n \times p$ probability assessment matrix $\mathbf{p}(\mathcal{M}_1) = [p_1(\mathcal{M}_1) \cdots p_n(\mathcal{M}_1)]^T$, which can be used to assess the probability of the true status. In this section we review a few familiar learning approaches that produce $\mathbf{p}(\mathcal{M}_1)$ for the sample observations.

1.4.1.1 Logistic regression

There are abundant research development to compute the vectors of class probability estimates. Among them, an approach resembling linear regression model is the multinomial logistic regression model by treating the multiple category indicator variable as the response and using the variables involved in Ω_1 as the predictor. From the fitted model we may then evaluate the model-based prediction on the probability scale.

Using the first category as the reference, a multinomial logistic regression model is given by

$$\log \frac{P(y_i = m|x_i)}{P(y_i = 1|x_i)} = \beta_{m0} + \beta_{m1}^T x_i, \qquad m = 2, ..., M, \tag{1.2}$$

where the regression coefficient β_{m1} characterizes the dependence of the outcome on the predictor variables. In general, for a response with *M* categories, we need $M - 1$ log-odds equations defined above to formulate the regression model. In fact, the $M - 1$ multinomial logit equations contrast each of categories $2, 3, \cdots$,*M* with the reference category 1. Thus, the model-based risk score can be computed by

$$p_{ik} = P(y_i = k|x_i) = \left\{ \sum_{m=1}^{M} \exp(\beta_m^T C_i) \right\}^{-1} \exp(\beta_k^T C_i), \qquad i = 1, \cdots, M, \tag{1.3}$$

where $C_i = (1, x_i^T)^T$, $\beta_k = (\beta_{k0}, \beta_{k1}^T)^T$ and $\beta_1 = \mathbf{0}$. In practice the regression coefficients β_k can be estimated from the maximum likelihood estimation. To implement the multinomial logistic regression, we may use the `R` package `nnet`. Suppose $\mathbf{y}$ is the n−dimensional vector of class labels and $\mathbf{d}$ is the $n \times p$ matrix of predictors. We can use the following code to obtain a fitted regression model $\mathcal{M}_1$.

```
> M1 <- nnet::multinom(y~d)
```

We may then use the command `nnet::predict(M1,type='probs')` to yield the probability assessment matrix $\mathbf{p}(\mathcal{M}_1)$ for the whole data set.

Logistic regression had a great success in medicine, marketing, political science and economics. It remains a routine method in many companies, for instance for mortgage default prediction within banks or for click-through rate prediction in marketing companies.

1.4.1.2 Support vector machine

The support vector machines (SVM) are based upon the idea of maximizing the margin, i.e., maximizing the minimum distance from the separating hyperplane to the nearest class member

[249, 241]. In addition to nice theoretical properties, SVMs give exceptionally good performance on classification tasks. The basic SVM supports only binary classification, but extensions [67, 157] have been proposed to handle the multi-category classification as well. In these extensions, additional parameters and constraints are added to the optimization problem to handle the separation of the different classes.

The solution of an SVM is usually obtained from minimizing a regularized hinge loss function where a tuning parameter λ determines the trade-off between increasing the margin-size and ensuring that the data features x_i lie on the correct side of the margin. The computation can be solved efficiently with existing software. However, there is a nonparametric form for the estimated risk score based on functional approximation. We usually only focus on the numerical output for risk prediction from the fitted model.

To implement SVM, we may use the `R` package `e1071`. Suppose **y** is the class label and **d** is the matrix of predictors. In SVM we usually have to specify a kernel function which defines an inner product for the original data. When a radial basis function (RBF) is chosen as the kernel, we may use the following code to obtain the fitted model.

```
> M1 <- e1071::svm(y~d,kernel="radial",probability = T)
```

We can then use `e1071::attr(predict(M1,d,probability = T),"probabilities")` to extract the probability assessment matrix $\mathbf{p}(\mathcal{M}_1)$.

SVM can be viewed as a penalization method where the slope coefficients of the basis expansions are shrunk toward zero. Suppose $y_i = \{1, -1\}$ and linear model $f(x_i) = w^T\phi(x_i) + b$, where $\phi(x_i)$ is a fixed transformation. SVM tries to minimize

$$\sum_{i=1}^{N} [1 - y_i f(x_i)]_+ + \lambda \|w\|^2 \tag{1.4}$$

where $[]_+$ represents the positive part. In general, other margin maximizing loss functions can be used to create similar classifiers. One can apply what we introduce in this book to carry out a similar investigation on their classification accuracy. One of the reasons for the popularity of SVMs was their ability to handle data of various types thanks to the notion of kernel.

1.4.1.3 Linear discriminant analysis

Linear discriminant analysis (LDA) is one of the oldest classification approaches, due to Ronald Fisher in 1930s, and still finds applications in plenty of real world problems. LDA seeks a linear combination of predictor variables that best separates two or more classes of subjects. The resulting combination may be used as a linear classifier.

Typically features x_i's in the mth class are assumed to be normally distributed with mean μ_m, covariance Σ_m, and probability density function $\phi_m(x_i)$. Under such an assumption, we may use the maximum likelihood estimation to estimate the parameters and then use the Bayes' theorem to derive the posterior probability for the ith subject:

$$p_{ik} = P(y_i = k|x_i) = \frac{\rho_k \phi_k(x_i)}{\sum_{m=1}^{M} \rho_m \phi_m(x_i)} \qquad k = 1, \cdots, M. \tag{1.5}$$

To implement LDA, we use the **R** package **MASS** in this tutorial. The following code gives the solution of an estimated LDA model.

```
> M1 <- MASS::lda(y~d)
```

We then use `MASS::predict(M1)$posterior` to obtain the probability assessment matrix.

Like other aforementioned learning approaches, LDA also has various generalized versions such as quadratic discriminant analysis (QDA) and robust parameter estimation. These methods are quite fundamental to the traditional ways classification was done, partially because they are relatively easy to understand and computationally straight forward. However, when p is large, the sample size required to estimate the parameters reasonably well can be prohibitive. Normality assumption is also needed for LDA.

In addition to the three classifiers we reviewed here, there are plenty of other strong methods such as random forests, gradient boosting, perceptron algorithm, deep neural networks, among others. In particular, we will also briefly mention the tree method in chapter 5.

1.4.2 Clustering

Clustering is an unsupervised learning problem. It means there is no outcome to be predicted, the algorithm tries to find pattern in data. It deals with identifying subgroups within the data such that a set of objects in a subgroup (cluster) are similar to each other compared to objects in another cluster.

The goal of cluster analysis is to find meaningful groups in data. Typically in the data these groups will be internally cohesive and separated from one another. The purpose is to find groups whose members have something in common that they do not share with members of other groups. One can use multivariate techniques to construct clusters. Historically cluster analysis was used in categorizing animals and plants, psychological analysis with latent factors, genetic data analysis and image data analysis. Recently one important area for clustering is community detection for network data analysis. A very common approach is the spectral clustering algorithm. Two popular clustering algorithms are K-means and .

K-means clustering algorithm iteratively groups a collection of data points into a fixed number of clusters (K) according to their similarity. The algorithm aims to reduce the distance between each data point and its corresponding cluster center, also called the centroid. The algorithm terminates when either the centroids remain stable, or a maximum number of iterations is achieved. K-means clustering has various applications, such as data analysis, image segmentation, anomaly detection, etc. In `R` we can use function `kmeans` to obtain the cluster results.

Hierarchical clustering algorithm organizes data points into a hierarchy of clusters based on their similarity or distance. There are two variants: agglomerative and divisive. Agglomerative clustering begins with each data point as a separate cluster and then combines the nearest clusters until there is only one cluster left. Divisive clustering begins with all data points in a single cluster and then divides the cluster repeatedly until each data point has its own cluster. Hierarchical clustering can be represented by a dendrogram, which is a graph that illustrates the nested arrangement of clusters and their distances. Hierarchical clustering has various uses, such as finding patterns, discovering hierarchies, or detecting outliers in data. In `R` we can use `hclust` to obtain the cluster results.

Most researchers agree that K-means clustering is faster and simpler, but requires choosing the number of clusters beforehand and may not capture complex structures. On the other hand, hierarchical clustering is more flexible and intuitive, but can be computationally expensive and sensitive to outliers. We do not delve further into this topic. One may refer to some excellent texts such as [88, 36].

1.4.3 Subgroup identification

Identifying subgroups is increasingly important across many applications. When there is no response variable, this can be achieved using a clustering algorithm. However, in many cases, such as medical studies, we often have information on how well patients respond to a treatment, making it possible to apply a tree-based algorithm to build subgroups. In Chapter 5, we review tree methods and distinguish them from our threshold model.

A central goal of precision medicine is to identify treatments that maximize individual health outcomes. This approach has garnered significant attention due to its promise and potential impact in medical practice. The rapid rise in interest in precision medicine can be attributed to several factors:

(i) the increasing recognition that the "one size fits all" strategy is ineffective for many serious diseases like cancer, where targeted therapies based on individual traits often yield better results; (ii) it is much more challenging to find a treatment that works universally across all patients; (iii) risk factors for diseases are likely to vary among different patient groups; and (iv) recent advancements in genomics, computational biology, medical imaging, and regenerative medicine have made targeted therapies more feasible. The overarching goal of precision medicine is to identify subgroups of patients that respond differently to various treatment regimens and to model the relationships between response variables and predictors uniquely across these subgroups.

There are two different goals for subgroup identification in precision medicine: namely, *prognostic* and *predictive* signature developments. Clark et al. [63] summarizes the two types as follows: a *prognostic* signature is a measurement associated with clinical outcome without therapy or with the application of a standard therapy that patients are likely to receive and it can be thought of as a measure of the natural history of the disease, and a *predictive* signature is a measurement that is associated with response to a particular therapy and that is best evaluated in a randomized clinical trial with a control group. Although our focus is on prognostic signature development, it is quite possible that the methods we develop here may also be applicable in the predictive setting.

There are many statistical methods for subgroup identification in the literature and some of the most popular ones are based on tree-like partitioning algorithms. Early work includes Automatic Interaction Detection (AID) [188] and theta automatic interaction detection (THAID) [186]. The regression tree (CART) [39] algorithm was particularly successful and tree-based methods become more widely used for subgroup identification. A tree recursively partitions the subjects into binary subgroups until certain stopping criterion is met. There are two approaches to find an optimal tree: *pre-pruning* and *post-pruning*. Pre-pruning relies on some internal stopping criterion to control the size of the tree, such as [278] and post-pruning prunes a very large tree to a smaller sized-tree based on some optimality criteria [174]. A classification tree is a tree method where the response takes on discrete or unordered values and a regression tree is a tree method where the response variable is continuous or has ordered discrete values [173]. There are many tree-based classification and regression subgroup identification methods, and they include generalized unbiased interaction detection and estimation (GUIDE) [175], model-based recursive partitioning (MOB) [278], interaction tree (IT) [239], subgroup identification based on difference effect search (SIDES) [171] and Virtual twins (VT) [98], among others. GUIDE and MOB can be applied to both prognostic and predictive signature developments, but IT, SIDES and VT are only for predictive signature development. Some researchers have also proposed subgroup identification methods for individualized treatment rules [152]. Doove et al. [84] and Lipkovich et al. [170] provide reviews and compare a few tree-based methods in this field.

Additionally, change point detection (CPD) methods have been applied in economic studies [139, 117] and may offer an alternative strategy for clustering subjects. Identifying a structural break for a covariate among observed subjects can effectively lead to subgrouping. However, implementing CPD presents two primary challenges: (i) determining the number of change points, and (ii) accurately estimating their locations. Earlier approaches, such as iterative cumulative-sum methods, are often computationally intensive. More recent techniques, such as penalization methods, have accelerated the process of change point detection. In this book, we explore a newly proposed method known as the two-stage multiple change-point detection (TSMCD) method [161], which has demonstrated solid theoretical advantages and practical performance. Furthermore, other methods for subgroup identification exist [129, 52], including Bayesian approaches [218, 231].

2

Single change point

Research on change points dates back to the 1950s [196, 194]. Over the next half century, many scholars focused on the standard single change point problems involving independent normal distributions and regression model change points. The primary methods used in this research field include the likelihood ratio tests, Bayesian approaches, and information-based methods. In addition to the frequentist vs Bayesian distinction, some researchers recently also classify change point detection problems into two categories: retrospective or a posterior tests (offline) and sequential or a priori tests (online) [13]. The offline approach assumes the existence of a training set that serves as a reference benchmark for normal data with no change. In another word, such a retrospective approach is based on observing all of the available data. On the other hand, online methods work on sequentially collected data and seek to declare changes as soon as possible after they have occurred. Nonetheless, the construction of test statistics and the development of asymptotic theories are quite similar for both methods. We do not differentiate the two approaches in this book.

In this chapter, we illustrate several important and fundamental problems related to change point detection. First, we address the change in population mean, assuming the observations follow a normal distribution. The second problem focuses on changes in the variance of a normal distribution or autoregressive models. The third issue concerns shifts in the underlying distribution itself. Finally, we explore changes in regression parameters. In each of these sections, we restrict our analysis to a single change point and provide detailed probabilistic theories, discussing both exact and asymptotic distributions.

2.1 Change in mean

Let $\boldsymbol{x}_1$, $\boldsymbol{x}_2$, ..., $\boldsymbol{x}_n$ be a sequence of independent m-dimensional normal random vectors with mean and covariance parameters $(\boldsymbol{\mu}_1, \boldsymbol{\Sigma}_1)$,$(\boldsymbol{\mu}_2, \boldsymbol{\Sigma}_2), \ldots, (\boldsymbol{\mu}_n, \boldsymbol{\Sigma}_n)$, respectively. In general, we assume $\boldsymbol{\Sigma}_1 = \boldsymbol{\Sigma}_2 = \ldots = \boldsymbol{\Sigma}_n = \boldsymbol{\Sigma}$ and $\boldsymbol{\Sigma}$ may be unknown. We intend to test the following hypothesis,

$$H_0 : \boldsymbol{\mu}_1 = \boldsymbol{\mu}_2 = \ldots = \boldsymbol{\mu}_n = \boldsymbol{\mu},$$

versus the alternative:

$$H_1 : \boldsymbol{\mu}_1 = \ldots = \boldsymbol{\mu}_k \neq \boldsymbol{\mu}_{k+1} = \ldots = \boldsymbol{\mu}_n,$$

where k represents the position of the single change point at each stage $1 \leq k \leq n-1$.
To simplify the presentation, let us first consider the special circumstance where the dimension $m = 1$. Now $x_1, x_2, \ldots, x_n$ is a sequence of independent one-dimensional normal random vectors with parameters (μ_1, σ_1),$(\mu_2, \sigma_2), \ldots, (\mu_n, \sigma_n)$ and $\sigma_1 = \sigma_2 = \ldots = \sigma_n = \sigma$. This special problem was first examined by [196, 194, 195]. Later, [29] and other authors also contributed to the study of this problem. Under this special case, the testing procedure depends on whether the parameter σ^2 is known or unknown. We will use the likelihood ratio method to test for the existence of a single change point.

DOI: 10.1201/9781032657073-2

2.1.1 Variance is known

In this section, we consider the variance known and assume $\sigma^2 = 1$. Without loss of generality, we assume that the mean before the k-th position is equal to μ_1 and after the k-th position is equal to μ_n. Let

$$\begin{aligned} V_k &= k(\overline{x}_k - \overline{x})^2 + (n-k)(\overline{x}_{n-k} - \overline{x})^2 \\ &= \frac{k(n-k)}{n}(\overline{x}_k - \overline{x}_{n-k})^2 \\ &= \frac{n}{k(n-k)}\left(\sum_{i=1}^{k}(x_i - \overline{x})\right)^2 \end{aligned} \tag{2.1}$$

$\overline{x}_k = \frac{1}{k}\Sigma_{i=1}^{k} x_i$ is the MLE of μ_1, $\overline{x}_{n-k} = \frac{1}{n-k}\Sigma_{i=k+1}^{n} x_i$ is the MLE of μ_n and $\overline{x}$ is the mean of $\boldsymbol{x} = (x_1, x_2, \ldots, x_n)^\top$.

We consider the following cumulative sum (CUSUM) test statistics

$$U_n = \sqrt{V_{k^*}} = \max_{1 \le k \le n-1} \sqrt{V_k} = \max_{1 \le k \le n-1} |T_k|, \tag{2.2}$$

where $k^* = \arg\max_{1 \le k \le n-1} |T_k|$, $T_k = \sqrt{\frac{n}{k(n-k)}}\left(\sum_{i=1}^{k}(x_i - \overline{x})\right)$ and $\{T_1, T_2, \ldots, T_{n-1}\}$ forms a Markov process (see [51]). The search for the optimal change point k^* usually requires users to scan the whole sample of $n-1$ points. While this exhaustive search may be computationally intensive for scenarios involving multiple change points, it is adequately efficient and practical for the simpler single change point setting discussed in this chapter.

We introduce the null distribution of the CUSUM statistic in the following theorem.

Theorem 2.1.1. *Assume $x_1, x_2, \ldots, x_n$ are independent and identically distributed (iid) one-dimensional standard normal random variables. Thus H_0 is correct. The probability density function (pdf) of U_n is given by*

$$f_{U_n}(x) = 2\phi(x, 0, 1)\sum_{k=1}^{n-1} g_k(x, x) g_{n-k}(x, x),$$

where $\phi(x, 0, 1)$ is the pdf of $N(0, 1)$, $g_1(x, s) = 1$ for $x, s \ge 0$, and

$$g_k(x, s) = P\left[|T_i| < s, i = 1, \ldots, k-1 \,\middle|\, |T_k| = x\right].$$

Proof of Theorem 2.1.1. Let A_k, B_k, and C_k be the following events,

$$A_k = \{|T_k| \in (x, x + dx)\},$$

$$B_k = \{|T_j| < |T_k|, j = 1, \ldots, k-1\},$$
$$C_k = \{|T_j| < |T_k|, j = k+1, \ldots, n-1\}.$$

Then,

$$\begin{aligned} P[U_n \in (x, x+dx)] &= P\left\{\bigcup_{k=1}^{n-1} [|T_k| \in (x, x+dx)] \bigcap [|T_k| > |T_j|, j \ne k]\right\} \\ &= \sum_{k=1}^{n-1} P(A_k \cap B_k \cap C_k) = \sum_{k=1}^{n-1} P(A_k) P[B_k | A_k] P[C_k | A_k \cap B_k]. \end{aligned}$$

Due to $T_k \sim N(0,1)$, $P(A_k) = 2\phi(x,0,1) + o(dx)$ and

$$P(B_k|A_k) = g_k(x,x) + o(dx).$$

From the previous conclusion that $\{T_1, \ldots, T_{n-1}\}$ is Markov process, we have that $\{T_1, \ldots, T_{k-1}\}$ and $\{T_{k+1}, \ldots, T_{n-1}\}$ are independent. Therefore, $P(C_k|B_k \cap A_k) = P(C_k|A_k)$. According to the probability symmetry between B_k and C_k, we obtain

$$P(C_k|A_k) = g_{n-k}(x,x) + o(dx).$$

Thus, we have

$$f_{u_n}(x) = \sum_{k=1}^{n-1} 2\phi(x,0,1) g_k(x,x) g_{n-k}(x,x).$$

Theorem 2.1.1 has been proven. □

Based on this theorem, after we obtain a sample value u_n from an empirical data set, the test p-value can be computed as $P(U_n > u_n) = \int_{u_n}^{\infty} f_{U_n}(x)dx$. A small p-value would indicate significant evidence to reject the null hypothesis of no change and lead to the support of existence of a change point. If we use the above distribution of U_n, we need to evaluate $g_k(x,s)$. The following theorem provides a computational method.

Theorem 2.1.2. *Under the assumption of Theorem 2.1.1, the function $g_k(x,s)$ is determined by the recursion*

$$g_k(x,s) = \int_0^s g_{k-1}(y,s)\left[\phi\left(y, \rho x, \tau^2\right) + \phi\left(y, -\rho x, \tau^2\right)\right] dy,$$

where $\rho = \rho_{k-1,k}$ is the correlation coefficient between T_{k-1} and T_k, and $\tau = \sqrt{1-\rho^2}$.

Proof of Theorem 2.1.2. Because $\{T_1, \ldots, T_{n-1}\}$ is a Markov process, $T_k \sim N(0,1)$ and $f(|T_{k-1}| = y||T_k| = x) = \phi\left(y, \rho x, \tau^2\right) + \phi\left(y, -\rho x, \tau^2\right)$, we have

$$\begin{aligned}
g_k(x,s) &= P\left(|T_j| < s, j = 1, \ldots, k-1 ||T_k| = x\right) \\
&= \int_0^s P\left(|T_j| < s, j = 1, \ldots, k-2 ||T_{k-1}| = y, |T_k| = x\right) dP\left(|T_{k-1}| < y ||T_k| = x\right) \\
&= \int_0^s P\left(|T_j| < s, j = 1, \ldots, k-2 ||T_{k-1}| = y\right) dP\left(|T_{k-1}| < y ||T_k| = x\right) \\
&= \int_0^s g_{k-1}(y,s)\left[\phi\left(y, \rho x, \tau^2\right) + \phi\left(y, -\rho x, \tau^2\right)\right] dy.
\end{aligned}$$

□

In addition to the distribution of U_n under the null hypothesis, we offer the following theorem which gives the distribution of the change point position k^*. This theorem only quantifies the random behavior of this estimate but has not yet led to a formula for confidence interval. We will explore the details of interval estimation for change points in Chapter 4.

Theorem 2.1.3. *Under the assumption of Theorem 2.1.1, if k^* is the position of the change point estimated by (2.2), then for $k = 1, 2, \ldots, n$*

$$P(k^* = k) = 2\int_0^{\infty} g_k(x,x) g_{n-k}(x,x) \phi(x,0,1) dx.$$

Proof of Theorem 2.1.3. From the fact that $\{T_1, \ldots, T_{n-1}\}$ is a Markov process, and $f(|T_k| = x) = 2\phi(x, 0, 1), x \geq 0$, we have

$$\begin{aligned} P(k^* = k) &= P(|T_j| < |T_k|, j \neq k) \\ &= \int_0^{\infty} P(|T_j| < x, j \neq k||T_k| = x) f(|T_k| = x) dx \\ &= 2\int_0^{\infty} g_k(x, x) g_{n-k}(x, x) \phi(x, 0, 1) dx. \end{aligned}$$

□

Through the above theorem, we have obtained the exact distribution of the statistic U_n. When n is large, we hope to obtain an asymptotic distribution of U_n. The following results also no longer require the normality assumption for the original data.

Let $W_k = x_1 + x_2 + \ldots + x_k, 1 \leq k \leq n$. Then, we can show easily that the CUSUM test can be written as

$$U_n = \max_{1 \leq k \leq n-1} \left| \frac{W_k}{\sqrt{n}} - \frac{k}{n} \frac{W_n}{\sqrt{n}} \right| / \left[\frac{k}{n} \left(1 - \frac{k}{n}\right) \right]^{1/2}.$$

Suppose $\{B(t); 0 \leq t < \infty\}$ is a standard Brownian motion, then under H_0, through simple algebraic transformations

$$U_n \stackrel{\mathrm{D}}{=} \max_{t=1/n, \ldots, (n-1)/n} \frac{|B_0(t)|}{[t(1-t)]^{\frac{1}{2}}},$$

where "$\stackrel{\mathrm{D}}{=}$" means "identically distributed as", $t = \frac{k}{n}, k = 1, \ldots, n-1, \{B(t), t \geq 0\}$ is a standard Brownian motion, $B_0(t) = B(t) - tB(1)$ is the Brownian bridge which is a Gaussian process and satisfies $E(B_0(x)) = 0, Cov(B_0(x_1), B_0(x_2)) = \min(x_1, x_2) - x_1 x_2, 0 \leq x_1, x_2 \leq 1$. The following theorem indicates that the asymptotic null distribution of U_n is a Gumbel distribution.

Theorem 2.1.4. *Under the assumption of Theorem 2.1.1, $-\infty < x < \infty$,*

$$\lim_{n \to \infty} P\left[a_n^{-1}(U_n - b_n) \leq x\right] = \exp\left\{-2\pi^{1/2} e^{-x}\right\},$$

where $a_n = (2 \log\log n)^{-1/2}$, $b_n = a_n^{-1} + \frac{1}{2} a_n \log\log\log n$.

To prove this theorem we need to use the properties of Brownian motion and the convergence rules. The following lemmas (see Lemma 2.8-2.9 of [51]) are needed to prove the theorem.

Lemma 2.1. *For $-\infty < x < \infty$,*

$$\lim_{n \to \infty} P\left[a_n^{-1}\left(\max_{1 \leq nt \leq [n/\log n]} \frac{|B(t)|}{\sqrt{t}} - b_n\right) \leq x\right] = \exp\left\{-\pi^{1/2} e^{-x}\right\}.$$

Lemma 2.2. *As $n \to \infty$,*

$$\max_{1 \leq nt \leq [n/2]} \frac{|B_0(t)|}{\sqrt{t(1-t)}} - \max_{1 \leq nt \leq [n/\log n]} \frac{|B(t)|}{\sqrt{t}} = o_p(a_n),$$

$$\max_{1 \leq n(1-t) \leq [n/2]} \frac{|B_0(t)|}{\sqrt{t(1-t)}} - \max_{1 \leq n(1-t) \leq [n/\log n]} \frac{|B(t) - B(1)|}{\sqrt{1-t}} = o_p(a_n).$$

Using the above lemmas, we can prove the Theorem 2.1.4.

Proof of Theorem 2.1.4. By Lemma 2.1, and Lemma 2.2, we derive

$$
\begin{aligned}
& P\left[a_n^{-1}(U_n - b_n) \leq x | H_0\right] \\
&= P\left[\max_{1\leq nt\leq [n/2]} \frac{|B_0(t)|}{\sqrt{t(1-t)}} \leq a_n x + b_n\right] \\
&\cdot P\left[\max_{1\leq n(1-t)\leq [n/2]} \frac{|B_0(t)|}{\sqrt{t(1-t)}} \leq a_n x + b_n\right] \\
&= P\left[\max_{1\leq nt\leq [n/\log n]} \frac{|B(t)|}{\sqrt{t}} \leq a_n x + b_n\right] \\
&\cdot P\left[\max_{1\leq n(1-t)\leq [n/\log n]} \frac{|B(t)-B(1)|}{\sqrt{1-t}} \leq a_n x + b_n\right] + o_p(1) \\
&\overset{n\to\infty}{\longrightarrow} \exp\left\{-\pi^{-1/2}e^{-x}\right\} \cdot \exp\left\{-\pi^{-1/2}e^{-x}\right\} \\
&= \exp\left\{-2\pi^{-1/2}e^{-x}\right\}.
\end{aligned}
$$

□

Using this theorem, we can build an α level test by first choosing the upper α quantile from the standard Gumbel distribution which is

$$z_\alpha = \log(-2\sqrt{\pi}/\log(\alpha)). \tag{2.3}$$

Then we reject the null hypothesis if $U_n > a_n z_\alpha + b_n$. This asymptotic test is much faster to implement in practice.

2.1.2 Variance is unknown

Likely, we do not know the population variance in practice. In this section, we relax the known variance assumption. Under H_0: $x_1, x_2, \ldots, x_n$ are independent and identically distributed (iid) normal random variables with mean μ and variance σ^2, the likelihood function is

$$L_0\left(\mu, \sigma^2\right) = \frac{1}{(\sqrt{2\pi}\sigma)^n} e^{-\sum_{i=1}^n (x_i-\mu)^2/2\sigma^2}$$

and the MLEs of μ and σ^2 are

$$\widehat{\mu} = \bar{x} = \frac{1}{n}\sum_{i=1}^n x_i, \quad \text{and} \quad n\widehat{\sigma}^2 = S = \sum_{i=1}^n (x_i - \bar{x})^2,$$

respectively.

Under H_1: $x_1, x_2, \ldots, x_n$ are independent, $x_1, x_2, \ldots, x_k$ have the identical normal distribution $N(\mu_1, \sigma_1^2)$ and $x_{k+1}, x_{k+2}, \ldots, x_n$ have the identical normal distribution $N(\mu_n, \sigma_1^2)$, the likelihood function is

$$L_1\left(\mu_1, \mu_n, \sigma_1^2\right) = \frac{1}{(\sqrt{2\pi}\sigma_1)^n} e^{-\sum_{i=1}^k (x_i-\mu_1)^2/2\sigma_1^2 - \sum_{i=k+1}^n (x_i-\mu_n)^2/2\sigma_1^2},$$

and the MLEs of μ_1, μ_n, and σ_1^2 are,

$$\widehat{\mu}_1 = \bar{x}_k = \frac{1}{k}\sum_{i=1}^k x_i, \quad \widehat{\mu}_n = \bar{x}_{n-k} = \frac{1}{n-k}\sum_{i=k+1}^n x_i,$$

and

$$n\widehat{\sigma}_1^2 = S_k = \sum_{i=1}^{k} (x_i - \bar{x}_k)^2 + \sum_{i=k+1}^{n} (x_i - \bar{x}_{n-k})^2,$$

respectively.

The log-likelihood ratio test statistic is

$$\begin{aligned}
& -2\log\left(\frac{L_0(\hat{\mu}, \hat{\sigma}^2)}{L_1(\hat{\mu}_1, \hat{\mu}_n, \hat{\sigma}_1^2)}\right) \\
&= -2\left(-\frac{n}{2}\log(\hat{\sigma}^2) + \frac{n}{2}\log(\hat{\sigma}_1^2)\right) \\
&= n\log\left(\frac{1}{n}\sum_{i=1}^{n}(x_i - \bar{x})^2\right) - n\log\left\{\frac{1}{n}\left[\sum_{i=1}^{k}(x_i - \bar{x}_k)^2 + \sum_{i=k+1}^{n}(x_i - \bar{x}_{n-k})^2\right]\right\} \\
&= n\log\left(\sum_{i=1}^{n}(x_i - \bar{x})^2\right) - n\log\left\{\sum_{i=1}^{k}[(x_i - \bar{x}) + (\bar{x} - \bar{x}_k)]^2 + \sum_{i=k+1}^{n}[(x_i - \bar{x}) + (\bar{x} - \bar{x}_{n-k})]^2\right\} \\
&= n\log\left(\sum_{i=1}^{n}(x_i - \bar{x})^2\right) - n\log\left(\sum_{i=1}^{n}(x_i - \bar{x})^2 - k(\bar{x}_k - \bar{x})^2 - (n-k)(\bar{x}_{n-k} - \bar{x})^2\right) \\
&= -n\log\left(1 - \frac{T_k^2}{S}\right),
\end{aligned}$$

where $V_k = T_k^2$ is defined the same as in (2.1), and $T_k = \sqrt{\frac{n}{k(n-k)}}\left(\sum_{i=1}^{k}(x_i - \bar{x})\right) = \sqrt{\frac{k(n-k)}{n}}\,(\bar{x}_k - \bar{x}_{n-k})$.

Under H_0, it is easy to show that $\frac{T_k^2}{S} = O_p(1/n)$, then the log-likelihood ratio test statistic $-n\log\left(1 - \frac{T_k^2}{S}\right) \approx \frac{nT_k^2}{S}$. Next, we will explain how to obtain the distribution of the test statistics

$$Z_n = \max_{1\le k\le n} \frac{|T_k|}{\sqrt{S}}$$

under the null hypothesis, based on [263].

Let $\boldsymbol{y} = (y_1, \ldots, y_n)^\top$,$y_i = (x_i - \bar{x})/\sqrt{S}$ and $\boldsymbol{x} = (x_1, \ldots, x_n)^\top$. Also, define

$$c_k = \sqrt{\frac{k}{n-k}} \quad \text{and} \quad b_{k_i} = \begin{cases} n^{-1/2}c_k^{-1} & i = 1, \ldots, k \\ -n^{1/2}c_k & i = k+1, \ldots, n \end{cases},$$

for $k = 1, 2, \ldots, n-1$. Let $\boldsymbol{b}_k = (b_{k_1}, \ldots, b_{k_n})^\top$. Because of $\boldsymbol{b}_k^\top \mathbf{1} = 0$ and $T_k = \boldsymbol{b}_k^\top \boldsymbol{x}$,

$$T_k/\sqrt{S} = \boldsymbol{b}_k^\top \boldsymbol{y}, \tag{2.4}$$

where $\mathbf{1}$ is the $n \times 1$ unit vector. Since $\boldsymbol{y}$ is n-dimensional and $\boldsymbol{y}^\top\boldsymbol{y} = 1$, thus $\boldsymbol{y}$ is uniformly distributed on the surface of the unit $(n-1)$-ball $C = \{\boldsymbol{y} : \boldsymbol{y}^\top\boldsymbol{y} = 1 \; and \; \boldsymbol{y}^\top\mathbf{1} = 0\}$. Define $D = \{\boldsymbol{y} : \mathbf{1}^\top\boldsymbol{y} = 0, |\boldsymbol{b}_k^\top\boldsymbol{y}| \le z, k = 1, \ldots, n-1\}$, then the event $\{Z_n \le z\} = \{\boldsymbol{y} \in C \bigcap D\}$. From the additive formula of probability and DeMorgan's law, we derive

$$\begin{aligned} P\{Z_n \le z\} &= P\left\{ \boldsymbol{y} \in \bigcap_{k=1}^{n-1} \left(A_k^+ \bigcup A_k^- \right)^c \right\} \\ &= 1 - \sum_{k=1}^{n} P\{\boldsymbol{y} \in A_k\} \\ &+ \ldots + (-1)^p \sum_{1 \le k_1 \le k_2 \le \cdots \le k_p \le n-1} P\left\{ \boldsymbol{y} \in \bigcap_{j=1}^{p} A_{k_j} \right\} \\ &+ \ldots + (-1)^{n-1} P\left\{ \boldsymbol{y} \in \bigcap_{j=1}^{n-1} A_{k_j} \right\}, \end{aligned} \tag{2.5}$$

where $A_k = A_k^+ \bigcup A_k^-$, $A_k^+ = \{\boldsymbol{y} : \boldsymbol{y} \in C, \boldsymbol{b}_k^\top \boldsymbol{y} > z\}$, and $A_k^- = \{\boldsymbol{y} : \boldsymbol{y} \in C, -\boldsymbol{b}_k^\top \boldsymbol{y} > z\}$.

First, consider the first term $P\{\boldsymbol{y} \in A_k\}$ in (2.5). Under H_0, $T_k \sim N(0, \sigma^2)$, $S_k/\sigma^2 \sim \mathcal{X}_{n-2}^2$. Because T_k is also independent of S_k, we have

$$\frac{T_k}{\sqrt{S_k}/\sqrt{n-2}} \sim t_{n-2}.$$

Then, by (2.4),

$$P\{\boldsymbol{y} \in A_k\} = P\{|\boldsymbol{b}_k^\top \boldsymbol{y}| > z\} = P\left\{|T_k/\sqrt{S}| > z\right\} \tag{2.6}$$

can be calculated through the t_{n-2} distribution since

$$\frac{|T_k|}{\sqrt{S}} = \left(\frac{S_k}{T_k^2} + 1\right)^{-1/2}.$$

Consider other terms in (2.5), we may investigate the general one such as $P\left\{\boldsymbol{y} \in \bigcap_{j=1}^{p} A_{kj}\right\}$ for $1 < p < n-1$, with $k_1 < \cdots < k_p$. Let $\boldsymbol{B}$ be a $p \times n$ matrix such that $\boldsymbol{B} = \left(\boldsymbol{b}_{k_1}^{*\top}, ..., \boldsymbol{b}_{k_p}^{*\top}\right)^\top$

$$\boldsymbol{b}_{k_j}^{*\top} = \begin{cases} \boldsymbol{b}_{k_j}, & \text{if } A_{kj} = A_{kj}^+, \\ -\boldsymbol{b}_{k_j}, & \text{if } A_{kj} = A_{kj}^-, \end{cases}$$

where $\boldsymbol{b}_{k_j} = (b_{k_{j_1}}, \ldots, b_{k_{j_n}})$. Define

$$b_{k_{j_i}} = \begin{cases} n^{-1/2} c_k^{-1}, & i = 1, \ldots, k_j, \\ -n^{1/2} c_k, & i = k_j + 1, \ldots, n, \end{cases}$$

Then, $P\left\{\boldsymbol{y} \in \bigcap_{j=1}^{p} A_{k_j}\right\} = P\{\boldsymbol{B}\boldsymbol{y} > z\boldsymbol{1}_p\}$. We just need to derive the null probability function of the random variable $\boldsymbol{Z} = \boldsymbol{B}\boldsymbol{y}$. To this end we employ the following theorem based on the results of [86].

Theorem 2.1.5. *Under H_0, the pdf of $\boldsymbol{Z} = \mathbf{B}\mathbf{y}$ at $\mathbf{z}$ is given by*

$$f_p(\mathbf{z}) = \begin{cases} \frac{\Gamma\left[\frac{n-1}{2}\right]}{\pi^{p/2}\Gamma\left[\frac{n-1-p}{2}\right]} |\boldsymbol{\Sigma}|^{-1/2} \left[1 - \mathbf{z}^\top \boldsymbol{\Sigma}^{-1} \mathbf{z}\right]^{(n-3-p)/2}, & \text{if } \mathbf{z}^\top \boldsymbol{\Sigma}^{-1} \mathbf{z} < 1, \\ 0, & \text{otherwise}, \end{cases}$$

where $\boldsymbol{\Sigma} = \boldsymbol{B}\boldsymbol{B}^\top$.

Thus, after we obtain a sample value of the test statistic z_n from an empirical data set, the test p-value can be easily computed as $P(Z_n > z_n) = 1 - P(Z_n \leq z_n)$ using (2.5) under the null hypothesis by applying the elliptic distribution in Theorem 2.1.5 and the t_{n-2} distribution. We can also obtain Bonferroni approximations for the distribution of Z_n by ignoring the higher-order terms in (2.5) directly (cf. [263]).

When testing for a mean shift, [224] proposed a self-normalization (SN) based Kolmogorov–Smirnov test. Let $\boldsymbol{x}_1, \boldsymbol{x}_2, \ldots, \boldsymbol{x}_n$ be independent d-dimensional random variables with different mean $E\boldsymbol{x}_i = \boldsymbol{\mu}_i$ and the same variance $Var(\boldsymbol{x}_i) = \Sigma$. Supposed to test the mean shift of $\{\boldsymbol{x}_1, \ldots, \boldsymbol{x}_n\}$. Let

$$\mathbf{W}_n(\lfloor nr \rfloor) = n^{-1/2} \sum_{t=1}^{\lfloor nr \rfloor} (\boldsymbol{x}_t - \bar{\boldsymbol{x}}_n), \quad r \in [0, 1],$$

where $\bar{\boldsymbol{x}}_n = \frac{1}{n}\sum_{i=1}^{n} \boldsymbol{x}_i$. Let $S_{t_1,t_2} = \sum_{j=t_1}^{t_2} \boldsymbol{x}_j$ if $t_1 \leq t_2$ and 0 otherwise. For $k = 1, \ldots, n-1$, define

$$\mathbf{D}_n(k) = n^{-2}\left\{\sum_{t=1}^{k} [S_{1,t} - (t/k)S_{1,k}]^2 + \sum_{t=k+1}^{n} [S_{t,n} - (n-t+1)/(n-k)S_{k+1,n}]^2\right\}.$$

The SN-type test statistic is

$$SN_n = \sup_{k=1,\ldots,n-1} \mathbf{W}_n(k)^\top \mathbf{D}_n(k)^{-1} \mathbf{W}_n(k).$$

Denote "$\Rightarrow$" to be weak convergence in $D^d[0,1]$ and $\{\mathbf{B}(r), r \geq 0\}$ to be a d-dimensional Brownian motion. Under the null hypothesis, we have

$$\frac{1}{\sqrt{n}} \sum_{t=1}^{\lfloor nr \rfloor} (\boldsymbol{x}_t - \boldsymbol{\mu}) \Rightarrow \Sigma^{1/2}\mathbf{B}(r).$$

The continuous mapping theorem implies that the following theorem holds (see [224]).

Theorem 2.1.6. *Under* H_0,

$$SN_n \Rightarrow \sup_{r\in[0,1]} [\mathbf{B}(r) - r\mathbf{B}(1)]^\top \mathbf{D}^{-1}(r)[\mathbf{B}(r) - r\mathbf{B}(1)],$$

where

$$\mathbf{D}(r) = \int_0^r [\mathbf{B}(s) - (s/r)\mathbf{B}(r)][\mathbf{B}(s) - (s/r)\mathbf{B}(r)]^\top ds + \int_r^1 \left\{\mathbf{B}(1) - \mathbf{B}(s) - \frac{1-s}{1-r}[\mathbf{B}(1) - \mathbf{B}(r)]\right\}\left\{\mathbf{B}(1) - \mathbf{B}(s) - \frac{1-s}{1-r}[\mathbf{B}(1) - \mathbf{B}(r)]\right\}^\top ds.$$

Significance tests can thus be constructed based on the theorem. We refer to [224] for more computational details.

2.2 Change in variance

In many practical scenarios, the mean parameter may remain constant while the variance undergoes a structural change at a certain threshold. Some authors refer to mean change as an additive change while variance change as non-additive change. In comparison, mean change is a shift in the central tendency of the data. This type of change is additive because it involves altering the expected value of the data, effectively adding or subtracting a constant from the mean without directly modifying the variability or spread of the data around the mean. On the other hand, variance change refers to a shift in the variability or dispersion of the data. This is termed non-additive because it does not directly shift the central location but instead modifies how data points spread around the mean. Variance changes can lead to heteroscedasticity (non-constant variance), making the data distribution broader or narrower. See Figure 2.1 for an illustration.

In this section, we begin by addressing the variance change problem within the Gaussian model, introducing methods based on the likelihood ratio test to detect such changes. Following this, we extend our discussion to detecting variance changes in autoregressive time series models where the model order is unknown. Unlike the mean change scenario, where it is possible to derive exact small-sample distributions for test statistics under the null hypothesis, most of the distributions we consider for variance change detection are asymptotic, providing approximations that are more accurate with larger sample sizes.

2.2.1 Likelihood ratio test

Let $x_1, x_2, ..., x_n$ be a sequence of independent normal random variables with parameters $(\mu, \sigma_1^2), (\mu, \sigma_2^2), \ldots, (\mu, \sigma_n^2)$, respectively. Assume that μ is known in this section. We want to test the hypothesis

$$H_0 : \sigma_1^2 = \sigma_2^2 = \ldots = \sigma_n^2 = \sigma^2(\text{unknown}) \tag{2.7}$$

versus

$$H_1 : \sigma_1^2 = \ldots = \sigma_k^2 \neq \sigma_{k+1}^2 = \ldots = \sigma_n^2, \tag{2.8}$$

where $1 < k < n$ is the unknown position of the change point.

Under H_0, the maximum likelihood is

$$\log L_0(\hat{\sigma}^2) = -\frac{n}{2}\log(2\pi) - \frac{n}{2}\log(\hat{\sigma}^2) - \frac{n}{2},$$

where $\hat{\sigma}^2 = \frac{1}{n}\sum_{i=1}^{n}(x_i - \mu)^2$. Under H_1, the maximum likelihood is

$$\log L_0(\hat{\sigma}^2) = -\frac{n}{2}\log(2\pi) - \frac{k}{2}\log(\hat{\sigma}_1^2) - \frac{n-k}{2}\log(\hat{\sigma}_2^2) - \frac{n}{2},$$

where $\hat{\sigma}_1^2 = \frac{1}{k}\sum_{i=1}^{k}(x_i - \mu)^2$ and $\hat{\sigma}_2^2 = \frac{1}{n-k}\sum_{i=k+1}^{n}(x_i - \mu)^2$.

Calculating the maximum likelihood estimation, we obtain the likelihood ratio statistic as

$$\lambda_n = \left\{\max_{1<k<n-1}\left[n\log\widehat{\sigma}^2 - k\log\widehat{\sigma}_1^2 - (n-k)\log\widehat{\sigma}_n^2\right]\right\}^{1/2}. \tag{2.9}$$

According to the principle of the maximum likelihood procedure, we estimate the position k of the change point by $\hat{k}$ such that (2.9) attains its maximum, that is,

$$\begin{aligned}\hat{k} &= \arg\max_{1<k<n-1}\left[n\log\widehat{\sigma}^2 - k\log\widehat{\sigma}_1^2 - (n-k)\log\widehat{\sigma}_n^2\right]\\ &= \arg\min_{1<k<n-1}\left[k\log\widehat{\sigma}_1^2 + (n-k)\log\widehat{\sigma}_n^2\right].\end{aligned}$$

We now derive the asymptotic null distribution of λ_n:

$$\lambda_n \overset{D}{=} \max_{1<k<n-1}\left[n\log\frac{\sum_{i=1}^{n} y_i^2}{n} - k\log\frac{\sum_{i=1}^{k} y_i^2}{k} - (n-k)\log\frac{\sum_{i=k+1}^{n} y_i^2}{n-k}\right]^{1/2},$$

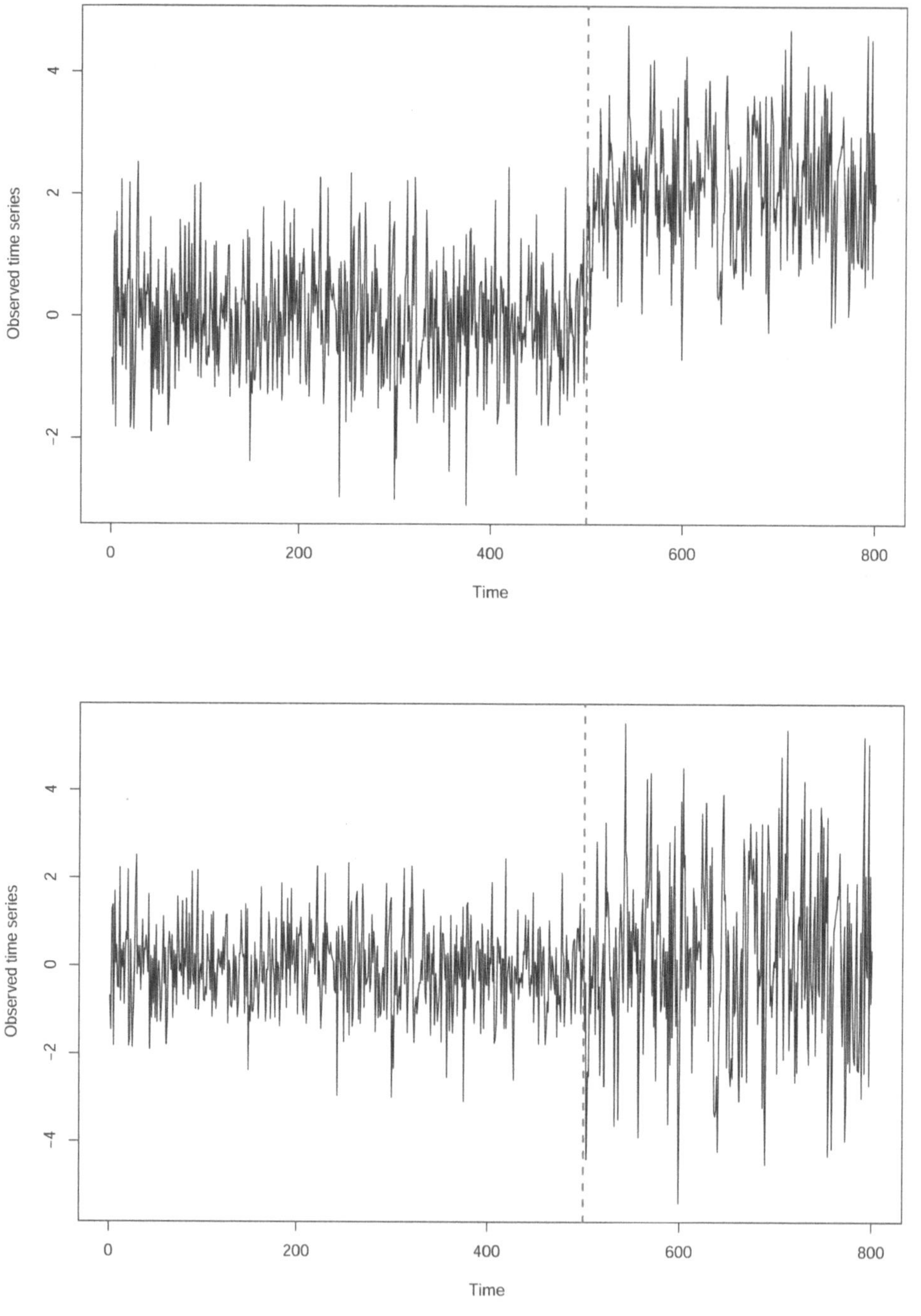

Figure 2.1: Comparison between changes in mean (top row) and variance (bottom row) of time series.

where $y_i = (x_i - \mu)/\sigma$. We apply Taylor's third-order expansion to transform the objective function on the right-hand side of the above equation and use simple algebraic operations, to arrive at

$$n \log \frac{\sum_{i=1}^n y_i^2}{n} - k \log \frac{\sum_{i=1}^k y_i^2}{k} - (n-k) \log \frac{\sum_{i=k+1}^n y_i^2}{n-k} \equiv W_k + Q_k + R_k \equiv L_k,$$

where

$$W_k = -\frac{1}{2n}\left[\sum_{i=1}^n (y_i^2 - 1)\right]^2 + \frac{1}{2k}\left[\sum_{i=1}^k (y_i^2 - 1)\right]^2 + \frac{1}{2(n-k)}\left[\sum_{i=k+1}^n (y_i^2 - 1)\right]^2,$$

$$Q_k = \frac{n}{3}\left(\xi_n^{(1)}\right)^{-3}\left(\frac{\sum_{i=1}^n y_i^2}{n} - 1\right)^3 - \frac{k}{3}\left(\xi_n^{(2)}\right)^{-3}\left(\frac{\sum_{i=1}^k y_i^2}{k} - 1\right)^3,$$

$$R_k = -\frac{n-k}{3}\left(\xi_n^{(3)}\right)^{-3}\left(\frac{\sum_{i=k+1}^n y_i^2}{n-k} - 1\right)^3,$$

$\xi_1^{(1)}$ is between $\frac{\sum_{i=1}^n y_i}{n}$ and 1, $\xi_1^{(2)}$ is between $\frac{\sum_{i=1}^k y_i}{n}$ and 1, and $\xi_1^{(3)}$ is between $\frac{\sum_{i=k+1}^n y_i}{n}$ and 1. This decomposition is helpful to establish the following theorem.

Theorem 2.2.1. *Under the null hypothesis H_0, as $n \to \infty, k \to \infty$, and $(k/n) \to \infty$; then for all $x \in R$,*

$$\lim_{n\to\infty} P\left[a(\log n)\lambda_n - b(\log n) \le x\right] = \exp\left\{-2e^{-x}\right\},$$

where $a(\log n) = (2 \log\log n)^{1/2}$, $b(\log n) = 2\log\log n + \frac{1}{2}\log\log\log n - \log\Gamma\left(\frac{1}{2}\right)$.

Using this limiting distribution we can evaluate the test P-value easily for practical data. We need the following lemmas to prove the main theorem (see Lemma 2.14–2.18 of [51]).

Lemma 2.3. (i) $\max_{1<k<n} k^{1/2}(\log\log k)^{-3/2}\left|Q_k\right| = O_p(1)$.

(ii) $\max_{1<k<n}(n-k)^{1/2}(\log\log(n-k))^{-3/2}\left|R_k\right| = O_p(1)$.

Lemma 2.4. *For all $x \in R$, as $n \to \infty$, the following hold.*

(i) $a^2(\log n)^2 \max_{1<k<\log n} W_k - (x + b(\log n))^2 \xrightarrow{P} -\infty$.

(ii) $a^2(\log n)^2 \max_{1<k<\log n} L_k - (x + b(\log n))^2 \xrightarrow{P} -\infty$.

(iii) $a^2(\log n)^2 \max_{n-\log n<k<n} W_k - (x + b(\log n))^2 \xrightarrow{P} -\infty$.

(iv) $a^2(\log n)^2 \max_{n-\log n<k<n} L_k - (x + b(\log n))^2 \xrightarrow{P} -\infty$.

Lemma 2.5. *As $n \to \infty$, the following hold.*

(i) $a^2(\log n)^2 \max_{\log n<k<n-\log n}\left|L_k - W_k\right| = o_p(1)$.

(ii) $a^2(\log n)^2 \max_{1<k<n/\log n}\left|\frac{1}{(n-k)}\left[\sum_{i=k+1}^n (y_i^2 - 1)\right]^2 - \frac{1}{n}\left[\sum_{i=1}^n (y_i^2 - 1)\right]^2\right| = o_p(1)$.

Lemma 2.6. *As $n \to \infty$, for all $x \in R$,*

$$a^2(\log n)^2 \max_{n/\log n<k<n-n/\log n} W_k - (x + b(\log n))^2 \xrightarrow{P} -\infty.$$

Lemma 2.7. *Suppose* $\left\{\left(z_i^{(1)},\ldots,z_i^{(d)}\right), 1\le i<\infty\right\}$ *be independent and identically distributed random vectors, and define* $S^{(j)}(k)=\sum_{i=1}^{k} z_i^{(i)}, 1\le i\le d$. *Assume* $E\left[z_i^{(1)}\right]=E\left[z_i^{(2)}\right]=\ldots=E\left[z_i^{(d)}\right]=0$, *covariance matrix of* $\left(z_i^{(1)},\ldots,z_i^{(d)}\right)$ *is the identity matrix, and* $\max_{1\le j\le d} E\left|z_i^{(j)}\right|^r<\infty$ *for some* $r>2$. *As* $n\to\infty$,

$$a(\log n)\max_{1\le j\le d}\left(\sum_{j=1}^{d}\left[k^{-1/2}S^{(j)}(k)\right]^2\right)^{1/2}-b(\log n)\xrightarrow{d} y^*,$$

where y^* *is a random variable with a distribution function* $F_{y^*}(x)=\exp\left\{-e^{-x}\right\}$, *and "*$\xrightarrow{d}$*" means "convergence in distribution".*

Next, based on these lemmas, we will complete the proof of Theorem 2.2.1.

Proof of Theorem 2.2.1. We have

$$\begin{aligned}&\lim_{n\to\infty} P\left[a(\log n)\lambda_n-b(\log n)\le x\right]\\ &=\lim_{n\to\infty} P\left[a(\log n)\max_{1<k<n-1} L_k^{1/2}-b(\log n)\le x\right]\\ &=\lim_{n\to\infty} P\left[a^2(\log n)^2\max_{1<k<n-1} L_k\le[x+b(\log n)]^2\right].\end{aligned}$$

Using Lemma 2.4 (ii), (iv), we obtain

$$\max_{1<k<n} L_k\overset{D}{=}\max_{\log n\le k\le n-\log n} L_k.$$

Then,

$$\begin{aligned}&\lim_{n\to\infty} P\left[a^2(\log n)^2\max_{1<k<n-1} L_k\le[x+b(\log n)]^2\right]\\ &=\lim_{n\to\infty} P\left[a^2(\log n)^2\max_{\log n<k<n-\log n} L_k\le[x+b(\log n)]^2\right].\end{aligned}$$

Then, by using Lemma 2.3 and Lemma 2.5 (i), we obtain

$$\max_{\log n\le k\le n-\log n} L_k\overset{D}{=}\max_{\log n\le k\le n-\log n} W_k.$$

Therefore,

$$\begin{aligned}&\lim_{n\to\infty} P\left[a^2(\log n)^2\max_{\log n<k<n-\log n} L_k\le[x+b(\log n)]^2\right]\\ &=\lim_{n\to\infty} P\left[a^2(\log n)^2\max_{\log n\le k\le n-\log n} W_k\le[x+b(\log n)]^2\right].\end{aligned}$$

Next, we divide $\log n<k<n-\log n$ into three parts $\log n<k<n/\log n$, $n/\log n<k<n-n/\log n$, and $n/\log n<k<n-n/\log n<k<n-\log n$. From Lemma 2.5(ii), we obtain

$$\max_{\log n\le k\le n/\log n} W_k\overset{D}{=}\max_{\log n\le k\le n/\log n}\frac{1}{2k}\left[\sum_{i=1}^{k}\left(y_i^2-1\right)\right]^2,$$

$$\max_{n-n/\log n\le k\le n-\log n} W_k\overset{D}{=}\max_{n-n/\log n\le k\le n-\log n}\frac{1}{2(n-k)}\left[\sum_{i=1}^{k}\left(y_i^2-1\right)\right]^2.$$

And from Lemma 2.6, we obtain

$$\max_{\log n\le k\le n-\log n} W_k \overset{D}{=} \max\left\{\left[\max_{\log n\le k\le n/\log n} W_k\right],\left[\max_{n-n/\log n\le k\le n-\log n} W_k\right]\right\}.$$

We derive

$$\begin{aligned}
&\lim_{n\to\infty} P\left[a^2(\log n)^2 \max_{\log n\le k\le n-\log n} W_k \le [x+b(\log n)]^2\right]\\
&=\lim_{n\to\infty} P\left[a^2(\log n)^2 \max\left\{\max_{1<k\le n/\log n}\frac{1}{2k}\left[\sum_{i=1}^{k}(y_i^2-1)\right]^2,\right.\right.\\
&\left.\left.\max_{n-n/\log n\le k<n}\frac{1}{2(n-k)}\left[\sum_{i=k+1}^{n}(y_i^2-1)\right]^2\right\}\le [x+b(\log n)]^2\right].
\end{aligned}$$

Since $\{y_i, 1\le i\le k, \log n\le k\le n/\log n\}$ and $\{y_j, k+1\le j\le n, n-n/\log n\le k\le n-\log n\}$ are independent, we can simplify the above formula as

$$\begin{aligned}
&\lim_{n\to\infty} P\left[a(\log n)\max_{1<k\le n/\log n}\frac{1}{\sqrt{2k}}\left|\sum_{i=1}^{k}(y_i^2-1)\right|-b(\log n)\le x\right]\times\\
&\lim_{n\to\infty} P\left[a(\log n)\max_{1<k\le n/\log n}\frac{1}{\sqrt{2(n-k)}}\left|\sum_{i=k+1}^{n}(y_i^2-1)\right|-b(\log n)\le x\right].
\end{aligned}\tag{2.10}$$

We consider the first term of (2.10). Define $t_i=(y_i^2-1)/\sqrt{2}$, we observe that $E(t_i)=0, Var(t_i)=1, E|t_i|^r<\infty$. Because $t_1,t_2,\ldots,t_k$ are independent, covariance matrix of $(t_1,t_2,\ldots,t_k)$ is the identity matrix. From Lemma 2.7, we derive

$$a(\log n)\max_{\log n<k\le n/\log n}\frac{1}{\sqrt{2k}}\left|\sum_{i=1}^{k}(y_i^2-1)\right|-b(\log n)\xrightarrow{d} y^*.$$

For the second item, the same method can be used to obtain

$$a(\log n)\max_{n-n\log n<k\le n-\log n}\frac{1}{\sqrt{2(n-k)}}\left|\sum_{i=1}^{k}(y_i^2-1)\right|-b(\log n)\xrightarrow{d} y^*.$$

Finally, (2.10) equals $\exp\{-2e^{-x}\}$, which completes the proof of the theorem. □

We may further consider the multivariate version of the above test. Let $\boldsymbol{x}_1,\boldsymbol{x}_2,\ldots,\boldsymbol{x}_n$ be a sequence of independent m-dimensional normal random vectors with parameters $(\boldsymbol{\mu}_1,\boldsymbol{\Sigma}_1),(\boldsymbol{\mu}_2,\boldsymbol{\Sigma}_2),\ldots,(\boldsymbol{\mu}_n,\boldsymbol{\Sigma}_n)$, respectively. In this section, we assume $\boldsymbol{\mu}_1=\boldsymbol{\mu}_2=\ldots=\boldsymbol{\mu}_n=\boldsymbol{\mu}$ and $\boldsymbol{\mu}$ is known. Without loss of generality, assume $\boldsymbol{\mu}=\mathbf{0}$. We desire to test the following hypothesis

$$H_0:\boldsymbol{\Sigma}_1=\boldsymbol{\Sigma}_2=\cdots=\boldsymbol{\Sigma}_n=\boldsymbol{\Sigma}\tag{2.11}$$

versus

$$H_1:\boldsymbol{\Sigma}_1=\cdots=\boldsymbol{\Sigma}_k\ne\boldsymbol{\Sigma}_{k+1}=\cdots=\boldsymbol{\Sigma}_n\ne\boldsymbol{\Sigma}.\tag{2.12}$$

Under H_0, the log-likelihood function is

$$\log L_0(\Sigma)=-\frac{1}{2}mn\log 2\pi-\frac{n}{2}\log|\Sigma|-\frac{1}{2}\sum_{i=1}^{n}\mathbf{x}_i'\Sigma^{-1}\mathbf{x}_i,$$

and the maximized log-likelihood with the unknown covariance replaced by the empirical estimate is

$$\log L_0(\widehat{\Sigma}) = -\frac{1}{2}mn\log 2\pi - \frac{n}{2}\log\left|\frac{1}{n}\sum_{i=1}^{n}\mathbf{x}_i\mathbf{x}_i'\right| - \frac{n}{2}.$$

where $\widehat{\Sigma} = (1/n)\sum_{i=1}^{n}\mathbf{x}_i\mathbf{x}_i'$.

Under H_1, with a fixed change position k, the log-likelihood function is

$$\begin{aligned}\log L_1(\Sigma_1, \Sigma_n) = & -\frac{mn}{2}\log 2\pi - \frac{k}{2}\log|\Sigma_1| - \frac{n-k}{2}\log|\Sigma_n| \\ & -\frac{1}{2}\left[\sum_{i=1}^{k}\mathbf{x}_i'\Sigma_1^{-1}\mathbf{x}_i + \sum_{i=k+1}^{n}\mathbf{x}_i'\Sigma_n^{-1}\mathbf{x}_i\right],\end{aligned}$$

and the maximum log-likelihood with the unknown covariances replaced by the empirical estimates is:

$$\begin{aligned}\log L_1\left(\widehat{\Sigma}_1, \widehat{\Sigma}_n\right) = & -\frac{mn}{2}\log 2\pi - \frac{k}{2}\log\left|\frac{1}{k}\sum_{i=1}^{k}\mathbf{x}_i\mathbf{x}_i'\right| \\ & -\frac{n-k}{2}\log\left|\frac{1}{n-k}\sum_{i=k+1}^{n}\mathbf{x}_i\mathbf{x}_i'\right| - \frac{n}{2}.\end{aligned}$$

where

$$\widehat{\Sigma}_1 = \frac{1}{k}\sum_{i=1}^{k}\mathbf{x}_i\mathbf{x}_i' \quad \text{and} \quad \widehat{\Sigma}_n = \frac{1}{n-k}\sum_{i=k+1}^{n}\mathbf{x}_i\mathbf{x}_i'.$$

Then the likelihood ratio test statistic is given by

$$\lambda_n = \max_{m<k<n-m}\left(\log\frac{\left|\frac{1}{n}\sum_{i=1}^{n}\boldsymbol{x}_i\boldsymbol{x}_i^\top\right|^n}{\left|\frac{1}{k}\sum_{i=1}^{k}\boldsymbol{x}_i\boldsymbol{x}_i^\top\right|^k\left|\frac{1}{n-k}\sum_{i=k+1}^{n}\boldsymbol{x}_i\boldsymbol{x}_i^\top\right|^{n-k}}\right)^{1/2}.$$

Through a similar calculation as for the univariate case, we can obtain the asymptotic null distribution of the test statistics in the following theorem.

Theorem 2.2.2. *Under the null hypothesis H_0, when $n \to \infty$,*

$$\lim_{n\to\infty} P\left\{a(\log n)\lambda_n - b_m(\log n) \le x\right\} = \exp\left\{-2e^{-x}\right\},$$

for all $x \in R$ where $a(\log n) = (2\log\log n)^{1/2}$ and

$$b_m(\log n) = 2\log\log n + (m/2)\log\log\log n - \log\Gamma(m/2).$$

The proof of Theorem 2.2.2 is very similar to the one-dimensional case, and the detailed proof can refer to [51].

2.2.2 Autoregressive models

Autoregressive (AR) models are the most popular statistical models for analyzing time series data [199]. The AR model relies only on past period values to predict current ones. It is used for forecasting when there is some correlation between values in a random time series. We only use past data to model their behavior, hence the name autoregressive (the Greek prefix "auto", meaning "self"). The process is basically a linear regression of the data in the current series against one or more past values in the same series. The model is often denoted as AR(q), where q indicates the order of the model, representing the number of lags included.

In our study, we address the challenge of determining if a change has occurred in the parameters or the order of an AR model. Leveraging the approach in [75], we propose an algorithm designed to detect variance changes in autoregressive settings, facilitating the identification of shifts in the model structure over time.

Let $X_1, \ldots, X_n$ be n consecutive observations from the following AR model with a change of the model structure at $t = k$:

$$\begin{aligned} X_t &= \phi_0 + \phi_1 X_{t-1} + \ldots + \phi_q X_{t-q} + \varepsilon_t,\ t \leq k, \\ &= \alpha_0 + \alpha_1 X_{t-1} + \ldots + \alpha_{q_1} X_{t-q_1} + \delta\varepsilon_t,\ t \geq k+1, \end{aligned} \tag{2.13}$$

where $\{\varepsilon_t\}$ is a sequence of independent, identically distributed random variables with the common mean 0 and variance σ^2 and $k \in (q, n]$. In addition to having different regression coefficients, the orders of the two AR models may not be the same, i.e. $q \neq q_1$. In general, the two segments defined in (2.13) would imply two different variance structures for the series $\{X_t\}$. It is often of interest to examine whether there exists a change point for an observed time series. The change point problem for this AR setting can be thus described as follows

$$H_0 : k = n \longleftrightarrow H_1 : k < n.$$

When H_0 is true the variance of the process $\{X_t\}$ would remain constant throughout the whole time period. On the other hand, the variance would experience a change at a finite time point $t = k$, due to the change in ϕ (into α), q (into q_1) and/or $\delta \neq 1$. To test this hypothesis, we may construct a CUSUM type test statistic. First, we define

$$Q(k) = \min_{\phi} \sum_{t=p+1}^{k} \left(X_t - \phi_0 - \phi_1 X_{t-1} - \ldots - \phi_q X_{t-q}\right)^2,$$

then it is easy to see the following matrix representation

$$Q(k) = \boldsymbol{X}_k^\top \boldsymbol{X}_k - \boldsymbol{X}_k^\top \boldsymbol{M}_k \left(\boldsymbol{M}_k^\top \boldsymbol{M}_k\right)^{-1} \boldsymbol{M}_k^\top \boldsymbol{X}_k = \boldsymbol{\varepsilon}_k^\top \boldsymbol{\varepsilon}_k - \boldsymbol{\varepsilon}_k^\top \boldsymbol{M}_k \left(\boldsymbol{M}_k^\top \boldsymbol{M}_k\right)^{-1} \boldsymbol{M}_k^\top \boldsymbol{\varepsilon}_k,$$

where $\boldsymbol{\phi} = (\phi_0, \phi_1, \ldots, \phi_q)^\top, \boldsymbol{X}_k = (X_{q+1}, \ldots, X_k)^\top, \boldsymbol{\varepsilon}_k = (\varepsilon_{q+1}, \ldots, \varepsilon_k)^\top$, and

$$\boldsymbol{M}_k = \begin{pmatrix} 1 & X_p & X_{q-1} & \ldots & X_1 \\ 1 & X_{q+1} & X_q & \ldots & X_2 \\ \vdots & \vdots & \vdots & \vdots & \vdots \\ 1 & X_{k-1} & X_{k-2} & \ldots & X_{k-q}, \end{pmatrix}.$$

Based on the above notations, the test statistic $\max_k D_k(q)$ will be provided in the following theorem along with its asymptotic property.

Theorem 2.2.3. *Under H_0, if $E\varepsilon_t^4 \equiv \eta_4 < \infty$, let*

$$D_k(q) = \sqrt{(n-2q-1)/2}\left|\frac{Q(k)}{Q(n)} - \frac{k-2q-1}{n-2q-1}\right|,$$

Then

$$\max_k D_k(q) \Rightarrow \sqrt{\frac{\eta_4 - \sigma^4}{2\sigma^4}} \sup_r |B_0(r)|,$$

where $B_0(r)$ is a Brownian bridge and $\Rightarrow$ stands for the weak convergence of the associated probability measures.

Proof of Theorem 2.2.3. Let $\Lambda_n(k) = \boldsymbol{\varepsilon}_k^\top \boldsymbol{M}_k \left(\boldsymbol{M}_k^\top \boldsymbol{M}_k\right)^{-1} \boldsymbol{M}_k^\top \boldsymbol{\varepsilon}_k$. Setting $\boldsymbol{S}_k = \boldsymbol{M}_k^\top \boldsymbol{\varepsilon}_k$ and $\boldsymbol{P}_k = \left(\boldsymbol{M}_k^\top \boldsymbol{M}_k\right)^{-1}$, we have

$$\Lambda_n(k) = \boldsymbol{S}_k^\top \boldsymbol{P}_k \boldsymbol{S}_k.$$

To derive the asymptotic behavior of the likelihood ratio statistic under H_0, we consider the process $Q_n(t) := \Lambda_n([nt]), t \in [0,1]$.

We assume $\sup_t E\left|\varepsilon_t\right|^{4+\delta} < \infty$ for some $0 < \delta \leq 1$. The asymptotic of $Q_n(t)$ is essentially governed by $\boldsymbol{S}_{[nt]}$ and $\boldsymbol{P}_{[nt]}$. Using Theorem 4 in [151] to the sequence $\boldsymbol{\xi}_n = \boldsymbol{S}_n - \boldsymbol{S}_{n-1}$, there exists an i.i.d. Gaussian random vector sequence $\{\boldsymbol{Z}_n\}$, which may be defined as a new probability space with a mean of zero and covariance matrix $\boldsymbol{\Gamma}_{q+1}$ such that

$$\boldsymbol{S}_k - \boldsymbol{U}_k \overset{\text{a.s.}}{=} O\left(k^{1/2-\lambda}\right) \quad \text{as } k \to \infty,$$

where $\boldsymbol{\Gamma}_{q+1} = [\gamma_{ij}]_{i,j=1,\ldots,q+1}$,

$$\Gamma_{ij} = \begin{cases} 1, & \text{if } i = j = 1, \\ \mu, & \text{if } i = 1, j > 1 \text{ or } i > 1, j = 1, \\ E\left(X_i X_j\right), & \text{if } i > 1, j > 1, \end{cases}$$

and for some $\lambda > 0$, $\boldsymbol{U}_k = \sum_{i=q+1}^k \boldsymbol{Z}_i$. It is immediate that

$$\boldsymbol{\Gamma}_{q+1}^{-1/2} \boldsymbol{S}_{[n\cdot]}/\sqrt{n} \Rightarrow \mathbf{B}(\cdot) \tag{2.14}$$

in $D^{p+1}[0,1]$.

Considering $\boldsymbol{P}_n$, the (i,j) $(i > 1$ and $j > 1)$ component of $\boldsymbol{P}_n^{-1}$ is $\sum_{s=1}^{n-q} X_{q+1-i+s} X_{q+1-j+s}$. Using Theorem 4 of [151] mentioned earlier again to the strongly mixing sequence $\{X_{q+1-i+s} X_{q+1-j+s} - E\left(X_i X_j\right), s = 1, 2, \ldots\}$, we have

$$\sum_{s=1}^{n-q} \left(X_{q+1-i+s} X_{q+1-j+s} - E\left(X_i X_j\right)\right) - \mathbf{V}(n-q) \overset{\text{a.s.}}{=} O\left(n^{1/2-\lambda}\right),$$

as $n \to \infty$, where $\mathbf{V}(t)$ is Brownian motion with variance

$$\operatorname{Var}\left(X_i X_j\right) + 2 \sum_{s=1}^{\infty} \operatorname{Cov}\left(X_{-i} X_{-j}, X_{s-i} X_{s-j}\right).$$

Because this argument can be used for all of the components of $\boldsymbol{P}_n^{-1}$, it follows easily that

$$n \boldsymbol{P}_{[nt]} \xrightarrow{p} t^{-1} \boldsymbol{\Gamma}_{q+1}^{-1}. \tag{2.15}$$

We conclude from (2.14) and (2.15), derive $Q_n(t) \Rightarrow \|\mathbf{B}(t)\|^2/t$ in $D[t_1, t_2]$, $0 < t_1 < t_2 \leq 1$, where $\mathbf{B}(\cdot)$ is $(q+1)$-dimensional Brownian motion with covariance matrix $\boldsymbol{I}_{q+1}$. As $n \to \infty$, $\Lambda_n(k)/\sqrt{n-q} \xrightarrow{p} 0, k = 1, \ldots, n$. Then, we have

$$\max_k \sqrt{(n-q)/2} \left| \frac{Q(k)}{Q(n)} - \frac{k-q}{n-q} \right| \Rightarrow \sqrt{\frac{\eta_4 - \sigma^4}{2\sigma^4}} \sup_r \left|B_0(r)\right|. \tag{2.16}$$

When n is small, $\Lambda_n(k)$ can't be omitted. As

$$\operatorname{tr}\left(\boldsymbol{I}_{k-p} - \boldsymbol{M}_k \left(\boldsymbol{M}_k^\top \boldsymbol{M}_k\right)^{-1} \boldsymbol{M}_k^\top\right) = k - 2q - 1 \quad \text{and} \quad q/n \to 0,$$

we transform statistic (2.16) to

$$\max_k \sqrt{(n-2q-1)/2}\left|\frac{Q(k)}{Q(n)} - \frac{k-2q-1}{n-2q-1}\right|, \tag{2.17}$$

which converges weakly to the supremum of a Brownian bridge. □

Let k^* be the value which optimizes $D_k(q)$ over k. If $\max_k D_k(q)$ exceeds a predetermined boundary, we may then declare that there is a change point near k^* and take k^* as an estimate of the change point.

In Theorem 2.2.3, η_4 and σ^4 are unknown parameters and can be consistently estimated by sample moments. In particular, when $\{\varepsilon_t\}$ is a sequence of random variable with normal distribution, we have $\sqrt{(\eta_4 - \sigma^4)/(2\sigma^4)} = 1$.

Table 2.1: Empirical 95th quantiles of $\max_k D_k(q)$.

		$q=1$			$q=2$
n	$q=0$	$\phi_1 = 0.2$	$\phi_1 = 0.5$	$\phi_1 = 0.7$	$\phi_1 = 0.5$ $\phi_2 = 0.1$
50	1.2443 (0.0006)	1.2222 (0.0005)	1.2310 (0.0006)	1.2439 (0.0005)	1.2343 (0.0007)
100	1.2738 (0.0007)	1.2805 (0.0006)	1.2779 (0.0006)	1.3065 (0.0005)	1.2711 (0.0006)
200	1.3019 (0.0005)	1.2910 (0.0006)	1.2992 (0.0005)	1.3075 (0.0005)	1.2969 (0.0006)
400	1.3165 (0.0006)	1.3190 (0.0007)	1.3125 (0.0006)	1.3068 (0.0007)	1.3227 (0.0005)

Notes: Estimated from 10,000 replicates of series (1) where ϵ_t independent $N(0,1)$, $\phi_0 = 0.1$ and $\phi_1 = 0.5$; numbers in parentheses are standard errors.

In Table 2.1, we present the empirical 95th quantiles of $\max_k D_k(q)$, generated from Monte Carlo simulations across different series lengths n and autoregressive orders q, with $M = 10000$ repetitions. These results are compared to the theoretical asymptotic 95th quantile of $\sup_r |B_0(r)|$, where $P(\sup_r |B_0(r)| < 1.358) = 0.95$. The finite-dimensional 95th percentile of $\max_k D_k(q)$ consistently falls below the asymptotic threshold of 1.358, which translates into a smaller empirical type I error when testing for variance changes. Notably, for series lengths $n \geq 200$, the asymptotic approximation appears to align well with practical data, indicating its reliability for use in real-world scenarios.

When the order q is unknown, Davis, Huang and Yao (1995) [75] proposed the following procedure to search the change point by adopting a special BIC.

Algorithm 1 BIC-CSS algorithm

1: Transform the data to satisfy the normality assumption by the Box-Cox transformation or the Johnson transformation.

2: For each $q \in [0, q_c]$, calculate $D_k(q)$. Let k_q be a point at which $\max_k D_k(q)$ is obtained, and let

$$\mathrm{BIC}(q) = (k_q - q)\log\left(\frac{Q(k_q)}{k_q - q}\right) + (q+1)\log(k_q - q).$$

3: Let $\hat{q}$ be a point at which $\min_q \mathrm{BIC}(q)$ is obtained.

4: If $\max_k D_k(\hat{q}) > 1.358$, declare that there is a change point $\hat{\tau}$ at which $\max_k D_k(\hat{q})$ is obtained. If $\max_k D_k(\hat{q}) \leq 1.358$, there is no change point in the series.

2.3 Change in distribution

Consider an independent random process $\{x(t), 0 \leq t \leq 1\}$. Let $\tau \in (0,1)$ be a potential change point. When $0 < t \leq \tau$, $x(t)$ comes from a distribution F_1, and when $\tau < t < 1$, $x(t)$ comes from another distribution F_2. If F_1 and F_2 are not identical, then τ is indeed a change point of the process. To detect whether the process has a change point, we sequentially observe the process $\{x(t), 0 \leq t \leq 1\}$ at discrete points, say, $t = \frac{1}{n+1}, \frac{2}{n+1}, \cdots, \frac{n}{n+1}$ and obtain the observed series values $x\left(\frac{i}{n+1}\right) \equiv x_{ni}, i = 1, 2, \cdots, n$. The problem to be discussed in this section is: given $\{x_{ni}, i = 1, 2, \cdots, n\}$, how shall we determine whether the two distributions F_1 and F_2, separated by the point τ, are identical; in another word, how shall we estimate the location of the change point based on the observed values of the process?

If both pre- and post-change data are from the same family of distributions (e.g., normal distributions with differing means or variances), then the methods introduced in the previous sections can be applied. When the data changes to an entirely different family of distributions, such as shifting from a normal to a chi-squared or exponential distribution, parametric methods may struggle due to mis-specification of the underlying models. See Figure 2.2 for an illustration. In theory, there are plenty of sensible statistical distributions for modeling series. In real-world scenarios, data might not neatly fit into a known distribution class, or the distribution may evolve. If it is hard to pre-assume an appropriate parametric distribution for the data, we may consider a more general non-parametric approach based on empirical data.

We note that $x_{n1}, \cdots, x_{nn}$ are independently selected and should in general depend on the total length n. But for the sake of presentation simplicity, in the following, we write x_{ni} as $x_i, i = 1, \cdots, n$. Similar to the previous sections, our problem can again be regarded as a hypothesis-testing problem.

Let $x_1, x_2, \cdots, x_n$ be independent, and we have assumed that

$$x_1, \cdots, x_{[n\tau]} \text{ iid. } \sim F_1, \qquad x_{[n\tau]+1}, \cdots, x_n \text{ iid. } \sim F_2. \tag{2.18}$$

The null hypothesis is $H_0 : F_1 = F_2$, for any τ specified in $(0,1)$, and the alternative hypothesis is H_1 : there exists a point τ, such that $F_1 \neq F_2$.

First, we introduce some notations. Let

$$F = \tau F_1 + (1 - \tau) F_2.$$

Since F_1 and F_2 are continuous functions, F is also continuous. We further write the empirical distribution functions as

$$\hat{F}_k(x) = \frac{1}{k} \sum_{i=1}^{k} I\left(x_i \leq x\right), k = 1, \ldots, n,$$

$$\breve{F}_{n-k}(x) = \frac{1}{n-k} \sum_{i=k+1}^{n} I\left(x_i \leq x\right), k = 0, 1, \ldots, n-1.$$

where $I(\cdot)$ is indicative function. In fact, $\hat{F}_k(x)$ is the empirical distribution function of observations of the first k samples $x_1, \ldots, x_k$ while $\breve{F}_{n-k}(x)$ is the empirical distribution function of observations of the last $n-k$ samples $x_{k+1}, \ldots, x_n$. Their corresponding empirical quantile functions are

$$\hat{F}_k^{-1}(y) = \inf\left\{x : \hat{F}_k(x) \geq y\right\}, \quad 0 < y < 1,$$

$$\breve{F}_{n-k}^{-1}(y) = \inf\left\{x : \breve{F}_{n-k}(x) \geq y\right\}, \quad 0 < y < 1.$$

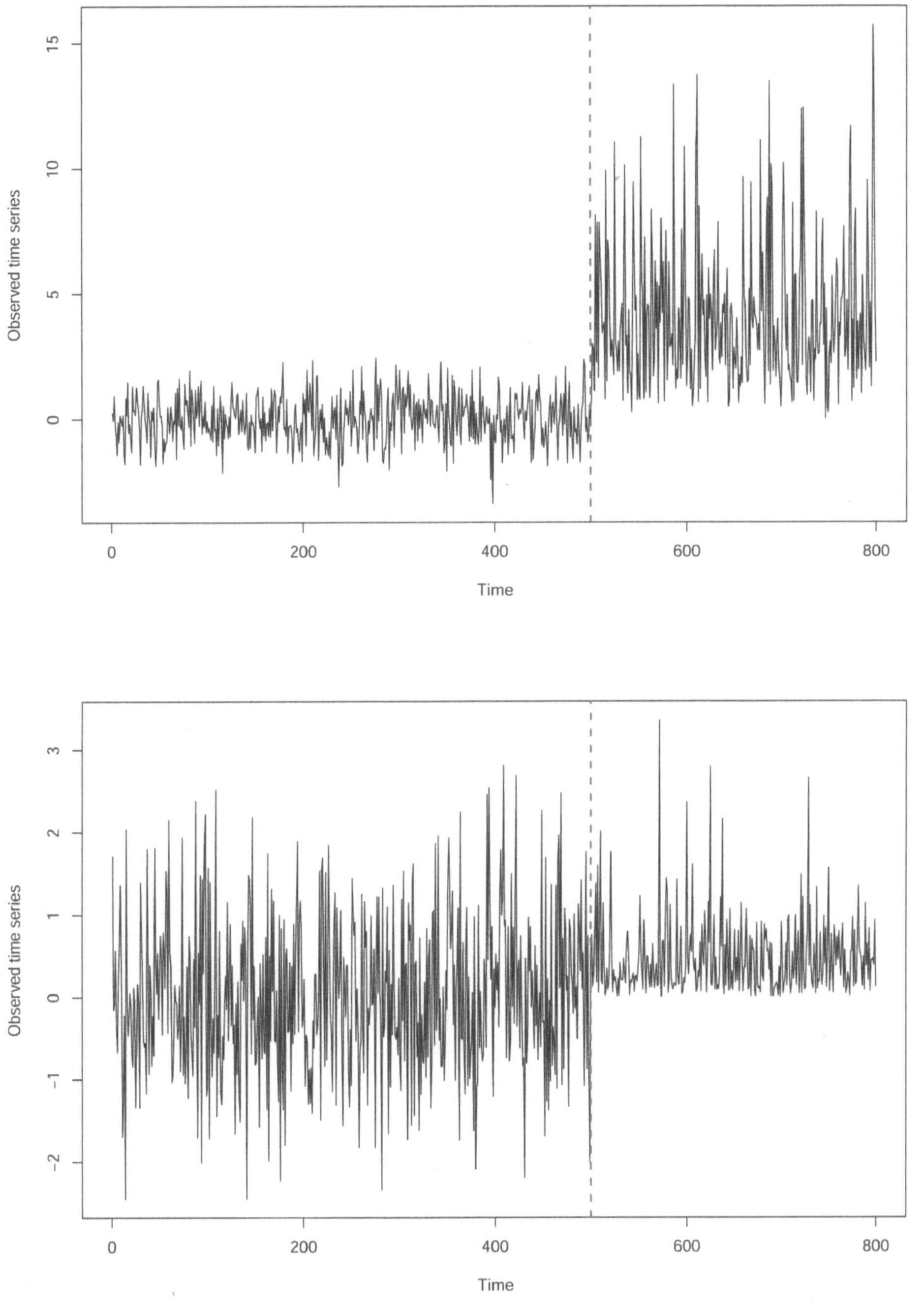

Figure 2.2: Comparison between changes from normal distribution to Chi-squared (top row) and exponential (bottom row) distributions.

2.3.1 Test the null hypothesis

To test H_0, when F_1 is known, define the test statistic to be

$$D_{nk}(x) = \sqrt{n-k}\left(\breve{F}_{(n-k)}(x) - F_1(x)\right), \quad k = 0, 1, \ldots, n-1,$$

and define the corresponding test process

$$D_n(t,x) = \frac{1}{\sqrt{n-[nt]}} \sum_{i=[nt]+1}^{n} \left\{I\left(x_i \le F_1^{-1}(x)\right) - x\right\}, \quad 0 < t < 1, 0 < x < 1.$$

In practice, the pre-change distribution and its density are often known while the post-change distribution may be unknown and could depend on the change point itself [243].

Remark 2.1. *When F_1 is unknown, we have to incorporate the uncertainty in estimating this function. In the empirical processes literature, [70] proposed a test based on a version of the Wilcoxon-type test process*

$$\tilde{D}_n(t,x) = \frac{1}{\sqrt{n}} \sum_{i=1}^{[nt]} \left\{I\left(x_i \le \hat{F}_n^{-1}(x)\right) - x\right\}, \quad 0 < t < 1, 0 < x < 1,$$

where $\hat{F}_n^{-1}(y) = \inf\left\{x : \hat{F}_n(x) \ge y\right\}$, $0 < y < 1$. If H_0 holds, $\tilde{D}_n(t,x)$ weakly converges to a Gaussian processes $\Gamma(t,x)$ with $E\Gamma(t,x) = 0$ and $E[\Gamma(t,u)\Gamma(s,v)] = (\min(s,t) - st)(\min(u,v) - uv)$. Such a result may facilitate the significance test even if we replace the true F_1 by the empirical function $\hat{F}_1$. To simplify our arguments, we assume F_1 is given in this section and leave out those complicated details.

Theorem 2.3.1. *If H_0 is true*

$$\sup_{0<t<1} \sup_{0<x<1} |D_n(t,x)| \Rightarrow \sup_{0 \le x \le 1} |B_0(x)|, \quad n \to \infty, \tag{2.19}$$

where $B_0(x)$ is a standard Brownian bridge.

Proof of Theorem 2.3.1. Due to the continuity of F, $U_i = F(x_i)$, $1 \le i \le n$ are independent uniform $(0,1)$ random variables. Let

$$\breve{E}_j(x) = \frac{1}{n-m} \sum_{i=n-j+1}^{n} I\left(U_i \le x\right), \quad 1 < j < n.$$

Under H_0,

$$\sum_{i=k+1}^{n} \left\{I\left(x_i \le F_1^{-1}(x)\right) - x\right\} = (n-k)\left(\breve{E}_{n-k}(x) - x\right),$$

derive

$$D_n(t,x) = \sqrt{n-k}\left(\breve{E}_{n-k}(x) - x\right).$$

According to Donsker's theorem in [72],

$$\sqrt{n-k}\left(\breve{E}_{n-k}(x) - x\right) \Rightarrow B_0(x),$$

where $\Rightarrow$ indicates weak convergence. Because $\sup_t \sup_x |\cdot|$ is continuous functional, (2.19) holds. □

Theorem 2.3.1 provides a method to test H_0. Similar to the limiting distribution of the two-sided Kolmogorov-Smirnov statistic ([146]),

$$P\left(\sup_{0\leq x\leq 1}|B_0(x)|>u\right)=\sum_{k\neq 0}(-1)^k\exp(-2k^2u^2),\quad u>0,$$

we can derive $1-\alpha$ quantile $C(\alpha)$ when the significance level of the test is $\alpha\in(0,1)$. If and only if $\sup_{0<t<1}\sup_{0<x<1}|D_n(t,x)|>C(\alpha)$, we reject the null hypothesis H_0.

To discuss the consistency of the test, we consider the asymptotic properties of test statistic $D_n(t,x)$ when H_1 is true.

Theorem 2.3.2. *Under H_1, let $x_0\in(0,1)$ and $F_1^{-1}(x_0)\neq F_2^{-1}(x_0)$, F_1^{-1} is continues at x_0,*

$$\lim_{n\to\infty}\frac{1}{\sqrt{n}}D_n(t,x_0)\overset{a.s.}{=}M_\tau(t,x_0)=\begin{cases}\frac{1-\tau}{\sqrt{1-t}}\left(F_2\left(F_1^{-1}(x_0)\right)-x_0\right),0<t\leq\tau,\\ \sqrt{1-t}\left(F_2\left(F_1^{-1}(x_0)\right)-x_0\right),\tau\leq t<1.\end{cases}\tag{2.20}$$

Proof of Theorem 2.3.2. Let

$$\hat{G}_k(x)=\frac{1}{k}\sum_{j=[n\tau]-k+1}^{[n\tau]}I(x_i\leq x),\quad 1\leq k\leq[n\tau].$$

From the Glivenko-Cantelli theorem,

$$\sup_{-\infty<x<+\infty}\left|\hat{G}_k(x)-F_1(x)\right|\overset{a.s.}{=}o(1),$$

$$\sup_{-\infty<x<+\infty}\left|\breve{F}_{n-k}(x)-F_2(x)\right|\overset{a.s.}{=}o(1),\quad 1\leq k\leq n-r.$$

Following the definition of $D_n(t,x)$, when $0<t\leq\tau$,

$$\begin{aligned}\frac{1}{\sqrt{n}}D_n(t,x)=&\frac{[n\tau]-[nt]}{\sqrt{n(n-[nt])}}\left(G_{[n\tau]-[nt]}\left(F_1^{-1}(x)\right)-x\right)\\&+\frac{n-[n\tau]}{\sqrt{n(n-[nt])}}\left(\breve{F}_{n-[n\tau]}\left(F_1^{-1}(x)\right)-x\right),\end{aligned}\tag{2.21}$$

when $\tau<t<1$,

$$\frac{1}{\sqrt{n}}D_n(t,x)=\sqrt{\frac{n-[nt]}{n}}\left(\breve{F}_{n-[nt]}\left(F_1^{-1}(x)\right)-x\right).\tag{2.22}$$

Because F_1^{-1} is continuous at x_0, we derive $F_1(F_1^{-1}(x_0))=x_0$. From the Glivenko-Cantelli theorem, (2.21), and (2.22), we derive (2.20). □

2.3.2 Estimation of the change point

When the null hypothesis H_0 is rejected by the aforementioned significance test, the change point estimation can be obtained as

$$D_n(t)=\sup_{0<x<1}|D_n(t,x)|,$$
$$D_n=\sup_{0<t<1}D_n(t),$$
$$\hat{t}_0(n)=\inf\{s:D_n(s)=D_n\}.$$

The following theorem justifies the consistency of the change point estimation.

Theorem 2.3.3. *If H_1 is true, F_1 is strictly monotonic and* $\sup_{0<x<1} |F_2(F_1(x)) - x| > 0$,

$$\lim_{n\to\infty} \hat{t}_0(n) \overset{a.s.}{=} \tau.$$

Proof of Theorem 2.3.3. Because F_1 is strictly monotonic, F is strictly monotonic. From Theorem 2.3.2, we derive

$$\sup_{0<t<1} \sup_{0<x<1} \left| \frac{1}{\sqrt{n}} D_n(t,x) - M_\tau(t,x) \right| \overset{a.s.}{=} o(1). \tag{2.23}$$

Therefore,

$$\lim_{n\to\infty} \frac{1}{\sqrt{n}} D_n \overset{\text{a.s.}}{=} \sqrt{1-\tau} \sup_{0<x<1} \left| F_2\left(F_1^{-1}(x)\right) - x \right|. \tag{2.24}$$

From the definition of $\hat{t}_0(n)$,

$$D_n\left(\hat{t}_0(n)\right) = D_n. \tag{2.25}$$

We know that $\hat{t}_0(n)$ has a limit, by the definitions of $D_n(t,x)$ and $\hat{t}_0(n)$. If there exists a set $A, P(A) > 0$, such that (2.23) and (2.25) hold but $\hat{t}_0(n,\omega) \to \tau^* \neq \tau, \omega \in A$. Because $M_\tau(t,x)$ reaches its maximum value at τ, derive by (2.23), then

$$\begin{aligned}\lim_{n\to\infty} \frac{1}{\sqrt{n}} \sup_{0<x<1} \left| D_n\left(\hat{t}_0(n), x\right) \right| &\overset{a.s.}{=} \sup_{0<x<1} M_\tau\left(\tau^*, x\right) \\ &< \sqrt{1-\tau} \sup_{0<x<1} \left| F_2\left(F_1^{-1}(x)\right) - x \right|,\end{aligned}$$

which is contradicted with (2.24). Then $\hat{t}_0(n) \xrightarrow{a.s.} \tau$ when $n \to \infty$. The theorem has been proven. □

2.4 Change in regression model

Regression analysis is a foundational statistical technique used to understand the relationship between a dependent variable and one or more independent variables. By analyzing these relationships, it allows businesses and researchers to predict outcomes and assess the impact of changes in independent variables on the dependent variable. Key objectives of regression analysis include quantifying how much a unit change in one variable influences another and identifying patterns or trends within data. This process is akin to deriving a mathematical equation that best fits the data, enabling predictions and clarifying the impact of specific factors on an outcome. If the underlying model structure changes after a structure break due to external events, then a single constant slope linear regression model may be unsuitable to explain the observed data. We have presented a few examples in section 1.2 on this subject. Research on switching in regression models parameters has been an active area. An `R` package `cpop` was specifically developed by Daniel Grose and Paul Fearnhead for this purpose [96]. This package implements a dynamic programming algorithm to find the optimal set of changes that minimizes an L_0 penalized cost. The core cost function used in this implementation is a weighted residual sum of squares, which helps in accurately determining the piecewise-linear structure of the data. The package can analyze data that is unevenly spaced, allow for heterogeneous noise variance, and accommodate a grid of potential change locations to be different from the locations of the sample data points. In this section we introduce some basic tools for this problem.

2.4.1 Least squared method

Many types of business data are naturally collected over time, resulting in sequential observations that form a time series. This structure is prevalent in areas such as financial markets, sales forecasting, customer behavior tracking, and economic indicators. Sequential data recording allows analysts to track changes, detect trends, and make future predictions based on historical patterns. Consider the following linear regression with a change point at an index k_0:

$$\begin{aligned} y_t &= \boldsymbol{w}_t^\top \boldsymbol{\alpha} + \boldsymbol{z}_t^\top \boldsymbol{\delta}_1 + \epsilon_t, \quad t = 1, 2, \ldots, k_0, \\ y_t &= \boldsymbol{w}_t^\top \boldsymbol{\alpha} + \boldsymbol{z}_t^\top \boldsymbol{\delta}_2 + \epsilon_t, \quad t = k_0 + 1, \ldots, n, \end{aligned} \tag{2.26}$$

where y_t is an observation on the dependent variable, $\boldsymbol{w}_t$ and $\boldsymbol{z}_t$ are $p-$ and $q-$vectors of regressors, and ϵ_t is a random disturbance. $\boldsymbol{\alpha}$, $\boldsymbol{\delta}_1$, $\boldsymbol{\delta}_2$ are unknown parameters. We assume $\boldsymbol{\delta}_1 \neq \boldsymbol{\delta}_2$, so that there is indeed a change in slope. The change point location k_0 is unknown to users and we need to estimate this parameter along with other regression parameters. Let $\boldsymbol{x}_t = (\boldsymbol{w}_t^\top, \boldsymbol{z}_t^\top)^\top, \boldsymbol{\beta} = (\boldsymbol{\alpha}^\top, \boldsymbol{\delta}_1^\top)^\top$ a $p+q$-dimensional vector, and $\boldsymbol{\delta} = \boldsymbol{\delta}_2 - \boldsymbol{\delta}_1$. Then, (2.26) can be equivalently rewritten as

$$\begin{aligned} y_t &= \boldsymbol{x}_t^\top \boldsymbol{\beta} + \epsilon_t, & t &= 1, 2, \ldots, k_0 \\ y_t &= \boldsymbol{x}_t^\top \boldsymbol{\beta} + \boldsymbol{z}_t^\top \boldsymbol{\delta} + \epsilon_t, & t &= k_0 + 1, \ldots, n. \end{aligned} \tag{2.27}$$

Let $\boldsymbol{z}_t = \boldsymbol{R}^\top \boldsymbol{x}_t$, where $\boldsymbol{R} = (\boldsymbol{0}, \boldsymbol{I})^\top$ is a $p \times q$ known matrix with full column rank and $\boldsymbol{I}$ is an identity matrix.

Let $\boldsymbol{Y} = (y_1, ..., y_n)^\top$, $\boldsymbol{X} = (\boldsymbol{x}_1, ..., \boldsymbol{x}_n)^\top$, $\boldsymbol{\epsilon} = (\epsilon_1, ..., \epsilon_n)^\top$, $\boldsymbol{X}_1 = (\boldsymbol{x}_1, ..., \boldsymbol{x}_k, \boldsymbol{0}, ...\boldsymbol{0})^\top$, $\boldsymbol{X}_2 = (\boldsymbol{0}, ..., \boldsymbol{0}, \boldsymbol{x}_{k+1}, ..., \boldsymbol{x}_n)^\top$, and $\boldsymbol{X}_0 = (\boldsymbol{0}, ..., \boldsymbol{0}, \boldsymbol{x}_{k_0+1}, ..., \boldsymbol{x}_n)^\top$. The matrices $\boldsymbol{X}_1$ and $\boldsymbol{X}_2$ both depend on a candidate position k. Define $\boldsymbol{Z}_1 = \boldsymbol{X}_1\boldsymbol{R}$, $\boldsymbol{Z}_2 = \boldsymbol{X}_2\boldsymbol{R}$, and $\boldsymbol{Z}_0 = \boldsymbol{X}_0\boldsymbol{R}$. We can rewrite (2.27) as

$$\boldsymbol{Y} = \boldsymbol{X}\boldsymbol{\beta} + \boldsymbol{Z}_0\boldsymbol{\delta} + \boldsymbol{\epsilon}. \tag{2.28}$$

For a given k, we may use the least squares method to estimate the model. Let $S_n(k)$ be the sum of squared residuals when regressing $\boldsymbol{Y}$ on $\boldsymbol{X}$ and $\boldsymbol{Z}_2$ with $k_0 = k$. The estimated change point $\hat{k}$ is given by

$$\hat{k} = \arg \min_{1 \le k \le n} S_n(k).$$

The sup-Wald-type test statistic is $\sup_{k \in [cn, (1-c)n]} W_n(k)$ (see [16]), where

$$W_n(k) = \frac{\hat{\boldsymbol{\delta}}_k^\top \left(\boldsymbol{Z}_2^\top \boldsymbol{M} \boldsymbol{Z}_2\right) \hat{\boldsymbol{\delta}}_k}{\hat{\sigma}^2(k)}, \tag{2.29}$$

$\hat{\boldsymbol{\delta}}_k = \left(\boldsymbol{Z}_2^\top \boldsymbol{M} \boldsymbol{Z}_2\right)^{-1} \boldsymbol{Z}_2^\top \mathbf{M}\mathbf{Y}$, $\boldsymbol{M} = \boldsymbol{I} - \boldsymbol{X}(\boldsymbol{X}^\top\boldsymbol{X})^{-1}\boldsymbol{X}$, $\hat{\sigma}^2(k) = S_n(k)/(n-p-q)$, and $c \in (0, \frac{1}{2})$ is a small number.

Since $S_n(n) - S_n(k) = \hat{\boldsymbol{\delta}}_k^\top \left(\boldsymbol{Z}_2^\top \boldsymbol{M} \boldsymbol{Z}_2\right) \hat{\boldsymbol{\delta}}_k$ where $S_n(n)$ is the sum of squared residuals by regressing $\boldsymbol{Y}$ on $\boldsymbol{X}$ alone, we can derive that the Wald-type test statistic (2.29) is in fact a monotone transformation of $S_n(k)$, that is,

$$W_n(k) = \frac{S_n(n) - S_n(k)}{S_n(k)/(n-p-q)} = (n-p-q)(S_n(k)^{-1} - 1).$$

It follows that

$$\begin{aligned} \hat{k} &= \arg\min_k S_n(k) = \arg\max_k \hat{\boldsymbol{\delta}}_k^\top \left(\boldsymbol{Z}_2^\top \boldsymbol{M} \boldsymbol{Z}_2\right) \hat{\boldsymbol{\delta}}_k \\ &= \arg\max_k W_n(k). \end{aligned} \tag{2.30}$$

We can use the above statistics to detect the presence of the change point in the observed sample data. The specific properties of this statistic are discussed in Chapter 4.

2.4.2 Information criterion approach

In the following, we focus on

$$y_i = \boldsymbol{x}_i^\top \boldsymbol{\beta} + \varepsilon_i, \quad i = 1, \dots, n,$$

where $\boldsymbol{x}_i,\ i = 1, \dots, n$ is a nonstochastic $(p+1)$-vector variable, $\boldsymbol{\beta}$ is a $(p+1)$ unknown regression vector and $\varepsilon_i,\ i = 1, \dots, n$ is a random error which is distributed as $N(0, \sigma^2)$ with σ^2 unknown. The null hypothesis we want to test is:

$$H_0 : E(y_i) = \boldsymbol{x}_i^\top \boldsymbol{\beta} \quad \text{for } i = 1, \dots, n,$$

while the alternative hypothesis is:

$$\begin{aligned} H_1 : & E(y_i) = \boldsymbol{x}_i^\top \boldsymbol{\beta}_1 \quad \text{for } i = 1, \dots, k, \\ \text{and} \quad & E(y_i) = \boldsymbol{x}_i^\top \boldsymbol{\beta}_2 \quad \text{for } i = k+1, \dots, n, \end{aligned}$$

where $k, k = p+1, \dots, n-p$ is the location of the change point to be tested.

Let

$$\boldsymbol{y} = \begin{pmatrix} y_1 \\ y_2 \\ \vdots \\ y_n \end{pmatrix}, \quad \boldsymbol{X} = \begin{pmatrix} 1 & x_{11} & \cdots & x_{p1} \\ 1 & x_{12} & \cdots & x_{p2} \\ \vdots & \vdots & \vdots & \vdots \\ 1 & x_{1n} & \cdots & x_{pn} \end{pmatrix} \equiv \begin{pmatrix} \boldsymbol{x}_1^\top \\ \boldsymbol{x}_2^\top \\ \vdots \\ \boldsymbol{x}_n^\top \end{pmatrix}, \quad \text{and } \boldsymbol{\beta} = \begin{pmatrix} \beta_0 \\ \beta_1 \\ \vdots \\ \beta_p \end{pmatrix}.$$

The null hypothesis H_0 corresponds to the model is

$$E(\boldsymbol{y}) = \boldsymbol{X}\boldsymbol{\beta}.$$

The likelihood function under H_0 in matrix notation is

$$L_0(\boldsymbol{\beta}, \sigma^2) = (2\pi)^{-n/2} \left(\sigma^2\right)^{-n/2} \exp\left\{-(\boldsymbol{y} - \boldsymbol{X}\boldsymbol{\beta})^\top (\boldsymbol{y} - \boldsymbol{X}\boldsymbol{\beta})/2\sigma^2\right\}.$$

The MLEs of $\boldsymbol{\beta}$, σ^2 are

$$\begin{aligned} \widehat{\boldsymbol{\beta}} &= \left(\boldsymbol{X}^\top \boldsymbol{X}\right)^{-1} \boldsymbol{X}^\top \boldsymbol{y}, \\ \hat{\sigma}^2 &= \frac{1}{n}(\boldsymbol{y} - \boldsymbol{X}\hat{\boldsymbol{\beta}})^\top (\boldsymbol{y} - \boldsymbol{X}\hat{\boldsymbol{\beta}}). \end{aligned}$$

Then, the maximum likelihood under H_0 is

$$\begin{aligned} L_0(\hat{\boldsymbol{\beta}}, \hat{\sigma}^2) &= (2\pi)^{-n/2} \left[\frac{1}{n}(\boldsymbol{y} - \boldsymbol{X}\hat{\boldsymbol{\beta}})^\top (\boldsymbol{y} - \boldsymbol{X}\hat{\boldsymbol{\beta}})\right]^{-n/2} e^{-n/2} \\ &= (2\pi)^{-n/2} \left(\frac{1}{n}(\boldsymbol{y}^\top \mathbf{M}\boldsymbol{y}\right)^{-n/2} e^{-n/2}, \end{aligned}$$

where $\mathbf{M} = \mathbf{I} - \mathbf{X}(\mathbf{X}^\top \mathbf{X})^{-1}\mathbf{X}$. The BIC under H_0 is given by

$$\text{BIC}(n) = n \log\left(\boldsymbol{y}^\top \mathbf{M}\boldsymbol{y}\right) + n(\log 2\pi + 1) + (p + 2 - n) \log n.$$

Let

$$\boldsymbol{y}_1 = \begin{pmatrix} y_1 \\ y_2 \\ \vdots \\ y_k \end{pmatrix}, \quad \boldsymbol{y}_2 = \begin{pmatrix} y_{k+1} \\ y_{k+2} \\ \vdots \\ y_n \end{pmatrix},$$

$$\boldsymbol{X}_1 = \begin{pmatrix} 1 & x_{11} & \cdots & x_{p1} \\ 1 & x_{12} & \cdots & x_{p2} \\ \vdots & \vdots & \vdots & \vdots \\ 1 & x_{1k} & \cdots & x_{pk} \end{pmatrix} \equiv \begin{pmatrix} \boldsymbol{x}_1^\top \\ \boldsymbol{x}_2^\top \\ \vdots \\ \boldsymbol{x}_k^\top \end{pmatrix},$$

$$\boldsymbol{X}_2 = \begin{pmatrix} 1 & x_{11} & \cdots & x_{p(k+1)} \\ 1 & x_{12} & \cdots & x_{p(k+2)} \\ \vdots & \vdots & \vdots & \vdots \\ 1 & x_{1k} & \cdots & x_{pn} \end{pmatrix} \equiv \begin{pmatrix} \boldsymbol{x}_{k+1}^\top \\ \boldsymbol{x}_{k+2}^\top \\ \vdots \\ \boldsymbol{x}_n^\top \end{pmatrix},$$

$$\boldsymbol{\beta}_1 = \begin{pmatrix} \beta_{10} \\ \beta_{11} \\ \vdots \\ \beta_{1p} \end{pmatrix}, \quad \boldsymbol{\beta}_2 = \begin{pmatrix} \beta_{20} \\ \beta_{21} \\ \vdots \\ \beta_{2p} \end{pmatrix}, \quad \text{and } \boldsymbol{\beta} = \begin{pmatrix} \beta_0 \\ \beta_1 \\ \vdots \\ \beta_p \end{pmatrix},$$

for $k = p+1, ..., n-p$. Then, the alternative hypothesis H_1 corresponds to the following models.

$$E(\boldsymbol{y}_1) = \boldsymbol{X}_1\boldsymbol{\beta}_1 \quad \text{and} \quad E(\boldsymbol{y}_2) = \boldsymbol{X}_2\boldsymbol{\beta}_2, \qquad \text{for } k = p+1, \ldots, n-p. \tag{2.31}$$

The maximum likelihood is

$$\begin{aligned} L_1(\hat{\boldsymbol{\beta}}_1, \hat{\boldsymbol{\beta}}_2, \hat{\sigma}^2) =&(2\pi)^{-n/2} \left\{ 1/n \left[\left(\boldsymbol{y}_1 - \boldsymbol{X}_1\hat{\boldsymbol{\beta}}_1 \right)^\top \left(\boldsymbol{y}_1 - \boldsymbol{X}_1\hat{\boldsymbol{\beta}}_1 \right) \right.\right. \\ &\left.\left. + \left(\boldsymbol{y}_2 - \boldsymbol{X}_2\hat{\boldsymbol{\beta}}_2 \right)^\top \left(\boldsymbol{y}_2 - \boldsymbol{X}_2\hat{\boldsymbol{\beta}}_2 \right) \right] \right\}^{-n/2} e^{-n/2} \\ =&(2\pi)^{-n/2} \left[1/n \left(\boldsymbol{y}_1^\top \mathbf{M}_1 \boldsymbol{y}_1 + \boldsymbol{y}_2^\top \mathbf{M}_2 \boldsymbol{y}_2 \right) \right]^{-n/2} e^{-n/2}, \end{aligned}$$

where $\mathbf{M}_1 = \mathbf{I} - \mathbf{X}_1(\mathbf{X}_1^\top \mathbf{X}_1)^{-1}\mathbf{X}_1$ and $\mathbf{M}_2 = \mathbf{I} - \mathbf{X}_2(\mathbf{X}_2^\top \mathbf{X}_2)^{-1}\mathbf{X}_2$. The BIC under H_1 is

$$\begin{aligned} \text{BIC}(k) =& n \log \left[\boldsymbol{y}_1^\top \mathbf{M}_1 \boldsymbol{y}_1 + \boldsymbol{y}_2^\top \mathbf{M}_2 \boldsymbol{y}_2 \right] \\ &+ n(\log 2\pi + 1) + (2p + 3 - n) \log n. \end{aligned}$$

According to the principle of information criterion in model selection, H_0 will be accepted when $\text{BIC}(n) \leq \min_{p+1 \leq k \leq n-p} \text{BIC}(k)$. Otherwise if H_1 is accepted, the estimated position of change point $\hat{k}$ is given by

$$\hat{k} = \arg \min_{p+1 \leq k \leq n-p} \text{BIC}(k).$$

2.4.3 Bayesian approach

The Bayesian approach to statistical modeling begins by specifying a prior distribution for the model parameters. This prior reflects the researcher's beliefs or knowledge about the parameters before observing the data, incorporating any relevant background information or theoretical considerations. In the context of time series analysis, the prior can be formed by examining historical data that is not part of the current sample, often referred to as out-of-sample data. This historical data can inform the prior by indicating trends, seasonality, or other patterns that might influence the parameter estimates, providing a more informed starting point for estimation. The Bayesian framework then updates this prior distribution with the observed data through the likelihood function, producing a conditional posterior distribution that represents the updated belief about the parameters. This approach allows for incorporating both prior knowledge and the information contained in the data, providing a comprehensive method for parameter estimation and prediction [46].

The switching of multiple linear regression model was investigated by [124] from a Bayesian point of view. We still consider the problem (2.31).

Let

$$\boldsymbol{y}=\begin{pmatrix} \boldsymbol{y}_1 \\ \boldsymbol{y}_2 \end{pmatrix}, \quad \boldsymbol{\beta}=\begin{pmatrix} \boldsymbol{\beta}_1 \\ \boldsymbol{\beta}_2 \end{pmatrix}, \quad \text{and } R=\frac{1}{\sigma^2},$$

where $\boldsymbol{y}_1, \boldsymbol{y}_2, \boldsymbol{\beta}_1, \boldsymbol{\beta}_2$ are defined in the previous section. We assign a discrete uniform prior to the change point position k

$$\pi_0(k)=\begin{cases} \frac{1}{n-2p}, & k=p+1,\ldots,n-p, \\ 0, & \text{otherwise}. \end{cases}$$

We also assume that the $2(p+1)$ parameter vector $\boldsymbol{\beta}$ and the parameter R are jointly independent of the change point position k. The prior distribution of parameter R and the conditional distribution of $\boldsymbol{\beta}$ given $R=r$ are assumed as:

$$\pi_0(R)=\begin{cases} \frac{b^a}{\Gamma(b)} r^{a-1} e^{-br}, & r>0, \\ 0, & \text{otherwise}, \end{cases}$$

and

$$\pi_0(\boldsymbol{\beta} \mid R=r)=\frac{r^{p+1}|\Sigma^{-1}|^{p+1}}{(2\pi)^{p+1}} \exp\left\{-\frac{r}{2}(\boldsymbol{\beta}-\boldsymbol{\beta}^*)^\top \Sigma^{-1}(\boldsymbol{\beta}-\boldsymbol{\beta}^*)\right\},$$

where R satisfies a gamma distribution with parameters a and b, $\boldsymbol{\beta}$ given $R=r$ satisfies a $2(p+1)$-dimensional normal distribution with mean vector $\boldsymbol{\beta}^*$ and covariance matrix $(1/r)\Sigma$. Therefore, the joint prior of $\boldsymbol{\beta}$ and R can be written as

$$\begin{aligned} \pi_0(\boldsymbol{\beta}, R) &= \frac{b^a}{\Gamma(b)} r^{a-1} e^{-br} \cdot \frac{r^{p+1}|\Sigma^{-1}|^{p+1}}{(2\pi)^{p+1}} \exp\left\{-\frac{r}{2}(\boldsymbol{\beta}-\boldsymbol{\beta}^*)^\top \Sigma^{-1}(\boldsymbol{\beta}-\boldsymbol{\beta}^*)\right\} \\ &\propto r^{a+p} \exp\left\{(-r)\left[b+\frac{1}{2}(\boldsymbol{\beta}-\boldsymbol{\beta}^*)^\top \Sigma^{-1}(\boldsymbol{\beta}-\boldsymbol{\beta}^*)\right]\right\}. \end{aligned}$$

By defining the $n \times (2p+2)$ matrix $\boldsymbol{X}(k)$:

$$\boldsymbol{X}(k)=\begin{pmatrix} \boldsymbol{X}_1 & \mathbf{0}_1 \\ \mathbf{0}_2 & \boldsymbol{X}_2 \end{pmatrix},$$

where $\boldsymbol{X}_1$ and $\boldsymbol{X}_2$ are defined in the same way as in the previous section, $\mathbf{0}_1$ is a $k \times (p+1)$ zero matrix and $\mathbf{0}_2$ is a $(n-k) \times (p+1)$ zero matrix. The likelihood function is

$$\begin{aligned} L_1(\boldsymbol{\beta}, R, k) =& L_1(\boldsymbol{\beta}_1, \boldsymbol{\beta}_2, R, k) \\ =& (2\pi)^{-n/2} r^{n/2} \exp\left\{-r(\boldsymbol{y}_1-\boldsymbol{X}_1\boldsymbol{\beta}_1)^\top(\boldsymbol{y}_1-\boldsymbol{X}_1\boldsymbol{\beta}_1)/2\right\} \\ & \cdot \exp\left\{-r(\boldsymbol{y}_2-\boldsymbol{X}_2\boldsymbol{\beta}_2)^\top(\boldsymbol{y}_2-\boldsymbol{X}_2\boldsymbol{\beta}_2)/2\right\} \\ =& (2\pi)^{-n/2} r^{-n/2} \exp\left\{-r(\boldsymbol{y}-\boldsymbol{X}(k)\boldsymbol{\beta})^\top(\boldsymbol{y}-\boldsymbol{X}(k)\boldsymbol{\beta})/2\right\}. \end{aligned}$$

Therefore, the joint posterior density of the parameters is

$$\begin{aligned}\pi_1(\boldsymbol{\beta}, R, k) =& f\left(\boldsymbol{\beta}, R, k \mid y_1, y_2, \ldots, y_n\right)\\ \propto& L_1(\boldsymbol{\beta}, R, k)\pi_0(\boldsymbol{\beta}, R)\pi_0(k)\\ \propto& r^{a+p+n/2} \cdot \exp\left\{(-r)\left[b+\frac{1}{2}\left(\boldsymbol{\beta}-\boldsymbol{\beta}^*\right)^\top \Sigma^{-1}\left(\boldsymbol{\beta}-\boldsymbol{\beta}^*\right)\right.\right.\\ &\left.\left.+\frac{1}{2}(\boldsymbol{y}-\boldsymbol{X}(k)\boldsymbol{\beta})^\top(\boldsymbol{y}-\boldsymbol{X}(k)\boldsymbol{\beta})\right]\right\}.\end{aligned}$$

We obtain the posterior density of the change point k as

$$\begin{aligned}\pi_1(k) &= f\left(k \mid y_1, y_2, \ldots, y_n\right)\\ &\propto D(k)^{-a^*}\left|\boldsymbol{X}(k)^\top \boldsymbol{X}(k)+\Sigma^{-1}\right|^{-1/2},\end{aligned}$$

for $k = p+1, \ldots, n-p$, where

$$\begin{aligned}&a^* = a+1+n/2,\\ &D(k) = b+\frac{1}{2}\left\{[\boldsymbol{y}-\widehat{\boldsymbol{y}}(k)]^\top[\boldsymbol{y}-\widehat{\boldsymbol{y}}(k)]+\left[\widehat{\boldsymbol{\beta}}(k)-\boldsymbol{\beta}^*\right]^\top \boldsymbol{w}(k)\left[\widehat{\boldsymbol{\beta}}(k)-\boldsymbol{\beta}^*\right]\right\},\\ &\boldsymbol{w}(k) = \boldsymbol{X}(k)^\top \boldsymbol{X}(k)\left[\boldsymbol{X}(k)^\top \boldsymbol{X}(k)+\Sigma^{-1}\right]^{-1}\Sigma^{-1},\\ &\widehat{\boldsymbol{\beta}}(k) = \left[\boldsymbol{X}(k)^\top \boldsymbol{X}(k)\right]^{-1}\boldsymbol{X}(k)^\top \boldsymbol{y},\\ &\widehat{\boldsymbol{y}}(k) = \boldsymbol{X}(k)\widehat{\boldsymbol{\beta}}(k).\end{aligned}$$

The distribution mode can serve as a change point estimator. Other inference results can also be derived from the posterior distribution. Readers interested in more details can refer to [62], and [55].

3

Multiple change points

Many scientific and business data involve more than one change point. Detecting multiple change points (Figure 3.1) is more challenging since we must determine the total number of change points. Traditional parameter estimation techniques, such as gradient-based optimization, are not directly applicable in this context because the number of change points is not an Euclidean parameter. Estimating the number of change points requires balancing model complexity and goodness of fit, often leading to adopting model selection criteria like AIC, BIC, or penalties in optimization frameworks. As a result, indirect approaches are typically employed to estimate the locations and number of change points. Specifically, methods need to account for the type of data (e.g., time series, multivariate) and its distributional properties. Efficient algorithms or approximations are more appealing for high-dimensional or long-time series.

Three main classes of multiple change points detection approaches are introduced in this chapter, including binary segmentation, penalized estimation, and moving window estimation. To simplify the discussion and focus on foundational concepts, we explore these methods in the context of shifts in the mean of independent univariate Gaussian variables with constant variance over time—a well-studied model within the field of change point detection. Each approach offers unique advantages for identifying multiple change points and serves as a building block for more complex settings. Specifically, for $t = 1, \ldots, n$, we assume the model

$$x_t = \mu_t + \epsilon_t, \quad \text{with} \quad \epsilon_t \overset{iid}{\sim} \mathcal{N}(0, \sigma^2), \tag{3.1}$$

where x_t is the observed data value, μ_t is a fixed one-dimensional, piece-wise constant signal process with an unknown number m of change points at unknown locations $0 < k_1 < \cdots < k_N < n$. Without loss of generality, the variance parameter σ^2 for the error ϵ_t is assumed to be 1.

3.1 Binary segmentation

Binary segmentation (BS) is used for multiple change point detection as a generic technique. Similar to the methods introduced in Chapter 2, a CUSUM-like procedure is first applied to find an initial change point in the entire dataset. Once the first change point is identified, we can split the data into two subsegments and apply the same method to the segments. We iterate this procedure until a certain stopping criterion is satisfied.

3.1.1 Standard binary segmentation

The basic ingredient of the standard BS algorithm is the CUSUM statistic defined by the inner product between the vector $(x_s, \ldots, x_e)$ and a particular vector of "contrast" weights. Specifically,

DOI: 10.1201/9781032657073-3

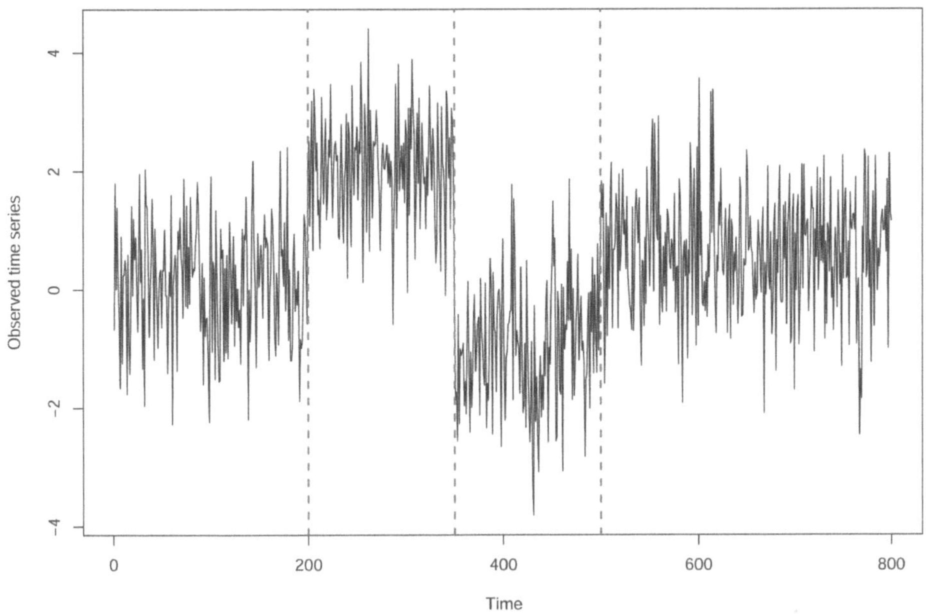

Figure 3.1: Multiple change points of a time series.

we consider

$$T_{s,e}^k = \sqrt{\frac{(e-k)(k-s+1)}{e-s+1}} \left(\frac{1}{k-s+1} \sum_{t=s}^{k} x_t - \frac{1}{e-k} \sum_{t=k+1}^{e} x_t \right), \tag{3.2}$$

where $s \leq k < e$. In the following BS methods, i.e., the standard BS, the wild BS, and the seeded BS, (3.2) is applied repeatedly with different segments to the model (3.1).

In the first step, $T_{s,e}^k$ is calculated for $k = 1, 2, \ldots, n-1$ in the BS algorithm, and the first estimated change point is denoted by $\hat{k}_1 = \arg\max_{1 \leq k < n} |T_{1,n}^k|$. If the value of $\max_{1 \leq k < n} |T_{1,n}^k|$ exceeds a threshold ζ_n, the change point test is considered significant, and the domain $[1, n]$ is split into two sub-intervals. The same procedure is continued by computing $T_{1,\hat{k}_1}^k$ and $T_{\hat{k}_1+1,n}^k$, which may result in further splits. When the procedure stops, we attain $\hat{N}$ estimated change points $\{\hat{k}_1, \cdots, \hat{k}_{\hat{N}}\}$. The BS algorithm is best defined recursively and hence described by the following pseudocode. The detail of BS is displayed in Algorithm 2.

To study the theoretical properties of BS, we first make some assumptions about the model and potential change points. The theoretical consistency result and its proof will be provided in the following:

Assumption 3.1. *(i) The model obeys (3.1) and* $\sigma^2 = 1$.
(ii) The sequence $\{\mu_t\}_{t=1}^n$ *is bounded, i.e.,* $|\mu_t| < \bar{\mu} < \infty$ *for a constant* $\bar{\mu}$, $t = 1, \ldots, n$.

Assumption 3.2. *The minimum distance between two adjacent change points satisfies* $\min_{i=1,\ldots,N+1} |k_i - k_{i-1}| \geq \delta_n$, *where* $\delta_n \geq Cn^{\Theta}$ *for* $C > 0$, *with* $\Theta \leq 1$. *In addition, the magnitudes* $\mu_i' = |\mu_{k_i} - \mu_{k_i - 1}|$ *of the jumps satisfy* $\min_{i=1,\ldots,N} \mu_i' \geq \underline{\mu}_n$, *where* $\underline{\mu}_n \geq Cn^{-\varpi}$, *with* $\varpi \geq 0$. *The parameters* Θ *and* ϖ *satisfy* $\Theta - \frac{\varpi}{2} > \frac{3}{4}$.

Algorithm 2 Standard BS

Require: Function BinSeg(s, e, ζ_n)
Ensure: Set of detected change points $\mathcal{T}_{s,e}$
 if $e - s < 1$ **then**
  STOP
 else
  $k_0 := \arg\max_{k\in\{s,\dots,e-1\}} |T^k_{s,e}|$
  if $|T^{k_0}_{s,e}| > \zeta_n$ **then**
   $\mathcal{T}_{s,e} = \mathcal{T}_{s,e} \cup \{k_0\}$
   BinSeg(s, k_0, ζ_n)
   BinSeg($k_0 + 1, e, \zeta_n$)
  else
   STOP
  end if
 end if

Theorem 3.1.1. *Suppose Assumption 3.1 and Assumption 3.2 hold. Let N and $k_1, \dots, k_m$ denote the number and locations of change points, respectively. Let $\hat{m}$ denote the number, and $\hat{k}_1, \dots, \hat{k}_{\hat{m}}$ the locations, sorted in increasing order, of the change point estimates obtained by the standard binary segmentation algorithm. Let the threshold parameter satisfy $\zeta_n = c_1 n^\theta$ where $\theta \in (1 - \Theta, \Theta - 1/2 - \varpi)$ if $\Theta \in (\frac{3}{4}, 1)$, or $\zeta_n \geq c_2 \log^p n$ $(p > 1/2)$ and $\zeta_n \leq c_3 n^\theta$ $(\theta < 1/2 - \varpi)$ if $\Theta = 1$, for any positive constants c_1, c_2, c_3. Then there exist positive constants C and C_1 such that $P(\mathcal{A}_n) \geq 1 - C_1 n^{-1}$, where*

$$\mathcal{A}_n = \left\{ \hat{N} = N; \max_{i=1,\dots,N} |\hat{k}_i - k_i| \leq C\epsilon_n \right\}$$

with $\epsilon_n = n^2 \delta_n^{-2} (\underline{\mu}_n)^{-2} \log n$.

Proof of Theorem 3.1.1. We will first define

$$\tilde{\mu}^k_{s,e} = \sqrt{\frac{(e-k)(k-s+1)}{(e-s+1)}} \left(\frac{1}{k-s+1} \sum_{t=s}^{k} \mu_t - \frac{1}{e-k} \sum_{t=k+1}^{e} \mu_t \right). \tag{3.3}$$

Let s, e satisfy

$$k_{p_0} \leq s < k_{p_0+1} < \cdots < k_{p_0+q} < e \leq k_{p_0+q+1} \tag{3.4}$$

with $0 \leq p_0 \leq N - q$, which will be the case at all stages of the algorithm while undetected change points remain. We also introduce a set A_n defined by

$$A_n = \left\{ \left| (e-k+1)^{-1/2} \sum_{i=k}^{e} \epsilon_i \right| < \lambda_2 \quad \forall 1 \leq k \leq e \leq n \right\}. \tag{3.5}$$

Note that because of Bonferroni's inequality, $P(A_n) \geq 1 - Cn^{-1}$ for $\lambda_2 \geq (6 \log n)^{1/2}$, where C is a positive constant. Let

$$B_n = \left\{ \max_{s,k,e:1\leq s\leq k<e\leq n} |T^k_{s,e} - \tilde{\mu}^k_{s,e}| \leq \lambda_1 \right\}, \tag{3.6}$$

where $\lambda_1 \geq \sqrt{8 \log n}$. By Lemma 3.1, we also have $P(B_n) \geq 1 - Cn^{-1}$, where C is a positive constant.

The following are on the event $B_n \cap A_n$, which has probability $P(B_n \cap A_n) \geq 1 - Cn^{-1}$. We impose the following conditions:

$$s < k_{p_0+r} - C\delta_n < k_{p_0+r} + C\delta_n < e \quad \text{for some } 1 \leq r \leq q, \tag{3.7}$$

$$\max\left(\min(k_{p_0+1} - s, s - k_{p_0}), \min(k_{p_0+q+1} - e, e - k_{p_0+q})\right) \leq C\epsilon_n. \tag{3.8}$$

Both (3.7) and (3.8) hold throughout the algorithm for all those segments starting at s and ending at e until every change-point has been identified, and we also find the detected change point within the distance of $C\epsilon_n$ from the true change point, by Lemmas 3.2–3.4. When one of the three conditions of Lemma 3.5 is satisfied, no further change points are detected.

The proof of the theorem is completed. □

Theorem 3.1.1 ensures that the correct number of change points can be identified with probability approaching one. The estimated change point locations are also uniformly consistent to the true locations. The lemmas used to prove the Theorem 3.1.1 are given below.

Lemma 3.1. *Let x_t follow model (3.1), and let the assumptions of Theorem 3.1.1 hold. Let $T_{s,e}^k$, $\tilde{\mu}_{s,e}^k$ and B_n be defined by (3.2), (3.3) and (3.6), respectively. We then have $P(B_n) \geq 1 - Cn^{-1}$, where $\lambda_1 \geq \sqrt{8 \log n}$, and C is a positive constant.*

Lemma 3.2. *Let x_t follow model (3.1) and let the assumptions of Theorem 3.1.1 hold. Let $T_{s,e}^k$ and $\tilde{\mu}_{s,e}^k$ be defined by (3.2) and (3.3), respectively. Assume (3.4), (3.7), and (3.8). The following holds on set B_n of Lemma 3.1. For $k = \arg\max_{t:s \leq t < e} |T_{s,e}^t|$, there exists $1 \leq r \leq q$ such that for large n, $|k - k_{p_0+r}| \leq C_1\gamma_n$ with $\gamma_n = n^{1/2}\lambda_1/\mu'_{p_0+r}$ (λ_1 as in Lemma 3.1). In addition, $|\tilde{\mu}_{s,e}^t|$ must then have a local maximum at $t = k_{p_0+r}$, and we must have*

$$\frac{|\tilde{\mu}_{s,e}^{k_{p_0+r}}|}{\max_{t:s \leq t < e} |\tilde{\mu}_{s,e}^t|} \geq C_2,$$

where C_1, C_2 are positive constants.

Lemma 3.3. *Let the conditions of Lemma 3.2 hold, and let the notation be the same as given in that lemma. On the set $B_n \cap A_n$, where B_n is defined in Lemma 3.1 and A_n in (3.5), we have for large n, $|k - k_{p_0+r}| \leq C\epsilon_n$, where $\epsilon_n = \lambda_2^2 n^2 \delta_n^{-2} (\mu'_{p_0+r})^2$ and C is a positive constant.*

Lemma 3.4. *Let x_t follow model (3.1), and let the assumptions of Theorem 3.1.1 hold. Let $T_{s,e}^k$ be defined by (3.2). Assume (3.4), (3.7), and (3.8). On the event B_n from Lemma 3.1, we have $|T_{s,e}^k| > Cn^{\Theta - 1/2 - \varpi}$, where $k = \arg\max_{t:s \leq t < e} |T_{s,e}^t|$.*

Lemma 3.5. *Let x_t follow model (3.1), and let the assumptions of Theorem 3.1.1 hold. Let $T_{s,e}^k$ be defined by (3.2). For some positive constant C, C', let s, e satisfy one of three conditions:*
(i) exists $1 \leq p \leq N$ such that $s \leq k_p \leq e$ and $(k_p - s + 1) \wedge (e - k_p) \leq C\epsilon_n$, or
(ii) exists $1 \leq p \leq N$ such that $s \leq k_p \leq k_{p+1} \leq e$ and $(k_p - s + 1) \vee (e - k_{p+1}) \leq C'\epsilon_n$, or
(iii) exists $1 \leq p \leq N$ such that $k_p < s < e \leq k_{p+1}$.
On the event $B_n \cap A_n$ from Lemma 3.3, we have $|T_{s,e}^k| < C\lambda_2 n^{1-\Theta} + \lambda_1$, where $k = \arg\max_{t:s \leq t < e} |T_{s,e}^t|$.

The proof of Lemma 3.1–3.5 can be found in [102].

3.1.2 Wild binary segmentation

When there are indeed more than one change points, the first step of the BS algorithm amounts to fitting a model with only a single change point, to data that actually contains multiple change points, effectively resulting in the fitting of a wrong model. The wild binary segmentation (WBS) [102] aims to randomly draw index pairs (s, e) and identify $\arg\max_{k:s\leq k<e} |T_{s,e}^k|$ for each draw. If the number of random draws is sufficiently large, we can guarantee, with high probability, the selection of a particularly favorable interval $[s, e]$ that is long enough and contains only one change point, located at a sufficient distance from its endpoints.

Denote by F_n^M a set of M random intervals $[s_m, e_m]$, $m = 1, \ldots, M$, whose start- and end-points have been drawn (independently with replacement) uniformly from the set $\{1, \ldots, n\}$. Guidance on how to select a suitable M will be given later. Using pseudocode, the main function of the wild binary segmentation (WBS) algorithm is described in Algorithm 3.

Algorithm 3 Wild BS

Require: Function WildBS(s, e, ζ_n)
Ensure: Set of detected change points $\mathcal{T}_{s,e}$
  if $e - s < 1$ **then**
    STOP
  else
    Denote $\mathcal{M}_{s,e}$ the set of those indices m for which $[s_m, e_m] \in F_n^M$ is such that $[s_m, e_m] \in [s, e]$
    (Optional augment: $\mathcal{M}_{s,e} := \mathcal{M}_{s,e} \cup \{0\}$, where $[s_0, e_0] = [s, e]$)
    $(m_0, k_0) := \arg\max_{m\in\mathcal{M}_{s,e}, k\in\{s_m,\ldots,e_m-1\}} |T_{s_m,e_m}^k|$
    if $|T_{s_{m_0},e_{m_0}}^{k_0}| > \zeta_n$ **then**
      $\mathcal{T}_{s,e} = \mathcal{T}_{s,e} \cup \{k_0\}$
      WildBS(s, k_0, ζ_n)
      WildBS($k_0 + 1, e, \zeta_n$)
    else
      STOP
    end if
  end if

The optional augmentation of $\mathcal{M}_{s,e}$ by $\{0\}$ is done to ensure that the algorithm also examines the entire current interval $[s, e]$, and not only its randomly drawn subintervals, in case $[s, e]$ only contains one change point and hence it is optimal to examine $[s, e]$ in its entirety.

We make the necessary technical assumption for the WBS, to establish the consistent result in the following theorem.

Assumption 3.3. *The minimum distance between two adjacent change points satisfies* $\min_{i=1,\ldots,N+1} |k_i - k_{i-1}| \geq \delta_n$, *and the magnitudes* $\mu_i' = |\mu_{k_i} - \mu_{k_i-1}|$ *of the jumps satisfy* $\min_{i=1,\ldots,N} \mu_i' \geq \underline{\mu}_n$, *where* δ_n *and* $\underline{\mu}_n$ *are linked by the requirement* $\delta_n^{1/2}\underline{\mu}_n \geq C\log^{1/2} n$ *for a large enough* C.

In what follows, denote $k_0 = 0$, $k_{N+1} = n$. The following theoretical results hold for the WBS algorithm.

Theorem 3.1.2. *Let* x_t *follow model 3.1, and suppose Assumption 3.1 and Assumption 3.3 hold. Let* N *and* $k_1, \ldots, k_N$ *denote the number and locations of change points, respectively. Let* $\hat{N}$ *denote the number, and* $\hat{k}_1, \ldots, \hat{k}_{\hat{N}}$ *the locations, sorted in increasing order, of the change point estimates obtained by the wild binary segmentation algorithm. There exist two constants* $\underline{C}, \overline{C}$ *such that if*

$\underline{C}\log^{1/2} n \leq \zeta_n \leq \overline{C}\delta_n^{1/2}\underline{\mu}_n$, *then* $P(\mathcal{A}) \geq 1 - C_1 n^{-1} - n\delta_n^{-1}(1-\delta_n^2 n^{-2}/9)^K$, *where*

$$\mathcal{A} = \left\{\hat{N} = N;\ \max_{i=1,\ldots,N} |\hat{k}_i - k_i| \leq C\log n(\underline{\mu}_n)^{-2}\right\}$$

for certain positive C, C_1.

Proof of Theorem 3.1.2. We start by defining intervals $\mathcal{I}_i$ between change points in such a way that their lengths are at least of order δ_n, and they are separated from the change points also by distances at least of order δ_n. To fix ideas, define $\mathcal{I}_i = [k_{i-1} + \frac{1}{3}(k_i - k_{i-1}), k_{i-1} + \frac{2}{3}(k_i - k_{i-1})], i = 1, \ldots, N+1$.

Each stage of the algorithm uses CUSUM statistics computed over M intervals (s_m, e_m), $k = 1, \ldots, M$, drawn uniformly (independently with replacement) from the set $\{(s_m, e_m) : s < e, 1 \leq s \leq n-1, 2 \leq e \leq n\}$. Define the event D_n^M as follows

$$D_n^M = \{\forall i = 1, \ldots, N \quad \exists m = 1, \ldots, M \quad (s_m, e_m) \in \mathcal{I}_i \times \mathcal{I}_{i+1}\}.$$

Note that

$$P((D_n^M)^c) \leq \sum_{i=1}^{N}\prod_{m=1}^{M}(1 - P((s_m, e_m) \in \mathcal{I}_i \times \mathcal{I}_{i+1})) \leq n\delta_n^{-1}(1 - \delta_n^2 n^{-2}/9)^M.$$

The remaining arguments will be valid on the set D_n^M. If an interval (s_m, e_m) is such that $(s_m, e_m) \in \mathcal{I}_i \times \mathcal{I}_{i+1}$, and thus (s_m, e_m) contains one change point only, k_i, then arguing as in Lemma 1 of [60], we have

$$|\tilde{\mu}_{s_m,e_m}^{k_i}| = \max_{t:s_k \leq t < e_k} |\tilde{\mu}_{s_m,e_m}^t| \geq C\delta_n^{1/2}\mu_i'.$$

Let (s, e) now be a generic interval satisfying (3.7) and (3.8), with ϵ_t and δ_n as in the statement of this theorem. The remaining arguments are valid on the set B_n. Consider

$$(m_0, k) = \arg\max_{(m,t):m\in\mathcal{M}_{s,e}, s_m\leq t<e_m} \left|T_{s_m,e_m}^t\right|, \tag{3.9}$$

where $\mathcal{M}_{s,e} = \{m : (s_m, e_m) \subseteq (s, e), 1 \leq m \leq M\}$. Imposing the condition

$$\delta_n \geq 3\epsilon_n, \tag{3.10}$$

we guarantee that both s and e are sufficiently bounded away from all the previously undetected change points $k_i \in (s, e)$ in the sense that $\mathcal{I}_i \cup \mathcal{I}_{i+1} \subset (s, e)$ for all such i. Denote the set of these i's by $\mathcal{J}_{s,e}$. For each $i \in \mathcal{J}_{s,e}$, there exists an $m_i \in \mathcal{M}_{s,e}$ such that $(s_{m_i}, e_{m_i}) \in \mathcal{I}_i \times \mathcal{I}_{i+1}$, and thus

$$\left|T_{s_{m_0},e_{m_0}}^k\right| \geq \max_{t:s_{m_i}\leq t<e_{m_i}} \left|T_{s_{m_i},e_{m_i}}^t\right| \geq \left|T_{s_{m_i},e_{m_i}}^{k_i}\right| \geq \left|\tilde{\mu}_{s_{m_i},e_{m_i}}^{k_i}\right| - \lambda_1 \geq C_1\delta_n^{1/2}\mu_i', \tag{3.11}$$

provided that

$$\delta_n \geq C_8\lambda_1^2(\underline{\mu}_n)^{-2}. \tag{3.12}$$

Therefore,

$$\left|\tilde{\mu}_{s_{m_0},e_{m_0}}^k\right| \geq \left|T_{s_{m_0},e_{m_0}}^k\right| - \lambda_1 \geq C_2\delta_n^{1/2}\max_{i\in\mathcal{J}_{s,e}} \mu_i'. \tag{3.13}$$

By Lemma 2.2 in [250], there exists a change point k_{p_0+r}, immediately to the left or to the right of k such that

$$|\tilde{\mu}_{s_{m_0},e_{m_0}}^{k_{p_0+r}}| > |\tilde{\mu}_{s_{m_0},e_{m_0}}^k| \geq C_2\delta_n^{1/2}\max_{i\in\mathcal{J}_{s,e}} \mu_i'. \tag{3.14}$$

Now, the following two situations are impossible:

(1) (s_{m_0}, e_{m_0}) contains one change point only, k_{p_0+r} and either $k_{p_0+r} - s_{m_0}$ or $e_{m_0} - k_{p_0+r}$ is not bounded from below by $C_3\delta_n$;

(2) (s_{m_0}, e_{m_0}) contains two change points only, say k_{p_0+r} and k_{p_0+r+1}, and both $k_{p_0+r} - s_{k_0}$ and $e_{k_0} - k_{p_0+r+1}$ are not bounded from below by $C_3\delta_n$.

Indeed, if either situation were true, then using arguments such as in Lemma 3.5, we would obtain that $\max_{t:s_{m_0}\le t<e_{m_0}} |\tilde{\mu}^t_{s_{m_0},e_{m_0}}|$ were not bounded from below by $C_2\delta_n^{1/2}\max_{i\in\mathcal{J}_{s,e}}\mu'_i$, a contradiction to (3.13). This proves that the interval (s_{m_0}, e_{m_0}) satisfies condition (3.7) (with δ_n as in the statement of this theorem), and thus we can follow the argument from the proof of Lemma 2 in [60] to establish that if $|k' - k_{p_0+r}| = C\gamma_n$ for a certain C, with $\gamma_n = \delta_n^{1/2}\lambda_1/\mu'_{p_0+r}$, and if $\tilde{\mu}^{k_{p_0+r}}_{s_{m_0},e_{m_0}} > \tilde{\mu}^{k'}_{s_{m_0},e_{m_0}}$ (assuming w.l.o.g. $\tilde{\mu}^{k_{p_0+r}}_{s_{m_0},e_{m_0}} > 0$), then $\tilde{\mu}^{k_{p_0+r}}_{s_{m_0},e_{m_0}} \ge \tilde{\mu}^{k'}_{s_{m_0},e_{m_0}} + 2\lambda_1$.

With this result, it is then straightforward to proceed like in the proof of Lemma 3.2 to show that $|k - k_{p_0+r}| \le C_4\gamma_n$, and that $|\tilde{\mu}^t_{s_{m_0},e_{m_0}}|$ must have a local maximum at $t = k_{p_0+r}$.

To establish that $|k - k_{p_0+r}| \le C_7\epsilon_n$, we need to use the above results to obtain an improved version of Lemma 3.3. The arguments in the remainder of the proof are valid on the set A_n. Following the proof of Lemma 3.3 for the interval (s_{m_0}, e_{m_0}) with $\gamma_n = \delta_n^{1/2}\lambda_1/\mu'_{p_0+r}$, in the notation of that lemma and using an argument like in Lemma 2 of [60], we obtain

$$\left|\tilde{\mu}^{k_{p_0+r}}_{s_{m_0},e_{m_0}}\right| - \left|\tilde{\mu}^{k}_{s_{m_0},e_{m_0}}\right| \ge C|k - k_{p_0+r}|\delta_n^{-1/2}\mu'_{p_0+r}.$$

Additionally, by (3.14), $|\tilde{\mu}^{k_{p_0+r}}_{s_{m_0},e_{m_0}}| \ge C_5\delta_n^{1/2}\mu'_{p_0+r}$, which combined yields

$$\sum_{t=s_{k_0}}^{e_{k_0}} (\mu_t - \bar{\mu}^k_{s_{m_0},e_{m_0}})^2 - \sum_{t=s_{k_0}}^{e_{k_0}} (\mu_t - \bar{\mu}^{k_{p_0+r}}_{s_{m_0},e_{m_0}})^2 \ge C|k - k_{p_0+r}|(\mu'_{p_0+r})^2.$$

This in turn leads to the following replacement for the triple inequality in Lemma 3.3:

$$|k - k_{p_0+r}|(\mu'_{p_0+r})^2 \ge C\max(\lambda_2|k - k_{p_0+r}|\delta_n^{-1/2}\mu'_{p_0+r}, \lambda_2|k - k_{p_0+r}|^{1/2}\mu'_{p_0+r}, \lambda_2^2).$$

These three inequalities yield, respectively, $\delta_n \ge C^2\lambda_2^2/\left(\mu'_{p_0+r}\right)^2$, $|k - k_{p_0+r}| \ge C^2\lambda_2^2/\left(\mu'_{p_0+r}\right)^2$, $|k - k_{p_0+r}| \ge C\lambda_2^2/\left(\mu'_{p_0+r}\right)^2$. The second inequality and the requirement that $|k - k_{p_0+r}| \le C_6\gamma_n = C_6\lambda_1\delta_n^{1/2}/\mu'_{p_0+r}$ (see the proof of Lemma 3.3) together imply $\delta_n \ge C^4C_6^{-2}\lambda_1^{-2}\lambda_2^4 \left(\mu'_{p_0+r}\right)^{-2}$. Combining this with the first inequality, we obtain

$$\delta_n \ge C^2\lambda_2^2\left(\mu'_{p_0+r}\right)^{-2}\max\left(C^2C_6^{-2}\lambda_1^{-2}\lambda_2^2, 1\right). \tag{3.15}$$

By the second and third inequalities, ϵ_n can be taken to be $\max\left(1, C^2\right)\lambda_2^2/\left(\mu'_{p_0+r}\right)^2$. At this point, we recall the constraints (3.10) and (3.12). Taking λ_1 and λ_2 to be of the lowest permissible order of magnitude, that is $\sim \log^{1/2} n$, these constraints together with (3.15) stipulate that we must have $\delta_n \ge C_9\log n/(\underline{\mu}_n)^2$ for a large enough C_9.

With the use of the above results, the proof of the theorem proceeds as follows; the following occurs on the event $B_n \cap A_n \cap D_n^M$, which has probability $\ge 1 - C_1n^{-1} - n\delta_n^{-1}\left(1 - \delta_n^2n^{-2}/9\right)^M$. At the start of the algorithm, as $s = 0$ and $e = n - 1$, (3.7) and (3.8) (with δ_n and ϵ_n as in the statement of this theorem) are satisfied, and therefore, by formula (3.11), the algorithm detects a change point k on that interval, defined by formula (3.9). By the above discussion, k is within the distance of $C\epsilon_n$ from the change point. Then (3.9) and (3.10) (with δ_n and ϵ_n as in the statement of this theorem) are satisfied within each segment until every change point in μ_t has been identified. Once this has happened, we note that every subsequent interval (s_m, e_m) satisfies the assumptions on (s, e) from Lemma 3.5 and therefore $\left|T^k_{s_{m_0},e_{m_0}}\right| < C\lambda_2 + \lambda_1 \le \zeta_n$, which means that no further change points are detected. □

Remark 3.1 (Tuning parameter choice of M). *The number M of the random draws should be chosen to be "as large as possible" subject to computational constraints. Note that with the optional augmentation of $\mathcal{M}_{s,e}$ by $\{0\}$, the WBS reduces to the standard BS for $M = 0$. That is, even a relatively small value of M is likely to bring benefits in practical performance. Cho and Fryzlewicz (2012) [60] recommended $M = 5000$ for data sets of length n not exceeding a few thousands. Moreover, the larger the value of M, the more negligible the dependence of the solution on a particular random draw. For $M = 5000$, this dependence has been observed to be minimal in our numerical studies.*

Remark 3.2 (Tuning parameter choice of ζ_n). *Since it is challenging to propose a particular choice of ζ_n without having a specific cost function, one possibility is to use a large-scale simulation study to select a default value of ζ_n that works well across a range of signals. The theoretical results do not permit a data-dependent choice of the threshold $\zeta_n = \sigma C\sqrt{\log n}$, i.e., having a reliable default choice is essential. The comparative simulation study in [60] gives two threshold constants: the default value of $C = 1$ and a slightly inflated value of $C = 1.3$.*

Remark 3.3. *There are two computational methods to determine the number of change points $\hat{N}$. One may perform model selection either by choosing a suitable threshold ζ_n and hence selecting the associated model, or by considering the sequence of model candidates and choosing one that optimizes a certain criterion, thereby by-passing the question of threshold choice entirely. [102] propose to select a model out of the candidates by minimizing the so-called "strengthened Schwarz information criterion" (SSIC).*

3.1.3 Seeded binary segmentation

Seeded Binary Segmentation (SBS), proposed by [149], is a flexible, fast, and statistically efficient approach to change point detection in large-scale problems. Similar to WBS [102], SBS analyzes a large number of data intervals separately.

In comparison, while WBS relies on random intervals to identify change points, SBS employs a deterministic construction of search intervals, known as seeded intervals. The key advantages of these seeded intervals include significantly faster computations, simpler theoretical analysis, and more stable and reproducible results.

In WBS, when the start and end points are drawn uniformly at random, many long intervals are generated. These long intervals, which often incur higher computational costs, are less desirable since they are more likely to contain multiple change points. The focus should instead be on relatively shorter intervals that are more likely to contain a single change point. This is where the concept of seeded intervals comes into play. Let us define the seeded intervals here.

Definition 3.1 (Seeded Intervals). *Let $a \in [1/2, 1)$ denote a given decay parameter. Define $\mathcal{I}_1 = (0, n]$ and, for $k = 2, \ldots, \lceil \log_{1/a}(n) \rceil$, i.e., logarithm with base $1/a$, define the kth layer as the collection of n_k intervals of initial length l_k that are evenly shifted, up to rounding, by the deterministic shift s_k as $\mathcal{I}_k = \bigcup_{i=1}^{n_k} \{(\lfloor (i-1)s_k \rfloor, \lceil (i-1)s_k + l_k \rceil]\}$ and their collection $\mathcal{I} = \bigcup_{k=1}^{\lceil \log_{1/a}(n) \rceil} \mathcal{I}_k$ as the set of seeded intervals, where $n_k = 2\lceil (1/a)^{k-1} \rceil - 1, l_k = na^{k-1}$ and $s_k = (n - l_k) / (n_k - 1)$. If some intervals appear multiple times on small scales due to rounding, they can be eliminated.*

Example 3.1. *Let $a = 1/2$. By Definition 3.1, we have $n_k = 2^k - 1, l_k = 2n/2^k$, and $s_k = n/2^k$. Then $\mathcal{I}_k = \bigcup_{i=1}^{2^k-1} \left\{ \left(\left\lfloor \frac{(i-1)n}{2^k} \right\rfloor, \left\lceil \frac{(i+1)n}{2^k} \right\rceil \right] \right\}$, that is, $\mathcal{I}_1 = (0, n]$, $\mathcal{I}_2 = \{(0, \lceil n/2 \rceil], (\lfloor n/4 \rfloor, \lceil 3n/4 \rceil], (\lfloor n/2 \rfloor, n]\}$, ..., $\mathcal{I}_{log_2(n)} = \{(0, 2], (1, 3], \ldots, (n-2, n]\}$.*

In practice, a more common choice for the decay $a = 1/2^{1/2}$ is well demonstrated to lead to competitive performance in many scenarios, and may thus be the default choice. In challenging problems, choosing a slower decay might slightly improve the expected estimation performance at the price of somewhat higher computational costs. Limiting the minimal number of observations K, i.e., stopping before the last layer, is reasonable if a certain minimal segment length is desirable, e.g., high-dimensional scenarios, or when prior knowledge of the true minimal segment length is available. Choosing a larger value for K speeds up computations and reduces the risk of selecting false positives from very short intervals during the final selection. The algorithm is described below.

Algorithm 4 Seeded BS

Require: A decay parameter $a \in [1/2, 1)$, a minimal segment length $K \geq 2$ and a selection method
Ensure: Set of detected change points
  Create a collection $\mathcal{I}$ of seeded intervals (Definition 3.1) with decay a, which cover $\geq K$ observations.
  for $i = 1$ to $|\mathcal{I}|$ **do**
    Take the ith interval in $\mathcal{I}$ and denote its boundaries with s and e
    Calculate the statistic $|T_{s,e}^k|$ for $k = s+1, \dots, e-1$
  end for
  Apply the chosen selection method to $|T_{s,e}^k|$, $(s, e] \in \mathcal{I}, k = s+1, \dots, e-1$, to output the final change point estimates.

Each interval $(s, e] \in \mathcal{I}$ gives a candidate $\hat{k}_{s,e} = \arg\max_{k=s+1,\dots,e-1} |T_{s,e}^k|$ and a maximal gain $G_{s,e} = |T_{s,e}^{\hat{k}_{s,e}}|$. Narrowest-over-threshold selection [20] for a single fixed threshold or multiple thresholds will be applied in this book. Greedy selection can also be applied, as referenced in [149].

The narrowest-over-threshold selection with a fixed threshold gathers all intervals $(s, e] \in \mathcal{I}$ for which $G_{s,e}$ is over the threshold. It takes the shortest, i.e., narrowest, interval among these with corresponding candidate $\hat{k}$. It eliminates all intervals that contain the found change point $\hat{k}$ and repeats these two steps as long as intervals with gain above the predefined threshold remain. The algorithm is described below.

Algorithm 5 Narrowest over threshold algorithm

Require: A collection $\mathcal{I}$ and the statistic $|T_{s,e}^k|$, $(s, e] \in \mathcal{I}, k = s+1, \dots, e-1$. $\mathcal{T} = \emptyset$
Ensure: Set of detected change points $\mathcal{T} \in \{1, \dots, n\}$.
  To start the algorithm: NOT$((0, n], \zeta_n)$.
  Procedure NOT$((s, e], \zeta_n)$.
  if $e - s \leq 1$ **then**
    STOP
  else
    $\mathcal{O}_{(s,e]} = \{(s_m, e_m] \in \mathcal{I} \cap (s, e] : \max_{s_m < k < e_m} |T_{s,e}^k| > \zeta_n)\}$
    if $\mathcal{O}_{(s,e]} = \emptyset$. **then**
      STOP
    else
      $(s_{m^*}, e_{m^*}] = \arg\min_{(s_m, e_m] \in \mathcal{O}_{(s,e]}} |e_m - s_m|$.
      $k^* = \arg\min_{s_{m^*} < k < e_{m^*}} |T_{s_{m^*}, e_{m^*}}^k|$.
      $\mathcal{T} = \mathcal{T} \cup \{k^*\}$
      NOT$((s, k^*], \zeta_n)$.
      NOT$((k^*, e], \zeta_n)$.
    end if
  end if
  End procedure

In the implementation, we treat all intervals on the same layer of seeded intervals as equally long despite slight differences due to rounding and rank all the intervals on the same layer according to their gains to eliminate ties when they are equally long. In some scenarios, this selection method can be somewhat unstable, as even similar thresholds can lead to segmentations differing in both the number and location of the boundaries.

Theorem 3.1.3. *For model (3.1), seeded binary segmentation has the worst-case $O(n \log n)$ computational and $O(n)$ memory complexity if combined with any of the following selection methods:*

(i) greedy selection with an arbitrarily fixed single threshold,

(ii) narrowest-over-threshold selection with an arbitrarily fixed single threshold,

(iii) greedy selection with a complete solution path, provided that the information criterion of the form $\tilde{k}_1 < \cdots < \tilde{k}_N \mapsto \sum_{i=1}^{N} \sum_{t \in (\tilde{k}_i, \tilde{k}_{i+1}]} (\bar{x}_{(\tilde{k}_i, \tilde{k}_{i+1}]} - x_t)^2 + PEN(\{\tilde{k}_1, \ldots, \tilde{k}_N\})$ with $\bar{x}_{(i,j]} = \sum_t x_t/(j-i)$, $\tilde{k}_0 = 0$ and $\tilde{k}_{N+1} = n$; and under the additional assumption that, given the penalty $PEN(S)$, the computation of $PEN(S \cup \{\tilde{k}\})$ is $O(\log n)$ for every finite set of split points S and any arbitrary additional split point $\tilde{k}$.

The proof can be found in [149]. Since the computation complexity is not the main focus of this chapter, it is omitted.

Seeded intervals accelerate the computation (see Theorem 3.1.3), simplifying the statistical analysis compared to random intervals. The following strong consistency result is presented.

Theorem 3.1.4. *Assume model (3.1) holds, and let $\mu_* = \min_{i=1,\ldots,N} \mu_i'$ with $\mu_i' = |\mu_{k_i+1} - \mu_{k_i}|$, and $\lambda = \min_{i=0,\ldots,N} |k_{i+1} - k_i|$ with $k_0 = 0$ and $k_{N+1} = n$. We further assume that*

$$\mu_*^2 \lambda \geq C_0 \log n \quad \textit{for some constant } C_0.$$

Let $\hat{m}$ and $\hat{k}_1 < \cdots < \hat{k}_{\hat{N}}$ denote respectively the number and locations of estimated change points by seeded binary segmentation with the narrowest over threshold as the selection method. Then there exist constants C_1 and C_2, independent of n and the decay a for seeded intervals, such that, given the threshold for the selection method $\zeta_n = C_1 \log^{1/2} n$, as $n \to \infty$,

$$P\left\{\hat{N} = N, \max_{i=1,\ldots,N} \mu_i'^2 |\hat{k}_i - k_i| \leq C_2 \log n\right\} \to 1.$$

Proof of Theorem 3.1.4. We consider interval $(k_i - \lambda, k_i + \lambda]$ for each change point k_i. By construction of seeded intervals, we can find a seeded interval $(c_i - r_i, c_i + r_i]$ such that $(c_i - r_i, c_i + r_i] \subseteq (k_i - \lambda, k_i + \lambda]$, $r_i \geq a^2\lambda$ and $|c_i - k_i| \leq r_i(1 + a^2)/2$. Then the problem almost reduces to the detection of change point on the interval $(c_i - r_i, c_i + r_i]$. The rest of the proof follows nearly the same as the proof of Theorem 1 in [20], and is thus omitted. □

For any candidate model of change points $\mathcal{T}_t = \{\hat{k}_1 < \cdots < \hat{k}_t\}$, the strengthened Schwarz information criterion ([102]) is given by

$$\text{sSIC}(\mathcal{T}_t) = \frac{n}{2} \log \hat{\sigma}_t^2 + t \log^{\theta} n,$$

where

$$\hat{\sigma}_t^2 = \frac{1}{n} \sum_{i=1}^{t+1} \sum_{k=k_{i-1}+1}^{k_i} \left(x_k - \frac{1}{k_i - k_{i-1}} \sum_{j=k_{i-1}+1}^{k_i} x_i \right)^2,$$

$\hat{k}_0 = 0, \hat{k}_{t+1} = n$. The only parameter in the above procedure is the constant θ, which must satisfy $\theta > 1$ (for example, $\theta = 1.01$). This introduces a stronger penalty compared to the standard SIC ($\theta = 1$), hence the term "strengthened" SIC. Let

$$\{\hat{k}_1 < \cdots < \hat{k}_{\hat{N}}\} = \arg\max_{\mathcal{T}_t} \text{sSIC}(\mathcal{T}_t).$$

The consistency of seeded binary segmentation is presented below, when we adopt the strengthened SIC.

Theorem 3.1.5. *Assume notation from Theorem 3.1.4, and that* $\lambda \geq \log^{\theta_0} n$, *and* N, $\mu'_j, j = 1, \ldots, N$ *lie in* (c_1, c_2) *for some constants* $\theta_0 > 1$ *and* $0 < c_1 < c_2 < \infty$. *Assume further that* N, *the number of change points, does not increase with* n. *Let* $\hat{N}$ *and* $\hat{k}_1 < \cdots < \hat{k}_{\hat{N}}$ *be respectively the number and locations of estimated change points by the strengthened Schwarz information criterion using* $\theta \in (1, \theta_0)$. *Then there is a constant* C *independent of* n *and the decay* a *for seeded intervals such that, as* $n \to \infty$,

$$P\left\{\hat{N} = N, \max_{i=1,\ldots,N} \mu_i'^2 |\hat{k}_i - k_i| \leq C \log n\right\} \to 1.$$

Proof of Theorem 3.1.5. It follows the same arguments as the proof of Theorem 3 in [20]. □

3.2 Penalization method

When the total number of change points is unknown, it is common to determine this number by adding a penalty function, related to the number of change points, to the cost function. The goal is to strike a balance between goodness-of-fit and model complexity. If the penalty is too large, no change points may be detected, while too small a penalty can lead to overestimating the number of change points. Thus, selecting an appropriate penalty is crucial to the success of this approach.

To estimate model (3.1) with unknown number and locations of the change points, suppose we choose m potential splitting points, set $k_0 = 0$ and $k_{m+1} = n$, and denote $\boldsymbol{x}_{s:t} = (x_s, ..., x_t)$, for $s \leq t$, to be the set of observations from time s to time t. Based on a candidate segmentation $\boldsymbol{k}_{1:m} = (k_1, \ldots, k_m)$, we can define a cost function

$$\sum_{j=0}^{m} \mathcal{C}\left(\boldsymbol{x}_{k_j+1:k_{j+1}}\right),$$

where

$$\mathcal{C}\left(\boldsymbol{x}_{s+1:t}\right) = \sum_{i=s+1}^{t} \left(x_i - \frac{1}{t-s} \sum_{j=s+1}^{t} x_j\right)^2.$$

Twice the negative log-likelihood is a commonly used cost function in the change point literature. Then, we estimate the number and location of the change points by minimizing

$$\sum_{j=0}^{m} \mathcal{C}\left(\boldsymbol{x}_{k_j+1:k_{j+1}}\right) + \mathcal{P}(\boldsymbol{k}_{1:m}), \tag{3.16}$$

with respect to m and $\boldsymbol{k}_{1:m}$, where $\mathcal{P}(\boldsymbol{k}_{1:m})$ is a penalty function. Next, we will introduce different penalty functions in details.

3.2.1 The l_0 penalty

The l_0 (linear) penalty is a classical choice of penalty, including the familiar Bayesian information criterion (BIC)[220] and Akaike information criterion(AIC)[3]. We denote l_0 penalty as

$$\mathcal{P}_{l_0}(\boldsymbol{k}_{1:m}) = \beta m,$$

where $\beta > 0$ is a tuning parameter.

Given the cost function, the only parameter to calibrate is β. We always choose a value of β that optimizes a certain statistical criterion. BIC chooses β by taking a Bayesian approach with prior distributions that have nonzero prior probabilities on some lower dimensional subspace, and AIC chooses β by minimizing the Kullback-Leibler (KL) divergence of the fitted model from the true model. When the cost function is twice the negative log-likelihood, BIC suggests $\beta = \log n$, and AIC suggests $\beta = 2$. Finally, when no model or distribution is assumed, cross-validation may be an objective approach to tune β.

Here, we provide a brief account of the general principle of BIC, which is one of the most commonly used model selection criteria. Suppose we are interested in selecting the best model from the candidate models $M_1, \dots, M_r$, and assume that the sampling density is $f_k(x|\boldsymbol{\theta}_k)$ under each model M_k, $k = 1, \cdots, r$. Denote the prior density of p_k-dimensional parameter $\boldsymbol{\theta}_k$ under model M_k as $\pi_k(\boldsymbol{\theta}_k)$, then the posterior density of model M_k under n independent and identically distributed samples $\boldsymbol{x}_{1:n} = (x_1, ..., x_n)$ is

$$P(M_k|\boldsymbol{x}_{1:n}) = \frac{P(M_k)\int f_k(\boldsymbol{x}_{1:n}|\boldsymbol{\theta}_k)\pi_k(\boldsymbol{\theta}_k)d\boldsymbol{\theta}_k}{\sum_{j=1}^{r} P(M_j)\int f_j(\boldsymbol{x}_{1:n}|\boldsymbol{\theta}_j)\pi_j(\boldsymbol{\theta}_j)d\boldsymbol{\theta}_j},$$

where $f_k(\boldsymbol{x}_{1:n}|\boldsymbol{\theta}_k)$ is the likelihood function of samples $\boldsymbol{x}_{1:n}$ under model M_k, and $P(M_k)$ is the prior probability of model M_k. In BIC, we consider the following uniform prior distribution at $k = 1, \dots, r$,

$$P(M_k) = \frac{1}{r}.$$

The model M_k corresponding to the maximum $P(M_k|\boldsymbol{x}_{1:n})$ will be selected by BIC. Ignoring the constant, we aim to maximize

$$\int f_k(\boldsymbol{x}_{1:n}|\boldsymbol{\theta}_k)\pi_k(\boldsymbol{\theta}_k)d\boldsymbol{\theta}_k.$$

Based on the Laplacian approximation, we have

$$\begin{aligned}
\int f_k(\boldsymbol{x}_{1:n}|\boldsymbol{\theta}_k)\pi_k(\boldsymbol{\theta}_k)d\boldsymbol{\theta}_k &= \int \exp\{\log[f_k(\boldsymbol{x}_{1:n}|\boldsymbol{\theta}_k)\pi_k(\boldsymbol{\theta}_k)]\}d\boldsymbol{\theta}_k \\
&\approx \int \exp\{\log[f_k(\boldsymbol{x}_{1:n}|\hat{\boldsymbol{\theta}}_k)\pi_k(\hat{\boldsymbol{\theta}}_k)] - \frac{1}{2}(\boldsymbol{\theta}_k - \hat{\boldsymbol{\theta}}_k)^T(nI_{\hat{\boldsymbol{\theta}}_k})(\boldsymbol{\theta}_k - \hat{\boldsymbol{\theta}}_k)\}d\boldsymbol{\theta}_k \\
&= f_k(\boldsymbol{x}_{1:n}|\hat{\boldsymbol{\theta}}_k)\pi_k(\hat{\boldsymbol{\theta}}_k)\int \exp\{-\frac{1}{2}(\boldsymbol{\theta}_k - \hat{\boldsymbol{\theta}}_k)^T(nI_{\hat{\boldsymbol{\theta}}_k})(\boldsymbol{\theta}_k - \hat{\boldsymbol{\theta}}_k)\}d\boldsymbol{\theta}_k \\
&= f_k(\boldsymbol{x}_{1:n}|\hat{\boldsymbol{\theta}}_k)(2\pi)^{\frac{p_k}{2}} n^{-\frac{p_k}{2}} |I_{\hat{\boldsymbol{\theta}}_k}^{-1}|^{\frac{1}{2}},
\end{aligned}$$

where $\hat{\boldsymbol{\theta}}_k$ is the maximum likelihood estimate of $\boldsymbol{\theta}_k$, and $I_{\hat{\boldsymbol{\theta}}_k}$ is the expected Fisher information matrix at $\hat{\boldsymbol{\theta}}_k$. For large n, keeping the terms involving n and ignoring the rest, we can get

$$-2\log \int f_k(\boldsymbol{x}_{1:n}|\boldsymbol{\theta}_k)\pi_k(\boldsymbol{\theta}_k)d\boldsymbol{\theta}_k = -2\log f_k(\boldsymbol{x}_{1:n}|\hat{\boldsymbol{\theta}}_k) + p_k \log n. \tag{3.17}$$

The right-hand side of (3.17) is the BIC value for model M_k.

Based on the l_0 penalty, several efficient algorithms have been proposed to detect multiple change points, such as OPT [132], PELT [140], and FPOP [181], among others. Since it is impossible to cover all of these algorithms within the page limit of this book, we focus on introducing the Pruned Exact Linear Time (PELT) algorithm as an example. PELT improves the computational efficiency of OPT by incorporating a pruning mechanism. Denote the minimization from (3.16) for data $\boldsymbol{x}_{1:t}$ as $F(t)$, and let $\mathcal{T}_s = \{\boldsymbol{k} : 0 = k_0 < k_1 < \cdots < k_{m_1} < k_{m_1+1} = s\}$, where m_1 is the number of potential splitting points over the segment $[0, s]$. Given $F(0) = -\beta$, we can obtain a recursion for an objective function $F(\cdot)$ as

$$\begin{aligned}
F(s) &= \min_{\boldsymbol{k}\in\mathcal{T}_s} \left\{ \sum_{j=0}^{m} [\mathcal{C}\left(\boldsymbol{x}_{k_j+1:k_{j+1}}\right) + \beta] - \beta \right\} \\
&= \min_{t} \left\{ \min_{\boldsymbol{k}\in\mathcal{T}_t} \sum_{j=0}^{m-1} \left[\mathcal{C}\left(\boldsymbol{x}_{k_j+1:k_{j+1}}\right) + \beta\right] - \beta + \mathcal{C}\left(\boldsymbol{x}_{(t+1):s}\right) + \beta \right\} \\
&= \min_{t}\{F(t) + \mathcal{C}\left(\boldsymbol{x}_{(t+1):s}\right) + \beta\},
\end{aligned}$$

which serves as the basis of OPT and PELT. Then, we may specify the pruning rule as: for $t < s < n$, if the following inequality

$$F(t) + \mathcal{C}\left(\boldsymbol{x}_{(t+1):s}\right) \geq F(s)$$

holds, t cannot be the last change point before n. The details of PELT are described in the following table.

Algorithm 6 PELT

Require: Data $\boldsymbol{x}_{1:n} = (x_1, \ldots, x_n)$, penalty β, cost function $\mathcal{C}$.

1: Let $F(0) = -\beta$, $\mathcal{T} = \emptyset$, $R_1 = \{0\}$.

2: **for** $k^* = 1, \ldots, n$ **do**

3: Calculate $F(k^*) = \min_{k\in R_{k^*}} \left[F(k) + \mathcal{C}\left(\boldsymbol{x}_{(k+1):k^*}\right) + \beta\right]$.

4: Let $k^1 = \arg\left\{\min_{k\in R_{k^*}} \left[F(k) + \mathcal{C}\left(\boldsymbol{x}_{(k+1):k^*}\right) + \beta\right]\right\}$.

5: Set $\mathcal{T} = \mathcal{T} \cup \{k^1\}$.

6: Set $R_{k^*+1} = \left\{k^* \cup \left\{k \in R_{k^*} : F(k) + \mathcal{C}\left(\boldsymbol{x}_{(k+1):k^*}\right) < F(k^*)\right\}\right\}$.

7: **end for**

Ensure: the set of change points $\mathcal{T}$

The pruning idea in PELT is based on the following theorem. Under the settings in this chapter, the constant $K = 0$.

Theorem 3.2.1. *We assume that when introducing a change point into a sequence of observations, the cost, $\mathcal{C}$, of the sequence reduces. More formally, we assume there exists a constant K such that for all $t < s < n$,*

$$\mathcal{C}(\boldsymbol{x}_{(t+1):s}) + \mathcal{C}(\boldsymbol{x}_{(s+1):n}) + K \leq \mathcal{C}(\boldsymbol{x}_{(t+1):n}).$$

Then if

$$F(t) + \mathcal{C}(\boldsymbol{x}_{(t+1):s}) + K \geq F(s)$$

holds, at a future time $n > s$, t can never be the last change point before n.

Proof of Theorem 3.2.1. For $t < s < n$,

$$\begin{aligned}
&F(t) + \mathcal{C}\left(\boldsymbol{x}_{(t+1):s}\right) + K \geq F(s) \\
\Longrightarrow &F(t) + \mathcal{C}\left(\boldsymbol{x}_{(t+1):s}\right) + \mathcal{C}\left(\boldsymbol{x}_{(s+1):n}\right) + K + \beta \geq F(s) + \mathcal{C}\left(\boldsymbol{x}_{(s+1):n}\right) + \beta \\
\Longrightarrow &F(t) + \mathcal{C}\left(\boldsymbol{x}_{(t+1):n}\right) + \beta \geq F(s) + \mathcal{C}\left(\boldsymbol{x}_{(s+1):n}\right) + \beta.
\end{aligned}$$

Hence, it follows that t cannot be a future minimizer of the sets

$$S_n := \left\{F(k) + \mathcal{C}\left(\boldsymbol{x}_{(k+1):n}\right) + \beta, k = 0, 1, \ldots, n-1\right\}, n > s,$$

and can be removed from the set of change points for each future step. □

However, in many cases, the penalty function is not linear. Theorem 3.2.2 guarantees PELT can also be used for concave penalties. Here, the penalty function is $f(m)$, that is, the criteria is

$$\sum_{j=0}^{m} \mathcal{C}\left(\boldsymbol{x}_{k_j+1:k_{j+1}}\right) + f(m) \tag{3.18}$$

where $f(m)$ is concave and differentiable. For an appropriately chosen γ, Theorem 3.2.2 shows that the optimum segmentation based on such a penalty corresponds to minimizing

$$\sum_{j=0}^{m} \mathcal{C}\left(\boldsymbol{x}_{k_j+1:k_{j+1}}\right) + m\gamma.$$

Theorem 3.2.2. *Assume that f is concave and differentiable, with the derivative denoted as f'. Further, let $\hat{m}$ be the value of m for which the criteria (3.18) is minimized. Then the optimal segmentation under this set of penalties is the segmentation that minimizes*

$$\sum_{j=0}^{m} \mathcal{C}\left(\boldsymbol{x}_{k_j+1:k_{j+1}}\right) + mf'(\hat{m}).$$

Proof of Theorem 3.2.2. Since f is concave and differentiable, for $m \geq 0$ and $\hat{m} \geq 0$,

$$f(m) \leq f(\hat{m}) + (m - \hat{m})f'(\hat{m}).$$

For any given m we can minimize $\sum_{j=0}^{m} \mathcal{C}\left(\boldsymbol{x}_{k_j+1:k_{j+1}}\right)$ with respect to the m change-points $k_1, \ldots, k_m$. Denote this minimum by $C(m)$. Then we have that $\hat{m}$ satisfies

$$f(\hat{m}) + C(\hat{m}) \leq f(m) + C(m)$$

for any $m = 0, \ldots, n-1$.

Then, using the concavity of $f(\cdot)$,

$$f(\hat{m}) + C(\hat{m}) \leq f(m) + C(m) \leq f(\hat{m}) + (m - \hat{m})f'(\hat{m}) + C(m).$$

Rearranging this shows

$$\hat{m}f'(\hat{m}) + C(\hat{m}) \leq mf'(\hat{m}) + C(m),$$

which means $\hat{m}$ minimizes

$$\sum_{j=0}^{m} \mathcal{C}\left(\boldsymbol{x}_{k_j+1:k_{j+1}}\right) + mf'(\hat{m}).$$

□

This theorem suggests that we can minimize the objective functions based on $f(m)$ using PELT, and just need to apply the correct penalty constant. A simple approach is to run PELT with an arbitrary penalty constant, say $\gamma = f'(1)$. Let m_0 denote the resulting number of change points estimated. We then run PELT with penalty constant $\gamma = f'(m_0)$ and get a new estimate set of the number of change points m_1. If $m_0 = m_1$ we stop. Otherwise, we update the penalty constant and repeat until convergence.

[140] provides a theoretical investigation on the computational cost of the PELT method. In most cases, PELT greatly reduces computational costs through pruning. Under the assumption that regime lengths are randomly drawn from a uniform distribution, the complexity of PELT is $O(n)$. The worst case is where no pruning occurs and the computational cost is $O(n^2)$.

3.2.2 The l_1 penalty

The l_1 penalty is also known as the fused lasso. When the cost function is the quadratic error loss, the l_0 (linear) penalty can be relaxed to a l_1 penalty by using appropriate auxiliary variables [123, 251]. From a computational point of view, l_1 penalty is indeed superior.

For model (3.1), given the potential number of change points m, we can rewrite the multiple change points detection problem as

$$\begin{aligned} \min_{\boldsymbol{\mu}_{1:n}} \quad & \sum_{i=1}^{n} (x_i - \mu_i)^2 \\ \text{subject to} \quad & \sum_{i}^{n-1} \mathbf{1}\left\{\mu_{i+1} - \mu_i\right\} = m, \end{aligned}$$

where $\mathbf{1}$ is an indicator function. Then, we can relax the l_0 constraint into an l_1 constraint on the magnitude of the jumps as follows:

$$\begin{aligned} \min_{\boldsymbol{\mu}_{1:n}} \quad & \sum_{i=1}^{n} (x_i - \mu_i)^2 \\ \text{subject to} \quad & \sum_{i=1}^{n-1} |\mu_{i+1} - \mu_i| \leq m\Delta, \end{aligned}$$

where Δ is the maximum jump amplitude of the real change points. Therefore, we arrive at the l_1 penalty function given by

$$\mathcal{P}_{l_1}(\boldsymbol{k}_{1:m}) = \beta \sum_{i=1}^{m} |\bar{\boldsymbol{x}}_{k_{j-1}+1:k_j} - \bar{\boldsymbol{x}}_{k_j+1:k_{j+1}}|,$$

where

$$\bar{\boldsymbol{x}}_{s+1:t} = \frac{1}{t-s} \sum_{j=s+1}^{t} x_j,$$

and $\beta > 0$ is the smoothing parameter.

This kind of relaxation strategy has been practiced in many statistical research problems such as the sparse PCA [73], and the sparse LDA [187]. In numerical analysis and image denoising, this technique is also known as the total variation regularizer. By adopting the group LARS algorithm, the complexity of l_1 penalty is $O(n \log n)$ [245]. From a theoretical perspective, the estimated change point fractions are asymptotically consistent under model (3.1) [123]. We leave out additional details on how to implement the l_1 penalty since many packages are developed and available for this purpose.

3.2.3 Other penalties

Several other penalty functions are available in the literature, but they tend to be more complex, as optimizing the penalized sum of costs is not fully tractable. In some cases, these penalties can be approximated by an l_0 penalty, allowing the use of algorithms like PELT and FPOP. Here, we introduce four main examples.

The classic BIC statistic usually does not work well here because of irregularities in the likelihood function. By asymptotic approximation of the Bayes factor, modified BIC (mBIC) is derived as a criterion for the model of Brownian motion with changing drift [282]. The penalty function $\mathcal{P}_{mBIC}$ is given by

$$\mathcal{P}_{mBIC}(\boldsymbol{k}_{1:m}) = 3m \log n + \sum_{i=1}^{m+1} \log(\frac{k_i - k_{i-1}}{n}).$$

Lebarbier (2005) [155] contributed another penalty following the results of [182]. One aim of the study was to assess the heuristic method proposed by [182] for the penalty constant estimation, and to calibrate it in the particular Gaussian regression framework [155]. The penalty function $\mathcal{P}_{LE}$ is given by

$$\mathcal{P}_{LE}(\boldsymbol{k}_{1:m}) = \frac{m+1}{n}(\beta_1 \log(\frac{m+1}{n}) + \beta_2),$$

where β_1 and β_2 are positive parameters.

Variable selection is an important part of high-dimensional statistical modeling. Some well-known consistent variable selection methods can also be employed to solve the multiple change points detection problem, such as the smoothly clipped absolute deviation (SCAD) [91] penalty and the minimax concave penalty (MCP) [281]. Specifically, we may rewrite model 3.1 as a linear model

$$x_t = \boldsymbol{\alpha}^T \boldsymbol{A}_t + \epsilon_t, \quad \text{for} \quad t = 1, \ldots, n,$$

where $\boldsymbol{\alpha}$ and $\boldsymbol{A}_t$ are n-dimensional vectors, $\mu_t = \sum_{j=1}^{t} \alpha_j$, $\boldsymbol{A}_{t,j} = 1$ for $j \leq t$ and 0 otherwise. Then, detecting change points $k_1, \ldots, k_m$ is equivalent to estimating $\boldsymbol{\alpha}$. Based on the familiar variable selection principles, we can minimize

$$\frac{1}{n}\sum_{i=1}^{n} \left(x_i - \boldsymbol{\alpha}^T \boldsymbol{A}_i\right)^2 + \sum_{j=1}^{n} \mathcal{P}\left(|\boldsymbol{\alpha}_j|\right).$$

The SCAD penalty function proposed by Fan and Li (2001) [91] is

$$\mathcal{P}_{SCAD}(x) = \begin{cases} \lambda x, & \text{if } x \leqslant \lambda, \\ \frac{\gamma\lambda x - 0.5\left(x^2+\lambda^2\right)}{\gamma-1}, & \text{if } \lambda < x \leqslant \gamma\lambda, \\ \frac{\lambda^2(\gamma+1)}{2}, & \text{if } x > \gamma\lambda, \end{cases},$$

and the MCP penalty function proposed by Zhang (2010) [281] is

$$\mathcal{P}_{MCP}(x) = \begin{cases} \lambda x - \frac{x^2}{2\gamma}, & \text{if } x \leqslant \gamma\lambda, \\ \frac{1}{2}\gamma\lambda^2, & \text{if } x > \gamma\lambda, \end{cases},$$

where $\lambda > 0$ and $\gamma > 0$ are tuning parameters in both penalty functions. Under certain regularity conditions, the estimates of $\boldsymbol{\alpha}$ based on SCAD and MCP can achieve desirable asymptotic properties [91, 281]. Applications of these penalty functions in change point detection have become quite popular, owing to such nice theoretical quality and stable computational performance.

There are many penalty functions to choose from, but it is important to note that some are only applicable to specific cost functions. The choice of cost function is also diverse and crucial in different models. For example, for binary or categorical outcomes in a regression setting, the least squares cost is usually not ideal, and users may adopt zero-one loss or deviance, among other choices. However, this section does not focus on cost functions, so we will not delve into that topic in detail.

3.2.4 Real data example

Genetic alterations are crucial in the development of numerous diseases, including cancer and neurodevelopmental disorders. One common type of alteration, DNA copy number alteration (CNA), can be measured through array comparative genomic hybridization (aCGH), allowing researchers to identify important molecular markers associated with human genetics and disease. Analyzing aCGH profiles from groups of individuals with shared phenotypic traits helps in identifying recurrent CNA patterns, which can indicate strong links to the phenotype in question. For example, as Coe et al. (2006) [65] demonstrated, studying CNAs across various subtypes of lung cancer has revealed distinct, recurrent alterations specific to each subtype. Therefore, detecting change points in DNA copy number alterations is significant for human health research.

In this analysis, we examine the change point detection results obtained by applying different penalty functions to the aCGH dataset from Bleakley and Vert (2011) [30], which documents copy number variations for multiple individuals with bladder tumors. Given that all individuals in this dataset share the same disease, we would expect consistency in the locations of change points across each individual's micro-array data, as these points represent variations in copy number that are typically stable within each segment.

In addition to identifying change points, Stransky et al. (2006) [237] provide a detailed biological analysis of this dataset, identifying specific regions of transcriptome bias that operate independently of DNA copy number alterations. This study further differentiates between "nongenetic" regions, where expression changes occur without DNA copy number shifts, and "genetic" regions, where expression is directly influenced by such changes. Notably, they observed significant differences between these regions in terms of chromosomal location and the proportion of downregulated genes. Such insights allow for a deeper biological interpretation of the identified change points and may help explain the observed variations in copy number alterations across individuals.

The data can be accessed in the `R` package `ecp`, under the dataset `aCGH` (James and Matteson, 2013) [133]. The original dataset included missing values, but after preprocessing by James and Matteson, a sample of 43 individuals with data points up to $n = 2215$ remained available for analysis.

We analyze the results obtained by applying different penalty functions to the real dataset, ACGH, from [30]. This dataset consists of microarray data from 57 individuals diagnosed with bladder tumors. Since all individuals share the same disease, we expect the change point locations to be nearly identical across each microarray set. In this context, a change point corresponds to a change in copy number, which is assumed to remain constant within each segment. After cleaning the missing records [133], a sample of 43 individuals with a size of $n = 2215$ was retained.

In this analysis, we focus on detecting change points in the copy-number variations of the first individual. We compare the results using AIC, BIC, fused lasso, SCAD, and MCP in the following.

In Figure 3.2, we observe that AIC detects 9 change points, while BIC identifies only 2. This demonstrates that, for large n, BIC imposes a stronger penalty on the number of change points compared to AIC, leading to fewer detected points. The most significant change points are located near positions 1724 and 1907.

However, the variable selection methods—fused lasso, SCAD, and MCP—do not achieve convergence on this dataset, likely due to the presence of too many small perturbations and noises. For comparison, we repeated each method 100 times and averaged the number of detected change points. Among the three methods, fused lasso identified the most change points, around 90, while SCAD detected the fewest, approximately 20. All three methods identified more change points than AIC and BIC.

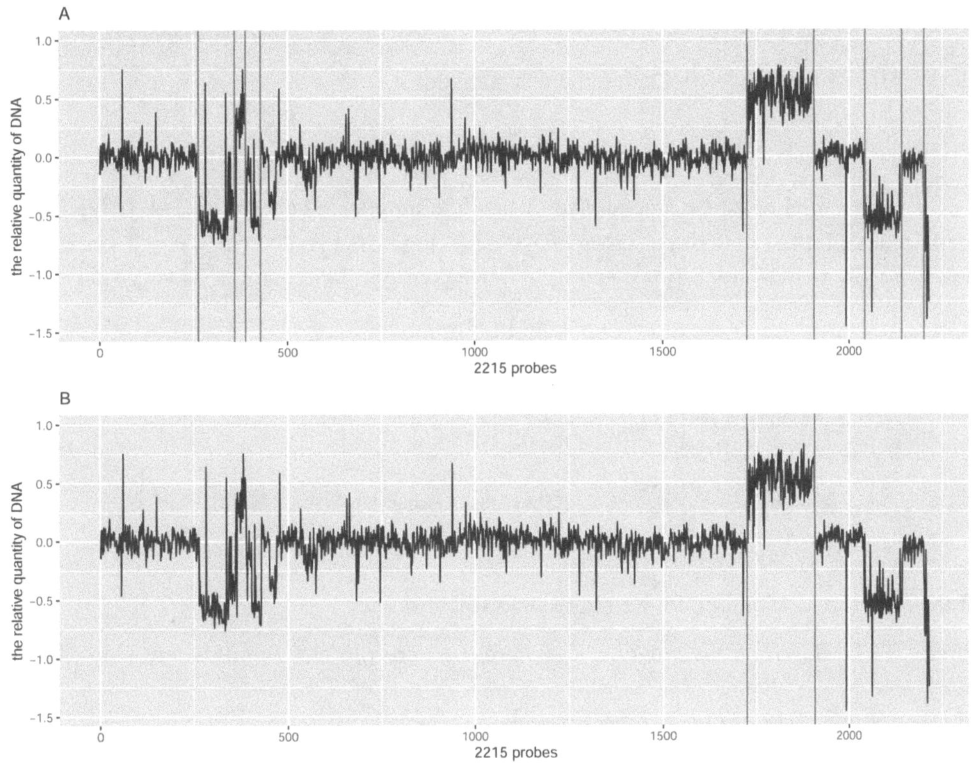

Figure 3.2: Locations of detected change points for the first individual in ACGH data by AIC (top panel) and BIC (bottom panel).

3.3 Moving windows

The concept of the moving window method is outlined as follows. This method requires selection of an appropriate window width. Its primary advantages include the ease of implementation and the low computational complexity.

Algorithm 7 Moving windows methods

1: Select a certain window width, and move the window.
2: Calculate the test statistics on the data in the windows, and get the curve corresponding to the test statistics.
3: Verify whether the local maximum of this curve exceeds the critical value of the test, and if so, it is a change point.

There is relatively little research on moving windows methods, comparing to the other methods. The moving windows method for detecting multiple change points proposed by [273] will be described in detail below.

Based on model (3.1), let

$$d_i = \mu_{k_i+1} - \mu_{k_i} \neq 0, \quad i = 1, \dots, m,$$

and suppose without loss of generality that $|d_1| \geq |d_2| \geq \cdots \geq |d_m|$, and if $|d_i| = |d_j|$, and $i < j$, then $k_i < k_j$.

Let $l = l(n) \uparrow \infty, l/n \to 0$. Define the test statistics

$$D_{nk} = \frac{x_{k+1} + \cdots + x_{k+l}}{l} - \frac{x_{k-1} + \cdots + x_{k-l}}{l},$$

for $l \leq k \leq n - l$. Let h_n be a sequence of positive numbers such that $h_n \to 0, l/(nh_n) \to 0$. For definiteness, let $h_n = c' \left(\log_3 n / \log_2 n\right)^{1/3}, l = \log n \left(\log_2 n\right)^{2/3} \left(\log_3 n\right)^{1/3}$ where $\log_1 n = 1 \wedge \log n, \log_k n = \log_1 \left(\log_{k-1} n\right)$, and $c' > 0$ is a constant. In general, c' is very small. Define

$$\begin{aligned}
I_1 &= \arg\max_k \left\{ |D_{nk}| - \frac{k}{n} h_n \right\} \\
I_2 &= \arg\max_k \left\{ |D_{nk}| - \frac{k}{n} h_n : |k - I_1| > 4l \right\}, \\
I_3 &= \arg\max_k \left\{ |D_{nk}| - \frac{k}{n} h_n : |k - I_1| > 4l \text{ and } |k - I_2| > 4l \right\}, \\
&\cdots.
\end{aligned}$$

If the definition is not unique, we choose the smallest one.

Theorem 3.3.1. *Based on the model 3.1, if there is a constant $c > 0$ that makes $(k_i - k_j)/n > c$ for any $1 < i, j < N$, then*

(1) $|I_i - k_i| \leq 2l$ for all large n, a. s, $i = 1, \ldots, N$.

(2) $D_{nI_i} \to d_i$, a. s., $i = 1, \ldots, N$.

(3) $D_{nI_i} = O\left(h_n\right)$, a. s. for $i > N$.

Theorem 3.3.1 gives strongly consistent estimators of the change points k_i and the jump size d_i, $i = 1, \ldots, m$. However, Theorem 3.3.1 does not supply a method to estimate the integer m, explicitly. But, based on D_{nI_i}, we can construct a strongly consistent estimator of m as follows:

Let

$$G_{nm} = \frac{1}{2^{m+1}} \left|D_{n,I_{m+1}}\right| + \frac{1}{2^{m+2}} \left|D_{n,I_{m+2}}\right| + \cdots + mc_n.$$

Here $c_n > 0$ with the properties $c_n \to 0$ and $h_n/c_n \to 0$. We can choose $c_n = c' \left(\log_2 n^{2/3} / \log_2 n\right)^{1/3}$, where $c' > 0$ has been defined in h_n. Let

$$\hat{N}_n = \arg\min_m G_{nm}.$$

After Theorem 3.3.1 is established, we may then further obtain the following theorem.

Theorem 3.3.2. *Based on the model 3.1, if there is a constant $c > 0$ that makes $(k_i - k_j)/n > c$ for any $1 < i, j < N$, then*

$$\hat{N}_n \to N, \text{ a. s. }.$$

The proofs of Theorem 3.3.1 and Theorem 3.3.2 are given below.

Proof of Theorem 3.3.1. In this part, we prove $|I_1 - k_1| \leq 2l$ for all large n, a. s. Introduce the following notations

$$\begin{aligned}
\Delta_k &= \frac{1}{l}(\mu_{k+1} + \cdots + \mu_{k+l}) - \frac{1}{l}(\mu_{k-1} + \cdots + \mu_{k-l}), \\
W_k^+ &= \frac{1}{l}(\epsilon_{k+1} + \cdots + \epsilon_{k+l}), \\
W_k^- &= \frac{1}{l}(\epsilon_{k-1} + \cdots + \epsilon_{k-l}).
\end{aligned}$$

At first, we note that for $|k-k_1| > 2l$,

$$P\left(|D_{nk_1}| - \frac{k_1}{n}h_n \le |D_{nk}| - \frac{k}{n}h_n\right)$$
$$\le P\left(|\Delta_{k_1}| - |\Delta_k| + \left(\frac{k-k_1}{n}\right)h_n \le |W_k^+| + |W_k^-| + |W_{k_1}^+| + |W_{k_1}^-|\right).$$

Because $|k-k_1| > 2l$, the points $k-l, \ldots, k-1, k+1, \ldots, k+l$ are all on the same side of k_1. If n is larger than some nonrandom number $N_1 > 0$, for any k with $|k-k_1| > 2l$, the interval $[k-l, k+l]$ can contain at most one discontinuity point k_i.

There are three possibilities: (a) No k_i in $[k-l, k+l]$, (b) $k_i \in [k-l, k+l]$, $|d_i| = |d_1|$, then $k_i > k_1$, (c) $k_i \in [k-l, k+l]$, $|d_i| < |d_1|$.

For case (a),

$$|\Delta_{k_1}| = |d_1|, \quad |\Delta_k| = 0,$$

so,

$$|\Delta_{k_1}| - |\Delta_k| + \frac{k-k_1}{n}h_n = |d_1| + \frac{k-k_1}{n}h_n > \frac{1}{2}|d_1|,$$

for $n \ge N_2 > 0$, where N_2 is a nonrandom constant.

For case (b),

$$|\Delta_{k_1}| = |d_1|, \quad |\Delta_k| = \frac{l - |k-k_i|}{l}|d_1|,$$

so, for a constant $c > 0$,

$$|\Delta_{k_1}| - |\Delta_k| + \frac{k-k_1}{n}h_n \ge \frac{k-k_1}{n}h_n \ge ch_n,$$

when $n \ge N_3 > 0$, where N_3 is nonrandom.

For case (c),

$$|\Delta_{k_1}| - |\Delta_k| + \frac{(k-k_1)}{n}h_n \ge |d_1| - |d_i| + o(1) \ge \frac{1}{2}(|d_1| - |d_i|) > 0,$$

when $n \ge N_4 > 0$, where N_4 is nonrandom.

Thus, for $n \ge \max(N_1, N_2, N_3, N_4)$, and for some constants $c_1 > 0, c_2 > 0$,

$$\begin{aligned}
P\left(|D_{nk_1}| - \frac{k_1}{n}h_n \le |D_{nk}| - \frac{k}{n}h_n\right)
&\le 4P\left(b_1 h_n \le \left|\frac{v_1 + \cdots + v_l}{l}\right|\right)\\
&\le 4P\left(\frac{b_1}{\sigma}h_n\sqrt{l} \le |z|\right)\\
&\le 8\frac{1}{\sqrt{2\pi}\frac{b_1}{\sigma}h_n\sqrt{l}}e^{-\frac{1}{2}\frac{b_1^2}{\sigma^2}h_n^2 l}\\
&\le b_2 e^{-b_3 \log n \log_3 n} = b_2 n^{-b_3 \log_3 n},
\end{aligned}$$

if we choose $h_n = c'(\log_3 n/\log_2 n)^{1/3}$, $l = \log n(\log_2 n)^{2/3}(\log_3 n)^{1/3}$. Here, $v_1, \ldots, v_l, z \overset{iid}{\sim} \mathcal{N}(0,1)$, and b_1, b_2, b_3 are positive constants, $\log_2 x = \log\log x, \log_3 x = \log\log_2 x$. Thus, the series

$$\sum_n P\left(|D_{nk_1}| - \frac{k_1}{n}h_n \le |D_{nk}| - \frac{k}{n}h_n, \text{ for some } k \text{ with } |k-k_1| > 2l\right)$$
$$\le \sum_n nP\left(|D_{nk_1}| - \frac{k_1}{n}h_n \le |D_{nk}| - \frac{k}{n}h_n\right) < \infty.$$

By the Borel-Cantelli Lemma,

$$P(\exists N, \text{ s.t. } \quad n \geq N, |k-k_1| > 2l \\ \Longrightarrow |D_{nk_1}| - \frac{k_1}{n}h_n > |D_{nk}| - \frac{k}{n}h_n) = 1,$$

or

$$P(\exists N, \text{ s.t. } \quad n \geq N, |D_{nk_1}| - \frac{k_1}{n}h_n \leq |D_{nk}| - \frac{k}{n}h_n \\ \Longrightarrow |k-k_1| \leq 2l) = 1.$$

Since $|D_{nI_1}| - \frac{I_1}{n}h_n \geq |D_{nk_1}| - \frac{k_1}{n}h_n$,

$$P(\exists N, \text{ s.t. } \quad n \geq N \Longrightarrow |I_1 - k_1| \leq 2l) = 1.$$

This proves (1) when $i = 1$.

By

$$|D_{nI_1}| - \frac{I_1}{n}h_n \geq |D_{nk_1}| - \frac{k_1}{n}h_n,$$

we have that

$$\liminf |D_{nI_1}| \geq |d_1|.$$

Besides, by the definition of D_{nI_1}, we have

$$|D_{nI_1}| \leq |d_1| + 2|z_l|,$$

where $z_l \sim \mathcal{N}(0, 1/l)$. So, we get that $D_{nI_1} \to d_1$ a.s.

2. In this part we prove $|I_i - k_i| \leq 2l$ for all large n, a.s., and $D_{nI_i} \to d_i$ a. s., for $2 \leq i \leq m$. But we carry out the proof only for the case $i = 2$.

Suppose k is such that $|k - k_1| > 2l$ and $|k - k_2| > 2l$. Suppose for these k, $|D_{nk_2}| - \frac{k_2}{n}h_n \leq |D_{nk}| - \frac{k}{n}h_n$, so that

$$|\Delta_{k_2}| - |\Delta_k| + \frac{k-k_2}{n}h_n \leq |W_k^+| + |W_k^-| + |W_{k_2}^+| + |W_{k_2}^-|.$$

Since $|k - k_1| > 2l$, $|k - k_2| > 2l$, k_1 and k_2 are not in the interval $[k - l, k + l]$, there exists $N_1 > 0$ such that as $n \geq N_1$, $[k - l, k + l]$ contains at most one discontinuity point.

There are three possibilities: (a) No k_i in $[k - l, k + l]$, (b) $k_i \in [k - l, k + l]$, $|d_i| = |d_2|$, then $i > 2$, $k_i > k_2$. (c) $k_i \in [k - l, k + l]$, $|d_i| < |d_2|$.

For (a), we have

$$|\Delta_{k_2}| = |d_2|, \quad |\Delta_k| = 0,$$

and then as $n \geq N_2, N_2$ nonrandom,

$$|\Delta_{k_2}| - |\Delta_k| + \frac{k-k_2}{n}h_n = |d_2| + \frac{k-k_2}{n}h_n > \frac{1}{2}|d_2|.$$

For (b),

$$|\Delta_{k_2}| = |d_2|, \quad |\Delta_k| = \frac{l - |k-k_i|}{l}|d_2|,$$

and as $n \geq N_3, N_3$ nonrandom, we have

$$|\Delta_{k_2}| - |\Delta_k| + \frac{k-k_2}{n}h_n \geq \frac{k-k_2}{n}h_n \geq ch_n,$$

for constant $c > 0$.

For (c), we can show in the same manner that as $n \geq N_4$, N_4 nonrandom,

$$|\Delta_{k_2}| - |\Delta_k| + \frac{k - k_2}{n} h_n > \frac{1}{2}(|d_2| - |d_i|) > 0.$$

Therefore, as $n \geq \max(N_1, N_2, N_3, N_4)$,

$$\begin{aligned} P\left(|D_{nk_2}| - \frac{k_2}{n} h_n \leq |D_{nk}| - \frac{k}{n} h_n\right) & \\ & \leq 4P\left(b_1 h_n \leq \left|\frac{v_1 + \cdots + v_l}{l}\right|\right) \\ & \leq 4P\left(b_2 h_n \sqrt{l} \leq |z|\right) \leq b_2 n^{-b_3 \log_3 n} \end{aligned}$$

for positive constants b_1, b_2, b_3. Here, $v_1, \ldots, v_l, z$ have the same significance as before. Therefore, the series

$$\sum_n P\left(|D_{nk_2}| - \frac{k_2}{n} h_n \leq |D_{nk}| - \frac{k}{n} h_n, \text{ for some } k \text{ with } |k - k_1| > 2l \text{ and } |k - k_2| > 2l\right) < \infty.$$

By the Borel-Cantelli Lemma,

$$\begin{aligned} P\Big(\exists N, \text{s.t.} \quad & n \geq N, |k - k_1| > 2l \text{ and } |k - k_2| > 2l \\ & \Longrightarrow |D_{nk_2}| - \frac{k_2}{n} h_n > |D_{nk}| - \frac{k}{n} h_n\Big) = 1, \end{aligned}$$

or

$$\begin{aligned} P\Big(\exists N, \text{s.t.} \quad & n \geq N, |D_{nk_2}| - \frac{k_2}{n} h_n \leq |D_{nk}| - \frac{k}{n} h_n \\ & \Longrightarrow |k - k_1| \leq 2l \text{ or } |k - k_2| \leq 2l\Big) = 1. \end{aligned}$$

Since

$$|k_2 - I_1| \geq |k_2 - k_1| - |k_1 - I_1| > 4l$$

for large n, by the definition of D_{nI_2}, we can get

$$|D_{nk_2}| - \frac{k_2}{n} h_n \leq |D_{nI_2}| - \frac{I_2}{n} h_n$$

for large n. Therefore,

$$P(\exists N, \text{ s.t.} \quad n \geq N \Longrightarrow |I_2 - k_1| \leq 2l \text{ or } |I_2 - k_2| \leq 2l) = 1.$$

However,

$$|I_2 - k_1| \geq |I_2 - I_1| - |I_1 - k_1| > 4l - 2l = 2l,$$

so $|I_2 - k_2| \leq 2l$ for large n a. s.. This proves (1) when $i = 2$.

By

$$|D_{nI_2}| - \frac{I_2}{n} h_n \geq |D_{nk_2}| - \frac{k_2}{n} h_n,$$

we have

$$\liminf |D_{nI_2}| \geq |d_2|.$$

Besides, by the definition of D_{nI_2}, we have

$$|D_{nI_2}| \leq |d_2| + 2|z_l|,$$

where $z_l \sim \mathcal{N}(0, 1/l)$. So, we get that $D_{nI_2} \to d_2$ a.s.

3. In this part, we prove that $D_{nI_{N+1}} = O(h_n)$, a. s.. By definition, $|I_{N+1} - I_i| > 4l, i = 1, \ldots, N$ for sufficiently large n, a. s. Thus, in this case,

$$|I_{N+1} - k_i| \geq |I_{N+1} - I_i| - |I_i - k_i| > 4l - 2l = 2l.$$

Therefore, in $[I_{N+1} - l, I_{N+1} + l]$, there are no discontinuity points. Then we can get

$$\left|D_{nI_{N+1}}\right| \leq \left|\Delta_{I_{N+1}}\right| + \left|W^{+}_{I_{N+1}}\right| + \left|W^{-}_{I_{N+1}}\right| = \left|W^{+}_{I_{N+1}}\right| + \left|W^{-}_{I_{N+1}}\right|,$$

and

$$\begin{aligned}
&\sum P\left(\left|D_{nI_{N+1}}\right| \geq h_n\right) \\
&\leq \sum P\left(h_n \leq \left|W^{+}_{I_{N+1}}\right| + \left|W^{-}_{I_{N+1}}\right|\right) \\
&\leq b \sum_n nP\left(b_1 h_n \leq \left|\frac{v_1 + \cdots + v_l}{l}\right|\right) \\
&\leq b \sum_n nP\left(b_2 \sqrt{l} h_n \leq |z|\right) < \infty,
\end{aligned}$$

where b, b_1, b_2 are positive constants, $v_1, \ldots, v_l, z$ have the same significance as before. Therefore,

$$\left|D_{nI_{N+1}}\right| = O(h_n), \quad \text{a.s.}.$$

□

Proof of Theorem 3.3.2. If $k < N$, then, almost surely, as $n \to \infty$

$$\begin{aligned}
G_{nk} - G_{nN} &\geq \frac{1}{2^{k+1}}\left|D_{nI_{k+1}}\right| + \cdots + \frac{1}{2^N}\left|D_{nI_N}\right| + (k - N)c_n \\
&\to \frac{1}{2^{k+1}}\left|d_{k+1}\right| + \cdots + \frac{1}{2^N}\left|d_N\right| > 0
\end{aligned}$$

by (2) of Theorem 3.3.1. That means that $\hat{N}_n \neq k$ for large n.

If $k > N$, then, noticing that $|D_{nI_i}| \downarrow$, as $i \uparrow$. By (3) of Theorem 3.3.1, we have, for a constant $C > 0$, that

$$\begin{aligned}
G_{nk} - G_{nN} &= -\frac{1}{2^{N+1}}\left|D_{nI_{N+1}}\right| - \cdots - \frac{1}{2^k}\left|D_{nI_k}\right| + (k - N)c_n \\
&\geq -\frac{1}{2^N}\left|D_{nI_{N+1}}\right| + (k - N)c_n \\
&\geq -\frac{C}{2^N} h_n + (k - N)c_n \\
&= c_n\left((k - N) - \frac{Ch_n}{2^N c_n}\right) \\
&\geq c_n \frac{1}{2}(k - N) > 0
\end{aligned}$$

for sufficiently large n, almost surely. In this case also $\hat{N}_n \neq k$. So, almost surely, for n sufficiently large, $\hat{N}_n = N$. □

To use this moving windows method, we need to determine c'. We propose to select the best value that minimizes the BIC:

$$BIC(i) = -2 \log L(\hat{\boldsymbol{\mu}}^i_{1:n}) + n_i \log n, \tag{3.19}$$

where n_i is the number of change points estimated by using the moving windows method with $c' = i$, and $\hat{\boldsymbol{\mu}}^i_{1:n}$ is the maximum likelihood estimation of $\boldsymbol{\mu}_{1:n}$ based on the change points estimated by using the moving windows method with $c' = i$.

This method can be used for more general stochastic processes, the details of which can be seen in [273].

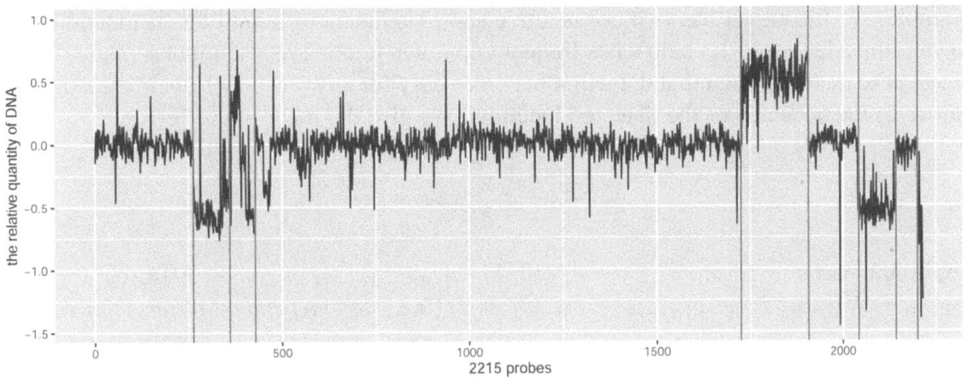

Figure 3.3: Alternate change points for the first individual in ACGH by the moving windows method.

Then, we use the moving windows method to analyze the copy-number variations of the first individual in ACGH [133]. By minimizing (3.19), we get $c' = 0.0007$. Therefore, we choose $h_n = 0.0007\left(\log_3 n/\log_2 n\right)^{1/3}$, $l = \log n\left(\log_2 n\right)^{2/3}\left(\log_3 n\right)^{1/3}$, and $c_n = 0.0007\left(\log_2 n^{2/3}/\log_2 n\right)^{1/3}$. We can get 8 change points which can be seen in Figure 3.3. This is similar to the change points found by AIC.

3.4 Comparison

This chapter introduces binary segmentation methods, penalization methods, and moving windows methods to detect multiple change points in Model 3.1. Here, we summarize and compare these three classes of approaches.

Among the penalization methods, using OPT, PELT, and FPOP to solve the l_0 penalty (AIC or BIC) is the most common choice. The computational complexity of OPT is $O(n^2)$. By pruning, the computational complexity of PELT can be $O(n)$ in some cases, but in the worst case where no pruning is involved, the complexity can remain at $O(n^2)$. FPOP combines ideas from PELT and pDPA, and FPOP always prunes more than PELT. Remarkably, we can obtain the exact solution by OPT, PELT, and FPOP. However, we can only get an approximate solution through binary segmentation and moving windows methods.

WBS is currently the most popular binary segmentation method. In the worst case, WBS needs $O(n^2)$ random intervals and thus an overall $O(n^3)$ computational complexity to achieve reasonable statistical performance. Seeded BS reduces computational complexity to $O(n \log n)$ by constructing a deterministic set of background intervals. NOT is a modification of WBS that focuses on the narrowest data intervals on which it is possible to locate the features of interest. NOT can estimate multiple change points consistently in piecewise linear and continuous signals as described in scenario 2 of [20], while WBS cannot.

Moving windows methods are often ignored and not commonly used when detecting multiple change points. The key to moving windows methods is to find a suitable window width m. The main benefits of moving windows methods are its low computational complexity $O(mn)$ and ease of implementation.

Most methods perform well when the minimal segment length and the minimal jump size are large enough. Here, we consider two special changing mean signals to compare these methods. In Example 3.2, the signal has two consecutive short segments of length ten at location $n/3$ with up/down jumps. Example 3.3 below has frequent up/down jumps every tenth observation. The jump sizes are set to be large, such that detection becomes easy for any reasonable estimation method. As a measure of the accuracy of the detected locations, we give the mean square error,

$$MSE = \frac{1}{n}\sum_{t=1}^{n}(\hat{\mu}_t - \mu_t)^2.$$

Example 3.2. *Let* $x_1, \ldots, x_{\lfloor n/3 \rfloor - 10} \sim \mathcal{N}(0,1)$, $x_{\lfloor n/3 \rfloor - 9}, \ldots, x_{\lfloor n/3 \rfloor} \sim \mathcal{N}(4,1)$, $x_{\lfloor n/3 \rfloor + 1}, \ldots, x_{\lfloor n/3 \rfloor + 10} \sim \mathcal{N}(-4,1)$ *and* $x_{\lfloor n/3 \rfloor + 11}, \ldots, x_T \sim \mathcal{N}(0,1)$ *be independent observations.*

Example 3.3. *Let* $x_1, \ldots, x_{10} \sim \mathcal{N}(4,1)$, $x_{11}, \ldots, x_{20} \sim \mathcal{N}(-4,1)$, $x_{21}, \ldots, x_{30} \sim \mathcal{N}(4,1)$, $x_{31}, \ldots, x_{40} \sim \mathcal{N}(-4,1)$, *etc. up to observation* n *be independent observations.*

Table 3.1 provides the average number of detected change points of five methods on 100 replications. Table 3.2 provides the average MSE of five methods on 100 replications. Table 3.3 provides the average running time of five methods on 100 replications.

Table 3.1: The average number of detected change points of five methods on 100 replications.

Estimated number		PELT	FPOP	WBS	NOT	Moving windows
Example 3.2	$n = 600$	3.00	3.04	3.08	3.02	3.18
	$n = 6000$	3.00	3.01	3.00	3.00	1.14
	$n = 60000$	3.00	3.00	2.97	2.56	1.00
Example 3.3	$n = 600$	59.0	59.2	59.0	0	1.95
	$n = 6000$	599	599	599	0	1.01
	$n = 60000$	6000	6000	6000	0	1.00

Table 3.2: The average MSE of five methods on 100 replications.

MSE		PELT	FPOP	WBS	NOT	Moving windows
Example 3.2	$n = 600$	$8.34e-3$	$9.22e-3$	$9.30e-3$	$9.30e-3$	$4.92e-1$
	$n = 6000$	$1.08e-3$	$1.11e-3$	$1.17e-3$	$1.08e-3$	$5.35e-2$
	$n = 60000$	$8.27e-5$	$8.51e-5$	$1.50e-4$	$1.52e-3$	$5.36e-3$
Example 3.3	$n = 600$	$1.00e-1$	$1.05e-1$	$1.01e-1$	$1.60e+1$	$1.58e+1$
	$n = 6000$	$1.00e-1$	$1.01e-1$	$1.69e-1$	$1.60e+1$	$1.60e+1$
	$n = 60000$	$1.01e-1$	$1.01e-1$	$2.09e-1$	$1.60e+1$	$1.60e+1$

As we can see, PELT performs the best in both examples, and WBS and FPOP also perform well. However, NOT and Moving windows method cannot work in Example 3.3, probably because the minimum segment length is too short and mutations are too frequent. From the point of view

Table 3.3: The average running time of five methods on 100 replications.

running time(s)		PELT	FPOP	WBS	NOT	Moving windows
Example 3.2	$n = 600$	$1.78e-3$	$2.35e-4$	$2.43e-2$	$3.11e-2$	$6.01e-4$
	$n = 6000$	$5.40e-3$	$1.52e-3$	$1.12e-1$	$6.82e-2$	$7.69e-3$
	$n = 60000$	$3.28e-2$	$1.60e-2$	$9.91e-1$	$4.47e-1$	$1.44e-1$
Example 3.3	$n = 600$	$2.13e-3$	$6.05e-4$	$2.52e-2$	$3.32e-2$	$5.41e-4$
	$n = 6000$	$6.49e-3$	$1.25e-3$	$1.11e-1$	$6.44e-2$	$6.55e-3$
	$n = 60000$	$5.40e-2$	$1.20e-2$	1.01	$4.41e-1$	$1.46e-1$

of computational efficiency, the running time of FPOP, PELT, and Moving windows method is relatively short. Especially when the sample size n is large, the computational efficiency of the pruning algorithm is faster. Therefore, when facing frequent mutations, we can give priority to PELT. More examples of these methods can be found in the book.

3.5 Other methods and R packages

3.5.1 A couple more methods

In addition to the well-known approaches mentioned above, there is also a significant body of research on multiple change point problems across various scenarios. This includes methods such as the Bayesian approach, genetic algorithms, and techniques for controlling the false discovery rate (FDR).

The Bayesian approach to multiple change point detection focuses on inferring the posterior distribution of the number and locations of change points. Barry and Hartigan (1993) [24] defined the partition $\rho = (U_1, U_2, \ldots, U_n)$. where the binary dummy variable $U_i = 1$ indicates the presence of a change point at position i with probability p_i and $U_i = 0$ indicates the absence of a change point with probability $1 - p_i$. Assume the total sum is $\sum_{i=1}^{n} U_i = b$, where b represents the number of change points. The estimators of the model parameters can be derived by maximizing the likelihood function of the observed data X_i given the partition

$$\prod_{i=1}^{n} f(X_i|\rho) f(\rho)$$

where the partition ρ follows a prior probability $f(\rho) = \frac{1}{p_0}\int_0^{p_0} p^{b-1}(1-p)^{n-b}dp$ and the tuning parameters p_0 takes value in the interval $(0, 1)$. The transition probability, p_i, which corresponds to the conditional probability of a change occurring at position i, can be obtained from the ratio

$$\frac{p_i}{1-p_i} = \frac{P(U_i = 1|X_i, i = 1, \ldots, n, U_j, j \neq i)}{P(U_i = 0|X_i, i = 1, \ldots, n, U_j, j \neq i)}.$$

A fast Markov Chain Monte Carlo (MCMC) implementation of Barry and Hartigan's method can be found in the `R` package `bcp` ([87]).

A genetic algorithm for detecting multiple change points in autoregressive time series was proposed by Davis et al. (2006) [76], referred to as Auto-PARM. This method applies the minimum description length (MDL) principle to evaluate different segmented AR models, viewing the structural break estimation as a model selection problem where candidate models may vary in the number of parameters. To optimize the MDL criterion, they utilized a genetic algorithm (GA), a widely used tool in statistical optimization that efficiently navigates complex solution spaces. Specifically, to implement GA, an initial population of chromosomes is generated at random, and then new chromosomes are generated by either a crossover operation or a mutation operation. This process continues for several generations. At the end of each migration, the overall best chromosome (i.e., the chromosome with the smallest MDL) is noted. If this best chromosome does not change for 10 consecutive migrations, or if the total number of migrations exceeds 20, then this best chromosome is taken as the solution to this optimization problem.

To automatically determine the number of change points from a sample, [50] introduces a data-adaptive selection procedure via false discovery rate (FDR) control based on order-preserving sample-splitting. The key idea is to construct a series of statistics that exhibit a symmetry property without a change point and yield positive values if a change point is present. Let $Z_o = \{X_{2i-1}, i = 1, \ldots, m\}$ and $Z_e = \{X_{2i}, i = 1, \ldots, m\}$ be the odd number and even number partitions of the original sample $\{X_i, i = 1, \cdots, n\}$ with $n = 2m$. Assume $\mathbb{T}_{p_n} = \{\hat{\tau}_1, \ldots, \hat{\tau}_{p_n}\}$ is the set of candidate change points identified on Z_o by a specific algorithm, such as BS or PELT, where p_n represents the number of detected change points. For some $h > 1$, $P(\max_{i=1,\ldots d} \min_{j=1,\ldots,p_n} |\tau_i - \hat{\tau}_j| \leq h) \to 1$ with $\{\tau_1, \ldots, \tau_d\}$ representing the true change points. Define $I_0 = \{j : \min_{i=1,\ldots,d} |\tau_i - \hat{\tau}_j| > h, j = 1, \ldots, p_n\}$ and $I_1 = \{j : \min_{i=1,\ldots,d} |\tau_i - \hat{\tau}_j| \leq h, j = 1, \ldots, p_n\}$. Consider a statistic $W_j = (S^0_{j+1} - S^o_j)\Omega_j(S^e_{j+1} - S^e_j)$ where S^o_j is a score statistic based on the data $\{X_{2i-1} : \hat{\tau}_{j-1} < 2i - 1 \leq \hat{\tau}_j\}$, S^e_j is based on the data $\{X_{2i} : \hat{\tau}_{j-1} < 2i \leq \hat{\tau}_j\}$ and Ω_j serves as a rough scale estimator for standardization. The statistic must satisfy that W_j is symmetric in the set I_0 and positive in the set I_1. Define

$$L = \inf\left\{t > 0, \frac{1 + \sharp\{j : W_j \leq -t\}}{\sharp\{j : W_j \geq t\} \vee 1} \leq \alpha\right\}$$

where α is the target FDR level. Then, the estimators of change points are given by $\{\hat{\tau}_j : W_j \geq L, j = 1, \ldots, p_n\}$.

3.5.2 R packages and an illustration

There are also abundant `R` packages which implement change point detection for various problems. We provide a brief survey in this section. In Table 3.4, we summarize the `R` packages and the main functions used in those packages. Packages `not` [20], `breakfast` [5, 103, 102, 104], `IDetect` [6], `trendsegmentR` [180], `mosum` [185], `ChangepointTesting` [44], `changepoint` [141, 142], `changepointsVar` [1, 189], `breakpoint` [204, 203, 202, 201], `Sementor3IsBack` [64] are all suitable for univariate time series variable. While packages `InsepectChangepoint` [260], `changepointsHD` [10] allow detection of change points for multiple variables.

The dataset used in this section is sourced from the Singapore Department of Statistics and can be directly download from the website: `https://www.singstat.gov.sg/find-data/search-by-theme/society/environment/latest-data`. It provides comprehensive climate data for Singapore from 1960 to 2023, including key meteorological indicators such as annual means of daily maximum and minimum air temperatures, extreme temperature values, total rainfall, number of rainy days, relative humidity, and sunshine duration. This dataset comprises 64 annual observations for each variable, providing a robust time series for analysis.

We study the change points in the weather conditions in Singapore from 1960 to 2023. The `changepointsHD` and `InspectChangepoint` packages can analyze multiple time series simultaneously. Applied to the Singapore climate dataset, they can jointly analyze variables such as

Table 3.4: R packages for change point analysis and their function names.

Packages	Functions
not	not()
breakfast	breakfast(idetect+ic)
	breakfast(idetect_seq+thresh)
	breakfast(wbs+ic)
	breakfast(wbs2+sdll)
	breakfast(not+ic)
	breakfast(tguh+lp)
	breakfast(wcm+gsc)
IDetect	ID()
trendsegmentR	trendsegment()
mosum	mosum()
ChangepointTesting	changePoint()
changepoint	cpt.meanvar()
changepointsVar	jumpointsVar()
breakpoint	CE.Normal.Mean()
Segmentor3IsBack	Segmentor()
InspectChangepoint	inspect()
changepointsHD	simulated_annealing()
	rank_one()

annual mean daily maximum and minimum air temperatures, total rainfall, highest daily rainfall total, number of rainy days, bright sunshine daily means, and 24-hour relative humidity means. Using these methods, we can identify significant shifts in Singapore's climate patterns, providing a comprehensive view of historical changes. To facilitate interpretation, we plot the detected change points in each weather variable's time series, allowing us to visualize and compare the structural changes identified by both packages across all variables.

Figure 3.4 illustrates the trend in annual mean daily maximum air temperatures. The vertical dashed lines represent potential change points, with different markers on these lines indicating the specific methods that identified changes at those times. It is evident that the trend in daily maximum air temperature shows an upward fluctuation, with most methods detecting change points around 1980, 1996, and 1998. Figure 3.5 displays the trend in annual mean daily minimum air temperatures. Similar to the maximum temperature trend, this series also exhibits an undulating rise. Change points for minimum air temperatures are most frequently identified around 1980, 1996, and 2013. Together, these observed temperature changes align with the global trend of rising temperatures over recent decades.

Figures 3.6–3.8 illustrate patterns in rainfall over time. In Figure 3.6, the annual mean of the highest daily rainfall in 1969 and 1978 stands out with significant peaks, reaching the highest recorded values. As expected, most methods identified change points around these years, reflecting these substantial increases. Interestingly, according to Figures 3.7 and 3.8, apart from `changepointHD` and `InpesctChangepoint`, only `Segmentor` identified change points in the annual total rainfall, while both `Segmentor` and `CE.Normal.Mean` obtain changes for the number of rainy days. Other methods did not indicate significant change points for these variables. Overall, we can conclude that there has been no significant change in the rainfall patterns in Singapore from 1960 to 2023.

For daily sunshine duration, we plot its time series in Figure 3.9. Majority of methods detected a clear change point in 2013, suggesting a significant alteration in sunshine patterns before and after this year. This change may also correlate with the slight shifts identified in the annual means of the highest daily rainfall dataset, hinting at a potential link between rainfall and sunshine dynamics.

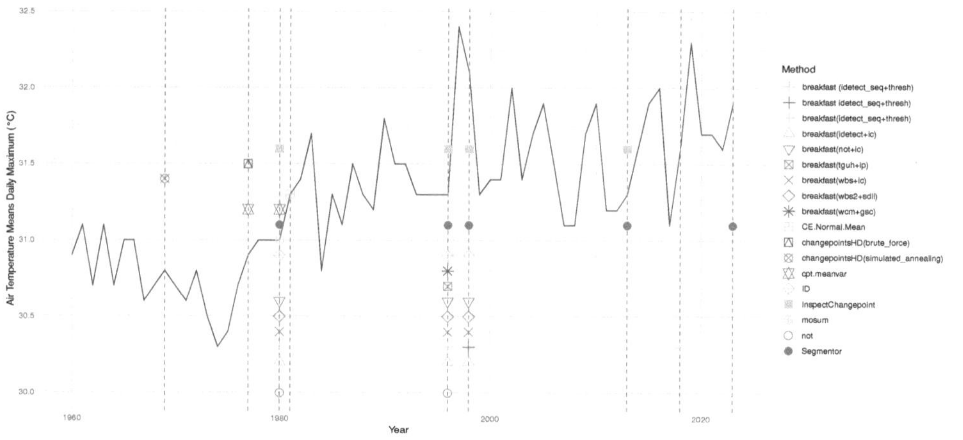

Figure 3.4: Annual air temperature means of daily maximum (Degree Celsius) from 1960 to 2023. Vertical dashed lines in the figure show all possible change points, dots with different shapes on these dashed lines indicate the methods that give change points in time.

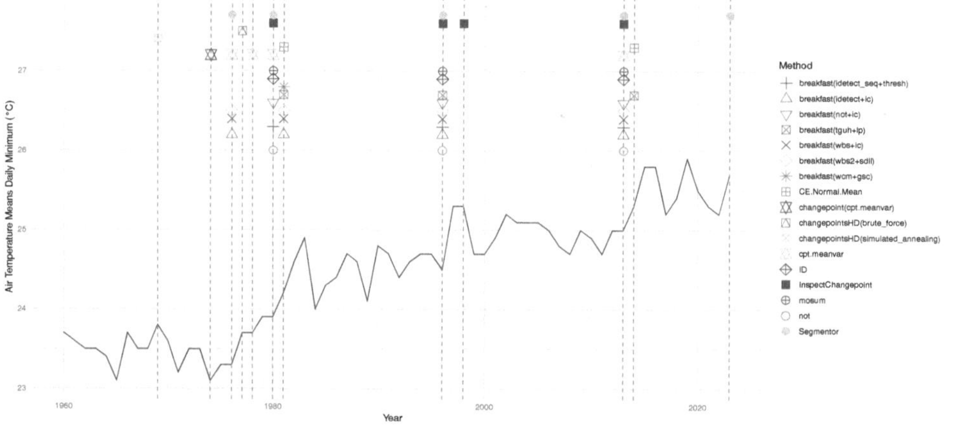

Figure 3.5: Annual air temperature means of daily minimum (Degree Celsius) from 1960 to 2023. Vertical dashed lines in the figure show all possible change points, dots with different shapes on these dashed lines indicate the methods that give change points in time.

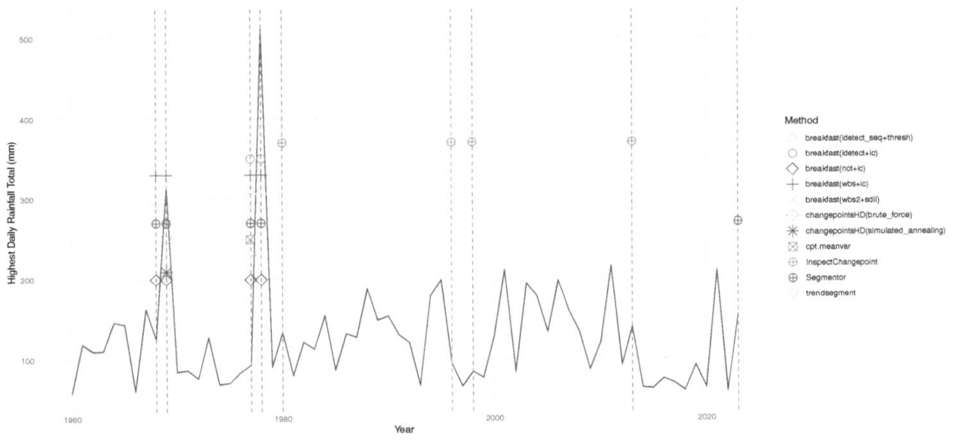

Figure 3.6: Annual means of highest daily rainfall from 1960 to 2023. Vertical dashed lines in the figure show all possible change points, dots with different shapes on these dashed lines indicate the methods that give change points in time.

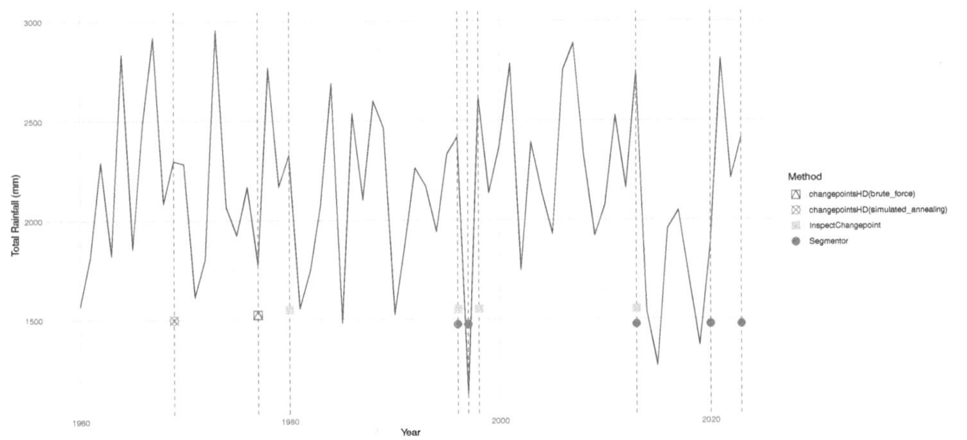

Figure 3.7: Annual total rainfall from 1960 to 2023. Vertical dashed lines in the figure show all possible change points, dots with different shapes on these dashed lines indicate the methods that give change points in time.

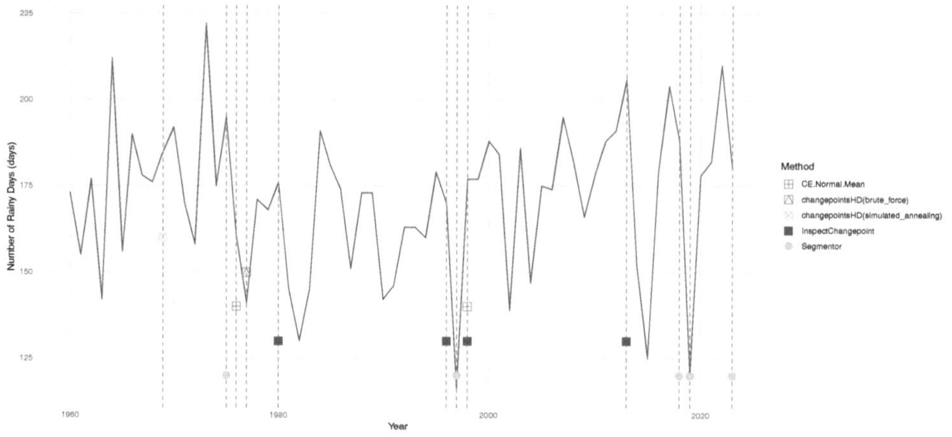

Figure 3.8: Number of rainy days from 1960 to 2023. Vertical dashed lines in the figure show all possible change points, dots with different shapes on these dashed lines indicate the methods that give change points in time.

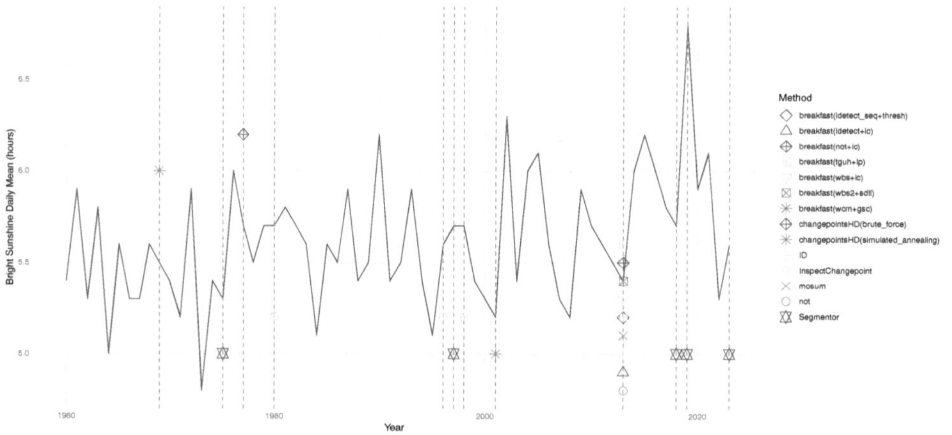

Figure 3.9: Annual means of duration of daily bright sunshine from 1960 to 2023. Vertical dashed lines in the figure show all possible change points, dots with different shapes on these dashed lines indicate the methods that give change points in time.

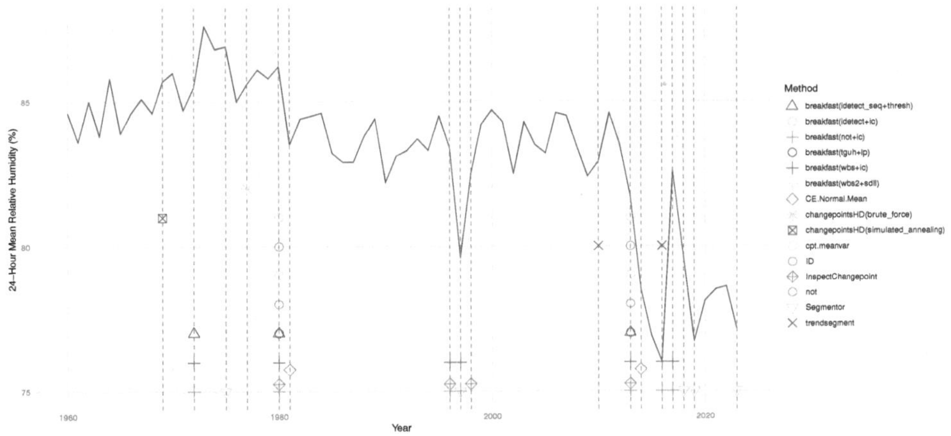

Figure 3.10: Annual means of 24-hours relative humidity from 1960 to 2023. Vertical dashed lines in the figure show all possible change points, dots with different shapes on these dashed lines indicate the methods that give change points in time.

As for daily relative humidity, we can find a clear trend of fluctuating downward after 1973 in Figure 3.10. And most methods detected change-points around 1980, 1997, and after 2013, with a few methods also identifying a shift in 1972. These periods coincide with the major transitions observed in temperature and rainfall, suggesting that humidity trends may have been influenced by broader climatic shifts, following a similar pattern of change over these key periods.

In conclusion, our analysis highlights three key periods of climatic change in Singapore: the late 1970s, the late 1990s, and around 2013. These periods mark significant transitions in temperature, rainfall, sunshine duration, and humidity, reflecting broader changes in Singapore's climate during these decades. The detailed computational methods for the packages covered in this section are beyond the scope of this monograph and we refer to the references provided above and the relevant documents in `R`.

4

Interval estimation

Interval estimation, also known as set estimation, is a form of statistical inference where we seek an interval of values that contains the true parameter with high probability. This interval is referred to as a confidence interval and plays a crucial role in statistical inference. While most research works on change points focus primarily on point estimation, the development of interval estimation methods in change point analysis has been relatively limited. If the researchers mainly aim at making decisions to group the data and quickly inform a jump position, they usually do not attempt to report the confidence interval for change points in practice. However, to quantify the uncertainty of the estimated change points we may adopt an appropriate confidence interval method to provide valuable information.

We explore several recent approaches to constructing confidence intervals for change points. As we will see, the primary challenge lies in the fact that the change point estimate does not converge to a normal distribution at the root n rate.

4.1 Interval estimation of a single change of mean

In this section, we investigate the simplest model of change point in which at most one change in the mean may occur. Consider the following model for a series $X(t)$ observed continuously over $t \in [0, 1]$:

$$X(t) = f(t) + e(t), \quad 0 \le t \le 1,$$

where $f(t)$ is a non-random function of the form

$$f(t) = \begin{cases} a, & 0 \le t \le t_0 \\ a + \theta, & t_0 < t \le 1 \end{cases},$$

where a, θ, and t_0 are unknown parameters and $0 \le t_0 \le 1$. We call t_0 the change point, and θ the magnitude of the jump at the change point. The error term $e(t)$ is a random variable whose distribution function F is independent of t. We assume that $E(e(t)) = 0$ for $t \in [0, 1]$, and $e(t)$ has a finite variance σ^2.

We will make inference on t_0 and θ by using observations made on $\{X(t)\}$. Suppose that we observe $X(i/n)$, $i = 1, \dots, n$. For simplicity of notation, we write X_i in place of $X(i/n)$ and assume that $X_1, X_2, \dots, X_n$ are independent in this section. Before we construct a confidence interval for the change point, we usually need to test if there is a significant jump in the mean. We consider the null hypothesis

$$H_0 : \quad \theta = 0 \text{ (i.e. no change point exists)}. \tag{4.1}$$

We will study the problem with different distribution assumptions on the error distribution F in the following subsections.

DOI: 10.1201/9781032657073-4

4.1.1 Normal error with known variance

We first consider the case that
$$F \sim \mathcal{N}(0, \sigma^2),$$
and σ^2 is known. The method is based upon the following theorem.

Theorem 4.1.1. *Suppose that $X_1, \ldots, X_n$ are independently and identically distributed (i.i.d.) with a common distribution $\mathcal{N}(a, \sigma^2)$. Let $\ell = \ell_n$ be a positive integer such that*
$$\lim_{n\to\infty} \ell/n = 0, \quad \lim_{n\to\infty} (\log n)^2/\ell = 0.$$
Define
$$Y_m = \frac{1}{\sqrt{2\ell}} \sum_{m-\ell+1}^{m} X_i - \sum_{m-2\ell+1}^{m-\ell} X_i, \quad m = 2\ell, 2\ell+1, \ldots,$$
$$\xi_{\mathrm{n}} = \max\{|Y_m| : m = 2\ell, 2\ell+1, \ldots, n\}, \tag{4.2}$$
$$A_n(x) = [2\log(3n/2\ell - 3)]^{-1/2}\left(x + 2\log(3n/2\ell-3) + \frac{1}{2}\log\log(3n/2\ell-3) - \frac{1}{2}\log\pi\right).$$
Then,
$$\lim_{n\to\infty} P(\xi_n/\sigma \le A_n(x)) = \exp(-e^{-x}), \quad -\infty < x < \infty. \tag{4.3}$$

Remark 4.1. *Theorem 4.1.1 is developed from Theorem 1 of [53], which presents a minor modification for the limiting distribution.*

Proof of Theorem 4.1.1. Construct a standard Brownian motion $\{W(t),\ t \ge 0\}$ such that
$$W\left(\frac{3m}{2\ell}\right) = \sqrt{3/2\ell}(x_1 + \ldots + x_m - ma)/\sigma, \quad m = 1, 2, \ldots,$$
and put
$$Z(t) = 3^{-1/2}[W(t) - 2W(t+3/2) + W(t+3)], \quad t \ge 0.$$
Then, it is seen $Y_m = \sigma Z(3m/2\ell - 3)$, $m = 2\ell, 2\ell+1, \ldots, n$. Put
$$\tilde{\xi}_n = \sup\{|Z(t)| : 0 \le t \le 3n/2\ell - 3\}, \quad \eta_n = \tilde{\xi}_n - \xi_n/\sigma. \tag{4.4}$$

We are going to show that
$$\lim_{n\to\infty} \eta_n \sqrt{\log n} = 0, \quad \text{a, s..} \tag{4.5}$$

In fact, we have
$$0 \le \eta_n \le \frac{4}{\sqrt{3}} \sup\{|W(t+s) - W(t)| : \quad 0 \le s \le 3/2\ell, \quad 0 \le t \le (3n-3)/2\ell - 3\}.$$

Insert $T = 3n/2\ell - 3,\ h = 3/2\ell,\ \varepsilon = 1$, and $\nu = (\sqrt{2}/4)\sqrt{\ell/\log n}\,\delta$ in Lemma 1.2.1 of [69], we can get
$$P(\eta_n\sqrt{\log n} \ge \delta) \le Cn\exp(-\delta^2\ell/24\log n). \tag{4.6}$$
Here, $\delta > 0$ is given, and C is a constant not depending on n. Since $\ell/(\log n)^2 \to \infty$, (4.5) follows from (4.4) and Borel-Cantelli Lemma.

$Z(t)$ is a stationary Gaussian process, with $E\left((Z(t))\right) = 0$, and the correlation coefficient function
$$\rho(\tau) = Cov\left(Z(t), Z(t+\tau)\right) = \begin{cases} 1 - |\tau|, & 0 \le |\tau| \le 3/2 \\ -1 + |\tau|/3, & 3/2 < |\tau| \le 3\,. \\ 0 & 3 < |\tau| < \infty \end{cases}$$

Next, we need to use Theorem 5.2 of [208]

$$\lim_{t\to\infty} P(a(t)^{-1}(\max_{0<s<t} Z(t) - b(t)) \leq x) = \exp(-e^{-x}). \tag{4.7}$$

where $a(t) = (2\log(t))^{-1/2}$,

$$b(t) = (2\log t)^{1/2} - (2\log t)^{-1/2} \log\left(2H_\alpha^{-1}(\pi \log t)^{1/2}\tilde{\sigma}^{-1}(1/(2\log t)^{1/2})\right). \tag{4.8}$$

In Theorem 5.2 of [208], $\rho(\tau) = 1 - |\tau|^\alpha H(|\tau|) + o(|\tau|^\alpha H(|\tau|))$ as $\tau \to 0$. Then $\alpha = 1$ in (4.8) and $H(|\tau|) = 1$, such that $\tilde{\sigma}^2(s) = 2|s|H(s) = 2|s|$ in [208], that is $\tilde{\sigma}^{-1}(1/(2\log t)^{1/2}) = 1/(4\log t)$ in (4.8). By Theorem 2.1 of [208],

$$H_\alpha = H_1 = \lim_{T\to\infty} T^{-1} \int_0^\infty e^s P(\sup_{0<t<T} Y(t) > s) ds,$$

where $Y(t) = W(t) - |t|/2$ is a drifted Wiener process. By Proposition 10.4 of [200], we can compute that $H_1 = 1/2$. Therefore,

$$b(t) = (2\log t)^{1/2} - (2\log t)^{-1/2} \left(1/2\log \pi - 1/2\log(\log t)\right), \tag{4.9}$$

and, combining (4.7) and (4.4), we have

$$\lim_{n\to\infty} P\left(\tilde{\xi}_n \leq A_n(x)\right) = \exp\{-e^{-x}\}, -\infty < x < \infty.$$

For large n, we have

$$A_n(x + \Delta x) - A_n(x) \geq \Delta x/\sqrt{2\log n}, \quad \Delta x > 0,$$

$$A_n(x + \Delta x) - A_n(x) \leq \Delta x/\sqrt{2\log n}, \quad \Delta x < 0.$$

Hence, for large n,

$$\begin{aligned} &P(\tilde{\xi}_n \leq A_n(x - |\Delta x|)) - P(\eta_n \geq |\Delta x|/\sqrt{2\log n}) \leq P(\xi_n/\sigma \leq A_n(x)) \\ &\leq P(\tilde{\xi}_n \leq A_n(x + |\Delta x|) + P(\eta_n \geq |\Delta x|/\sqrt{2\log n}). \end{aligned} \tag{4.10}$$

From (4.6)–(4.10), letting $n \to \infty$, and then $\Delta x \to 0$, we get (4.3). □

Remark 4.2. *Following Theorem 4.1.1, we can test the hypothesis (4.1) at a significance level α by solving the equation $\exp(-e^{-x}) = 1 - \alpha$. The solution to the equation is*

$$x(\alpha) = -\log(-\log(1 - \alpha)).$$

Calculate $d = 2\ell/n$, $C_n(\alpha, d) = A_n(x(\alpha))$, and reject the null hypothesis (4.1) when

$$\xi_n > \sigma C_n(\alpha, d). \tag{4.11}$$

Under the null hypothesis (4.1), this test has an asymptotic size α as the sample size n tends to infinity.

To investigate the effectiveness, we can study the power $\beta(\theta) = \beta_n(\theta, \sigma)$ of the test. Let r be the integer such that

$$\frac{r}{n} \leq t_0 < \frac{r+1}{n}.$$

Then, $Y_{r+\ell} \sim \mathcal{N}(\sqrt{\frac{\ell}{2}}\theta, \sigma^2)$. Therefore,

$$\begin{aligned}\beta(\theta) &\geq P(|Y_{r+\ell}| > \sigma \mathrm{C}_n(\alpha, d)) \\ &= \Phi\left(\frac{|\theta|\sqrt{\ell}}{\sqrt{2}\sigma} - \mathrm{C}_n(\alpha, d)\right) + \Phi\left(-\frac{|\theta|\sqrt{\ell}}{\sqrt{2}\sigma} - \mathrm{C}_n(\alpha, d)\right) \\ &> \Phi\left(\frac{|\theta|\sqrt{\ell}}{\sqrt{2}\sigma} - \mathrm{C}_n(\alpha, d)\right),\end{aligned} \tag{4.12}$$

where Φ is the standard normal distribution function. From the result of the inequality, it is obvious that to achieve a large power $\beta(\theta)$, a larger value of ℓ is preferred. Furthermore, (4.12) suggests that $\beta(\theta)$ increases steadily with $|\theta|/\sigma$. Practically, a change point is significant only when the ratio $|\theta|/\sigma$ is substantially different from 0.

After confirming the existence of a change point t_0, we consider the interval estimation of t_0. We may simply find the integer k such that $|Y_k| = \xi_n$. Take $[(k-2\ell)/n,\ k/n]$ as a candidate confidence interval of t_0.

It is easy to see that the length of this interval is $d = 2\ell/n$. Therefore, to improve the precision, one may want to choose a small ℓ. However, some other factors should also be considered. First, the theoretical result of testing the existence of a change point shows that a small ℓ increases the risk of false acceptance of the hypothesis (4.1). In addition, a small ℓ might lead to a small confidence coefficient of the interval. Specifically, we give an estimate of the confidence coefficient γ of this interval by

$$\gamma = P\left(\frac{k-2\ell}{n} \leq t_0 \leq \frac{k}{n}\right) \geq P\left(\left\{\sup_{m \notin (r, r+2\ell)} |Y_m| \leq \sigma \mathrm{C}_n(\alpha, d)\right\} \cap \{|Y_{r+\ell}| > \sigma \mathrm{C}_n(\alpha, d)\}\right).$$

Since $X_m, X_{m-1}, \ldots, X_{m-2\ell-1}$ are i.i.d. when $m \notin (r, r+2\ell)$, by Theorem 4.1.1, we have

$$P\left(\sup_{m \notin (r, r+2\ell)} |Y_m| \leq \sigma \mathrm{C}_n(\alpha, d)\right) \geq P\left(\xi_n \leq \sigma \mathrm{C}_n(\alpha, d)\right) = 1 - \alpha, \tag{4.13}$$

where the probability $P(\xi_n \leq \sigma \mathrm{C}_n(\alpha, d))$ is computed under the condition that $X_1, X_2, \ldots, X_n$ are i.i.d.. Hence, by (4.13) and (4.12), we have

$$\begin{aligned}\gamma &\geq (1-\alpha) + P(|Y_{r+\ell}| \geq \sigma \mathrm{C}_n(\alpha, d)) - 1 \\ &\geq \Phi\left(\frac{|\theta|\sqrt{\ell}}{\sqrt{2}\sigma} - \mathrm{C}_n(\alpha, d)\right) - \alpha,\end{aligned} \tag{4.14}$$

and it suggests that the confidence coefficient increases with $|\theta|/\sigma$.

By (4.14), we can provide a confidence interval of t_0 with prescribed length d_0 and confidence coefficient $1 - \alpha_0$. The main question is how to choose ℓ and n. We set α in (4.14) the value $\alpha_0/2$, and solve

$$\Phi\left(\frac{|\theta|\sqrt{\ell}}{\sqrt{2}\sigma} - \mathrm{C}_n(\alpha_0/2, d)\right) = 1 - \frac{\alpha_0}{2}$$

to get

$$\ell = 2\left(\frac{|\theta|}{\sigma}\right)^{-2} \left(\mathrm{C}_n(\alpha_0/2, d_0) + \mathrm{U}_{\alpha_0/2}\right)^2, \quad n = 2\ell/d_0. \tag{4.15}$$

Here, $\mathrm{U}_{\alpha_0/2}$ is the upper $\alpha_0/2$ quantile of $\mathcal{N}(0,1)$.

The above solution requires the specification of the unknown parameters θ and σ. In some cases, estimating these quantities with preliminary samples may be feasible. On the other hand, we recommend another sensible approach by taking a practical consideration of a constant M such that only when $|\theta|/\sigma \geq M$, the change point t_0 is of real importance. Thus we simply replace $|\theta|/\sigma$ in (4.15) by a given M.

For example, take $\alpha_0 = 0.05$, $d_0 = 0.1$, $M = 1.926$, (4.15) gives

$$\ell = 19.2, \quad n = 384.1.$$

Remark 4.3. *The result appears reasonable as it is comparable with the sample size needed in estimating the success probability p of the binomial distribution $B(n,p)$. The sample size required to guarantee a confidence interval of p not wider than 0.1 and with a confidence coefficient not smaller than 0.95 is roughly 384.*

4.1.2 Normal error with unknown variance

We now consider the more complicated case with unknown variance in the normal distribution. To apply the result of the preceding subsection, a natural idea is to estimate the unknown variance with the empirical sample $X_1, X_2, \ldots, X_n$. Denote by $\hat{\sigma}^2$ the estimator of σ^2, and replace σ in (4.11) to perform the hypothesis test. To attain an asymptotic size α, the estimator $\hat{\sigma}^2$ must satisfy certain conditions stated in the following theorem.

Theorem 4.1.2. *Suppose that the conditions of Theorem 4.1.1 hold, and $\hat{\sigma}^2$ is an estimator of σ^2 satisfying*

$$|\hat{\sigma}^2 - \sigma^2| \log n \xrightarrow{P} 0. \tag{4.16}$$

Then,

$$\lim_{n\to\infty} P(\xi_n/\hat{\sigma} \leq A_n(x)) = \exp(-e^{-x}).$$

Proof of Theorem 4.1.2. From (4.16), we can know that $|\hat{\sigma} - \sigma| \log n \xrightarrow{P} 0$. Thus, for $\forall \epsilon > 0$, we have

$$\begin{aligned} &P\left(\xi_n/\sigma \leq A_n(x)(1 - \epsilon/\log n)\right) - P\left(|\hat{\sigma} - \sigma| \log n \geq \epsilon\right) \\ &\leq P\left(\xi_n/\hat{\sigma} \leq A_n(x)\right) \\ &\leq P\left(\xi_n/\sigma \leq A_n(x)(1 + \epsilon/\log n)\right) + P\left(|\hat{\sigma} - \sigma| \log n \geq \epsilon\right). \end{aligned} \tag{4.17}$$

Given $\delta > 0$, For large n, we have

$$A_n(x - \delta) \leq A_n(x)(1 - \epsilon/\log n) \leq A_n(x)(1 + \epsilon/\log n) \leq A_n(x + \delta).$$

Therefore, in this condition we have

$$\begin{aligned} &P\left(\xi_n/\sigma \leq A_n(x)(x - \delta)\right) - P\left(|\hat{\sigma} - \sigma| \log n \geq \epsilon\right) \\ &\leq P\left(\xi_n/\hat{\sigma} \leq A_n(x)\right) \\ &\leq P\left(\xi_n/\sigma \leq A_n(x)(x + \delta)\right) + P\left(|\hat{\sigma} - \sigma| \log n \geq \epsilon\right). \end{aligned} \tag{4.18}$$

Letting $n \to \infty$, and then $\delta \to 0$, we finish the proof. □

The following theorem provides an estimator of σ^2 satisfies (4.16).

Theorem 4.1.3. *Suppose that*

$$X_i \sim F(x), \quad i = 1, \ldots, m,$$

$$X_i \sim F(x - \theta), \quad i = m + 1, \ldots, n,$$

where F is a distribution function whose moment of $(2+\delta)$-th order exists, and θ is a constant. Define

$$\overline{X}_{kc} = \sum_{i=k}^{c} X_i/(c-k+1), \quad \overline{X}_n = \sum_{i=1}^{n} X_i/n,$$

$$S_c^2 = \sum_{i=1}^{c}(X_i - \overline{X}_{1c})^2 + \sum_{i=c+1}^{n}(X_i - \overline{X}_{(c+1)n})^2, \quad c = 1, \ldots, n,$$

$$S_{\hat{m}}^2 = \min\{S_c^2 : c = 1, \ldots, n-1\}, \ \hat{m} = \arg\min_c\{S_c^2 : c = 1, \ldots, n-1\}$$

$$\hat{\sigma}_n^2 = \hat{\sigma}^2 = S_{\hat{c}}^2/n.$$

Then, we have

$$|\hat{\sigma}_n^2 - \sigma^2| \log n \xrightarrow{P} 0, \tag{4.19}$$

where σ^2 is the variance of F.

Proof of Theorem 4.1.3. Put $Y_i = X_i - EX_i, T_i = Y_1 + \ldots + Y_i$. For given $\epsilon > 0$, from Kolmogorov inequality, if $n/(\epsilon^2(\log n)^2) \le c \le n$ we have

$$P(|T_c/c| \le \epsilon/\log n) \ge 1 - D(\log n)^4/n, \tag{4.20}$$

and if $1 \le c \le n(1 - 1/(\epsilon^2(\log n)^2)$, we have

$$P(|(T_n - T_c)/(n-c)| \le \epsilon/\log n) \ge 1 - D(\log n)^4/n. \tag{4.21}$$

We shall use D to denote a constant not depending on n here and in the following, which may assume different values in each appearance. From (4.20) and (4.21), if $n/(\epsilon^2(\log n)^2) \le c \le n(1 - 1/(\epsilon^2(\log n)^2))$, we have

$$P\left(|T_c/c| \le \epsilon/\log n, |(T_n - T_c)/(n-c)| \le \epsilon/\log n\right) \ge 1 - 2D(\log n)^4/n. \tag{4.22}$$

First, we consider the case $\theta = 0$. Since

$$S_c^2 - S_n^2 = -\frac{c(n-c)}{n}\left(\overline{X}_{1c} - \overline{X}_{(c+1)n}\right)^2, \tag{4.23}$$

and $\overline{X}_{1c} - \overline{X}_{(c+1)n} = T_c/c - (T_n - T_c)/(n-c)$, when $\theta = 0$, by (4.22) and (4.23), if $n/(\epsilon^2(\log n)^2) \le c \le n(1 - 1/(\epsilon^2(\log n)^2))$, we have that

$$\begin{aligned}
&P(S_c^2 - S_n^2 \ge -n\epsilon^2/(\log n)^2) \\
&= P\left(-\frac{c(n-c)}{n}(T_c/c - (T_n - T_c)/(n-c))^2 \ge -n\epsilon^2/(\log n)^2\right) \\
&= P\left(|T_c/c - (T_n - T_c)/(n-c)| \le \frac{n}{\sqrt{c(n-c)}}\epsilon/(\log n)\right) \\
&\ge P\left(|T_c/c| \le \frac{n}{2\sqrt{c(n-c)}}\epsilon/(\log n), |(T_n - T_c)/(n-c)| \le \frac{n}{2\sqrt{c(n-c)}}\epsilon/(\log n)\right) \\
&\ge P\left(|T_c/c| \le \epsilon/\log n, |(T_n - T_c)/(n-c)| \le \epsilon/\log n\right) \ge 1 - 2D(\log n)^4/n.
\end{aligned} \tag{4.24}$$

For $c \le n/(\epsilon^2(\log n)^2)$, we have

$$\begin{aligned}
S_c^2 - S_n^2 &= -\frac{n}{c(n-c)}\left(\sum_{i=1}^{c}(X_i - \overline{X})\right)^2 \\
&\ge -\frac{n}{(n-c)}\sum_{i=1}^{c}(X_i - \overline{X})^2 \\
&\ge -2\left(\sum_{i=1}^{c} X_i^2 + \frac{c}{n}\sum_{i=1}^{n} X_i^2\right) \\
&\ge -2\left(\sum_{i=1}^{n/(\epsilon^2(\log n)^2)} X_i^2 + \frac{n/(\epsilon^2(\log n)^2)}{n}\sum_{i=1}^{n} X_i^2\right).
\end{aligned} \tag{4.25}$$

Similarly, for $c \ge n(1 - 1/(\epsilon^2(\log n)^2))$, we have

$$S_c^2 - S_n^2 \ge -2\left(\sum_{i=n(1-1/(\epsilon^2(\log n)^2))}^{n} X_i^2 + \frac{n/(\epsilon^2(\log n)^2)}{n}\sum_{i=1}^{n} X_i^2\right). \tag{4.26}$$

From (4.24)–(4.26), and putting

$$Q = \frac{\epsilon^2}{\log n} + \frac{4}{\epsilon^2 n \log n}\sum_{i=1}^{n} X_i^2 + \frac{2\log n}{n}\left(\sum_{i=1}^{n/(\epsilon^2(\log n)^2)} X_i^2 + \sum_{i=n(1-1/(\epsilon^2(\log n)^2))}^{n} X_i^2\right),$$

we have

$$P(|\hat{\sigma}^2 - S_n^2/n| \log n \ge Q) \le 2D(\log n)^4/n. \tag{4.27}$$

Denoting by the expectation of variable $EQ = \frac{\epsilon^2}{\log n} + \frac{8EX_1^2}{\epsilon^2 \log n}$, we have

$$Q \xrightarrow{p} 0. \tag{4.28}$$

Further, it can easily be shown that under the conditions of the present theorem, there exists $s > 0$ such that

$$n^s(\sigma^2 - S_n^2/n) \xrightarrow{p} 0, \tag{4.29}$$

From (4.27)–(4.29), then (4.19) follows.

Now, turning to the case of $\theta \ne 0$, we can easily verify that

$$S_c^2 - S_m^2 = \begin{cases} \frac{m(c-m)}{c}\left(\overline{X}_{1m} - \overline{X}_{(m+1)c}\right)^2 - \frac{(n-c)(c-m)}{n-m}\left(\overline{X}_{(c+1)n} - \overline{X}_{(m+1)c}\right)^2, & c > m \\ \frac{(n-m)(m-c)}{n-c}\left(\overline{X}_{(m+1)n} - \overline{X}_{(c+1)m}\right)^2 - \frac{c(m-c)}{m}\left(\overline{X}_{1c} - \overline{X}_{(c+1)m}\right)^2, & c < m. \end{cases}$$

Given $\epsilon > 0$ small enough, by $S_{\hat{c}}^2 - S_m^2 \le 0$ and $n^s(\sigma^2 - S_m^2/n) \xrightarrow{p} 0$ for some $s > 0$, we only need prove

$$P(S_{\hat{c}}^2 - S_m^2 \le -n\epsilon/(\log n)^2) \to 0, \tag{4.30}$$

that is, if $\hat{c} > m$, we need prove

$$P\left(\frac{m}{\hat{c}}\left(\overline{X}_{1m} - \overline{X}_{(m+1)\hat{c}}\right)^2 \le \frac{n-\hat{c}}{n-m}\left(\overline{X}_{(\hat{c}+1)n} - \overline{X}_{(m+1)\hat{c}}\right)^2 - \frac{n\epsilon}{(\hat{c}-m)(\log n)^2}\right) \to 0, \tag{4.31}$$

and if $\hat{c} < m$, we need to prove that

$$P\left(\frac{n-m}{n-\hat{c}}\left(\overline{X}_{(m+1)n} - \overline{X}_{(\hat{c}+1)m}\right)^2 \leq \frac{\hat{c}}{m}\left(\overline{X}_{1\hat{c}} - \overline{X}_{(\hat{c}+1)m}\right)^2 - \frac{n\epsilon}{(m-\hat{c})(\log n)^2}\right) \to 0. \tag{4.32}$$

We separate the following two cases for $\hat{c} > m$:

1. $\hat{c} > m$ and $(\hat{c}-m)/n > \delta$ for some $\delta > 0$. As $m/n \to t_0$, we have $t_0/(t_0+\delta) < m/\hat{c} \leq 1$ and $0 \leq (n-\hat{c})/(n-m) < 1$. By $\overline{X}_{1m} - \overline{X}_{(m+1)\hat{c}} \xrightarrow{P} \theta$ and $\overline{X}_{(\hat{c}+1)n} - \overline{X}_{(m+1)\hat{c}} \xrightarrow{P} 0$ and $n\epsilon/((\hat{c}-m)(\log n)^2) \to 0$, then (4.31) is held.

2. $\hat{c} > m$ and $(\hat{c}-m)/n \to 0$. As $m/n \to t_0 \in (0,1)$, we have $m/\hat{c} = 1/(1+(\hat{c}-m)/m) \to 1$ and $(n-\hat{c})/(n-m) = 1 + (\hat{c}-m)/(n-m) \to 1$. By $\frac{m(\hat{c}-m)}{\hat{c}}\left(\overline{X}_{1m} - \overline{X}_{(m+1)\hat{c}}\right)^2 = O_p(1)$, $\frac{(n-\hat{c})(\hat{c}-m)}{n-m}\left(\overline{X}_{(\hat{c}+1)n} - \overline{X}_{(m+1)\hat{c}}\right)^2 = O_p(1)$, and $n\epsilon/(\log n)^2 \to \infty$, and then (4.31) is held.

The similar inequality (4.32) obtained for the case of $\hat{c} \leq m$, and the theorem was proved. □

4.1.3 Bias of the variance estimator

Keep the same notations of Section 4.1.2. Since $S_{\hat{c}}^2 = \min\{S_c^2\} \leq S_m^2$, and $ES_m^2/(n-2) = \sigma^2$, we notice that even if we replace $\hat{\sigma}^2$ by $\tilde{\sigma}^2 = S_{\hat{c}}^2/(n-2)$, $\tilde{\sigma}^2$ still underestimates σ^2. Intuitively, the bias issue should be more severe if $|\theta|/\sigma$ is small. Thus we carry out some theoretical investigation of the bias in the case $\theta = 0$.

From (4.23), we have

$$\frac{1}{\sigma^2}\left(S_n^2 - S_{\hat{c}}^2\right) = \max_{1\leq c\leq n} \frac{n}{c(n-c)}\left(T_n - \frac{c}{n}T_n\right)^2 = \left\{\sup_{0<t<1} |Z_n(t)|/\sqrt{t(1-t)}\right\}^2 \triangleq U_n^2,$$

where $U_n = \sup_{0<t<1} |Z_n(t)|/\sqrt{t(1-t)}$, $T_c = \sum_{i=1}^{c}(X_i - a)/\sigma$, and

$$Z_n(t) = \begin{cases} (T_{[(n+1)t]} - [(n+1)t]T_n/n)/\sqrt{n}, & 0 \leq t < 1 \\ 0, & t = 1. \end{cases}$$

When X_1 is normally distributed, or more generally, when $E|X_1|^\delta < \infty$ for some $\delta > 2$, the following asymptotic distribution is valid in Theorem 2.1.4, $a_n^{-1}(U_n - b_n) \xrightarrow{D}$ Gummbel$(\log 2 + 1/2\log\pi, 1)$, where $a_n = (2\log\log n)^{-1/2}$, $b_n = a_n^{-1} + \frac{1}{2}a_n \log\log\log n$, that is,

$$\lim_{n\to\infty} P\left(a_n^{-1}(U_n - b_n) \leq x\right) = \exp\left\{-2\pi^{1/2}e^{-x}\right\}.$$

From this, using the mean and variance of the Gumbel distribution, we have

$$a_n^{-1}(EU_n - b_n) \to \log 2 + 1/2\log\pi + \gamma$$

where $\gamma = -\,\epsilon_0^\infty \log(y)e^{-y}dy \approx -0.577$ is the Euler–Mascheroni constant, and

$$a_n^{-2}Var(U_n) \to \frac{\pi^2}{6}.$$

Then, for large n,

$$EU_n^2 \approx a_n^2\left(\frac{\pi^2}{6} + \left(a_n^{-1}b_n + 2 + 1/2\log\pi + \gamma\right)^2\right).$$

As $a_n^{-1}b_n \to \infty$, we have

$$EU_n^2 - b_n^2 \to 0. \tag{4.33}$$

Notice that

$$E\frac{1}{\sigma^2}\left(S_n^2 - S_{\hat{c}}^2\right) = n - 1 - \frac{ES_{\hat{c}}^2}{\sigma^2} = EU_n^2.$$

Then

$$ES_{\hat{c}}^2 = \sigma^2\left(n - 1 - EU_n^2\right).$$

From (4.33), when n is large, we may take

$$S_{\hat{c}}^2/\left(n - 1 - b_n^2\right) \to \sigma^2.$$

as the estimator of σ^2 to bring down the bias.

In some applications, we might know in advance that there exist some constants λ_1, λ_2, $0 < \lambda_1 < \lambda_2 < 1$, such that the change point t_0 lies in the interval $[\lambda_1, \lambda_2]$. In this case, we may use

$$S^2(\lambda_1, \lambda_2) = \min\{S_c^2 : \lambda_1 n \le c \le \lambda_2 n\}$$

instead of S^2. We have

$$\frac{1}{\sigma^2}(S_n^2 - S^2(\lambda_1, \lambda_2)) = \{\sup_{\lambda_1 \le t \le \lambda_2}\{|Z_n(t)|/\sqrt{t(1-t)}\}\}^2 \triangleq v_n.$$

Put $\lambda = \min(\lambda_1, 1 - \lambda_2)$, we can have

$$\begin{aligned}\frac{1}{\sigma^2}(S_n^2 - S^2(\lambda_1, \lambda_2)) &\le \max\left\{\frac{n}{c(n-c)}(T_c - \frac{c}{n}T_n)^2 : \lambda n \le c \le (1-\lambda)n\right\} \\ &\le \frac{1}{\lambda(1-\lambda)}\left\{\max_{\lambda n \le c \le (1-\lambda)n} |\frac{1}{\sqrt{n}}T_c - \frac{1}{\sqrt{n}}\frac{c}{n}T_n|\right\}^2 \\ &\le \frac{1}{\lambda(1-\lambda)}\left\{\max_{\lambda n \le c \le (1-\lambda)n} |B(t)|\right\}^2 \\ &\le \frac{1}{\lambda(1-\lambda)}\left\{\max_{0 \le t \le 1} |B(t)|\right\}^2.\end{aligned}$$

where $\{B(t), 0 \le t \le 1\}$ is the Standard Brownian Bridge.

The distribution function of $\max_{0 \le t \le 1}|B(t)|$ is well-known as $1 - \sum_{k=1}^{\infty} 2(-1)^{k+1} \exp(-2k^2x^2)$. From which it is easy to get

$$E\left\{\max_{0 \le t \le 1}|B(t)|\right\}^2 = 1 - \frac{1}{2^2} + \frac{1}{3^2} - \frac{1}{4^2} + \cdots \approx 0.85.$$

Hence,

$$\frac{1}{\sigma^2}(S_n^2 - S^2(\lambda_1, \lambda_2)) \le 0.85\lambda^{-1}(1-\lambda)^{-1}.$$

Therefore, if we use

$$\hat{\sigma}^2(\lambda_1, \lambda_2) = S^2(\lambda_1, \lambda_2)/(n-2)$$

to estimate σ^2, the relative bias does not exceed

$$[\sigma^2 - E\left(\hat{\sigma}^2(\lambda_1, \lambda_2)\right)]/\sigma^2 = (0.85\lambda^{-1}(1-\lambda)^{-1} - 1)/(n-2). \tag{4.34}$$

For example, take $\lambda_1 = 1 - \lambda_2 = 0.1$. This value can perhaps be considered as small enough for many applications. Take $n = 100$. According to (4.34), the relative bias does not exceed 86%, which is reasonably good, considering the relatively small sample size 100 in estimating the variance of a complicated model.

4.1.4 Non-normal error

When the distribution of the random error $e(t)$ is not normal, we can use the theory of strong approximation of partial sums of i.i.d. random variables by the Brownian motion process to provide extensions of Theorem 4.1.1. In this way, the methods of preceding sections can still be applied.

Theorem 4.1.4. *Replacing the assumption of Theorem 4.1.1 that* $X_1 \sim \mathcal{N}(a, \sigma^2)$ *by*

$$E(\exp(tX_1)) < \infty \quad \textit{for } |t| \textit{ small enough.} \tag{4.35}$$

Then, the conclusion of Theorem 4.1.1 remains valid.

Proof of Theorem 4.1.4. Put

$$S_k = (X_1 + \ldots + X_k - ka)/\sigma, \quad k = 1, 2, \ldots$$

According to [147], there exists a Brownian Motion process $\{W(t), t \geq 0\}$ such that

$$\lim_{n\to\infty} \sup_{k\leq n} |S_k - W(k)|/\log n < \infty, \quad \text{a.s..} \tag{4.36}$$

Since

$$Y_m/\sigma = [(S_m - S_{m-\ell}) - (S_{m-\ell} - S_{m-2\ell})]/\sqrt{2\ell},$$

we have for $m \leq n$,

$$|Y_m/\sigma - [W(m) - 2W(m-\ell) + W(m-2\ell)]/\sqrt{2\ell}| \leq 4 \sup_{1\leq k\leq n} |S_k - W(k)|/\sqrt{2\ell}.$$

According to (4.36), noticing that $\log n/\sqrt{\ell} \to 0$ as $n \to \infty$, we get

$$\lim_{n\to\infty} \left\{ \sup_{m\leq n} |Y_m/\sigma - [W(m) - 2W(m-\ell) + W(m-2\ell)]/\sqrt{2\ell}| \right\} = 0, \quad \text{a.s..} \tag{4.37}$$

From Theorem 4.1.1, we have

$$\begin{aligned} \lim_{n\to\infty} P\left\{\sup([W(m) - 2W(m-\ell) + W(m-2\ell)]/\sqrt{2\ell} : m = 2\ell, \ldots, n) \leq A_n(x)\right\} \\ = \exp(-e^{-x}), \quad -\infty < x < \infty. \end{aligned} \tag{4.38}$$

Finally, (4.2) follows from (4.37) and (4.38). Theorem 4.1.4 is proved. □

Since under the assumption (4.35), the conclusion of Theorem 4.1.3 is still true, it follows that the method of the previous two sections applies to the case in which (4.35) holds.

The condition ensuring the asymptotic distribution can be further weakened.

Theorem 4.1.5. *Replacing the assumptions of Theorem 4.1.1 that* $X_1 \sim \mathcal{N}(a, \sigma^2)$ *and* $(\log n)^2/\ell \to 0$ *by: For some* $\delta > 2$

$$E|X_1|^\delta < \infty, \tag{4.39}$$

$$\lim_{n\to\infty} n^{2/\delta}/\ell = 0. \tag{4.40}$$

Then, the conclusion of Theorem 4.1.1 remains valid.

The proof parallels that of Theorem 4.1.4, using the result from [147], which establishes

$$\lim_{n\to\infty} \left\{ \sup_{k\leq n} |S_k - W(k)| n^{-1/\delta} \right\} = 0, \quad \text{a,s.}$$

under condition (4.39).

4.1.5 Jump estimation

To construct an estimate of the jump parameter θ at the change point t_0, we suggest the following procedure: First, find $\hat{k}$ such that $|Y_{\hat{k}}| = \xi_n$. Then calculate the jump size estimate as

$$\hat{\theta} = \hat{\theta}_n = \frac{1}{n-\hat{k}} \sum_{i=\hat{k}+1}^{n} X_i - \frac{1}{\hat{k}-2\ell} \sum_{i=1}^{\hat{k}-2\ell} X_i.$$

When $\hat{k} = 2\ell$, or $\hat{k} = n$, $\hat{\theta}$ is not defined. In general, if $\hat{k}$ is too near 2ℓ or n, this would imply that the change point t_0 is too near 0 or 1, and the samples at our disposal are perhaps not enough to give a reasonable estimate.

For an interval estimation of θ, we may use the following theorem about the asymptotic distribution of $\hat{\theta}$.

Theorem 4.1.6. *Supposed that the δ-th order moment of distribution F is finite for some $\delta > 2$. Further, $\theta \neq 0$ and ℓ satisfies (4.40) and $\ell/n \to 0$. Then, as $n \to \infty$, we have*

$$\left(\sqrt{nt_0(1-t_0)}/\sigma\right)(\hat{\theta}-\theta) \xrightarrow{d} \mathcal{N}(0,1). \tag{4.41}$$

Proof of Theorem 4.1.6. Without losing generality, we assume $a = 0$, $\sigma = 1$. Using Theorem 4.1.5 and slightly modifying the argument of Section 4.1.2, we easily show that

$$\lim_{n\to\infty} P(nt_0 \le \hat{k} \le nt_0 + 2\ell) = 1. \tag{4.42}$$

Define $n_1 = [nt_0 - 2\ell]$, $n_2 = n - [nt_0 + 2\ell] - 1$, and

$$T_n = \sqrt{nt_0(1-t_0)}\left(\frac{1}{n_2}\sum_{i=n-n_2+1}^{n}(X_i-\theta) - \frac{1}{n_1}\sum_{i=1}^{n_1} X_i\right).$$

Since $\ell/2 \to 0$, $0 < t_0 < 1$, we have as $n \to \infty$,

$$T_n \xrightarrow{d} \mathcal{N}(0,1). \tag{4.43}$$

Now,

$$T_n - \sqrt{nt_0(1-t_0)}(\hat{\theta}-\theta) = \sqrt{nt_0(1-t_0)}\left\{\frac{n-\hat{k}-n_2}{n_2(n-\hat{k})}\sum_{i=n-n_2+1}^{n}(X_i-\theta)\right.$$

$$\left. - \frac{1}{n-\hat{k}}\sum_{i=\hat{k}+1}^{n-n_2}(X_i-\theta) - \frac{\hat{k}-2\ell-n_1}{n_1(\hat{k}-2\ell)}\sum_{i=1}^{n_1} X_i + \frac{1}{\hat{k}-2\ell}\sum_{i=1}^{\hat{k}-2\ell} X_i\right\}.$$

Therefore,

$$\begin{aligned}
&\sup\left\{|T_n - \sqrt{nt_0(1-t_0)}(\hat\theta-\theta)| : nt_0 \le \hat k \le nt_0 + 2\ell\right\} \\
&\le C\left\{n^{-3/2}\ell \sup\left(|\sum_{i=j}^{n}(X_i-\theta)| : j \ge n-n_2+1\right)\right. \\
&+ n^{-1/2}\sup\left(|\sum_{i=j}^{n-n_2}(X_i-\theta)| : [nt_0]+1 \le j \le n-n_2\right) \\
&+ n^{-3/2}\ell \sup\left(|\sum_{i=1}^{j} X_i| : 1 \le j \le n_1\right) \\
&\left.+ n^{-1/2}\sup\left(|\sum_{i=n_1+1}^{j} X_i| : n_1+1 \le j \le [nt_0]\right)\right\} \triangleq \sum_{i=1}^{4} I_{ni},
\end{aligned}$$

where C is a constant not depending on n. Since $\ell/n \to 0$, $X_1, \ldots, X_n$ are independent with variance 1, $EX_i = \theta$ for $i \ge n - n_2 + 1$, it follows from Donsker's Theorem that

$$I_{n1} \xrightarrow{p} 0, \quad \text{as} \quad n \to \infty. \tag{4.44}$$

Similarly,

$$I_{n3} \xrightarrow{p} 0, \quad \text{as} \quad n \to \infty. \tag{4.45}$$

Now define $\tilde X_i = X_i + \theta$ for $1 \le i \le [nt_0]$, $\tilde X_i = X_i$ for $[nt_0]+1 \le i \le n$. Given arbitrarily $\epsilon > 0$, we have by Donsker's Theorem

$$\begin{aligned}
I_{n2} &\le n^{-1/2}\sup\left(|\sum_{i=j}^{n-n_2}(\tilde X_i - \theta)| : [n(t_0-\varepsilon)] \le j \le n-n_2\right) \\
&\xrightarrow{d} \sup\left(|W(t)| : t_0-\epsilon \le t \le t_0\right), \quad \text{as} \quad n \to \infty \\
&\to 0, \quad \text{a.s.}, \quad \text{as} \quad \epsilon \to 0.
\end{aligned}$$

Hence,

$$I_{n2} \xrightarrow{d} 0, \quad \text{as} \quad n \to \infty. \tag{4.46}$$

Similarly,

$$I_{n4} \xrightarrow{d} 0, \quad \text{as} \quad n \to \infty. \tag{4.47}$$

Summing up (4.44)–(4.47), we get

$$\sup\left(|T_n - \sqrt{nt_0(1-t_0)}(\hat\theta-\theta)| : nt_0 \le \hat k \le nt_0 + 2\ell\right) \xrightarrow{P} 0$$

as $n \to \infty$. From this and (4.42), (4.43), we obtain (4.41). The proof is concluded. □

It is easy to see that $\hat t_0 = (\hat k - \ell)/n$ is a consistent estimator of t_0 (when $\theta \ne 0$, t_0 is well-defined). Previously we have introduced a consistent estimate $\hat\sigma$ of σ. Substituting $\hat t_0$ for t_0 and $\hat\sigma$ for σ, we have the following theorem.

Theorem 4.1.7. *Suppose that the conditions of Theorem 4.1.6 are met, we have*

$$\left(\sqrt{n\hat t_0(1-\hat t_0)}/\hat\sigma\right)(\hat\theta-\theta) \xrightarrow{d} \mathcal{N}(0,1), \tag{4.48}$$

as $n \to \infty$.

Even when $\theta = 0$, making t_0 undefined, the statistic $\hat{t}_0$ remains well-defined. As a result, we cannot determine whether (4.48) holds true for $\theta = 0$. Consequently, Theorem 4.1.7 cannot be applied to test the hypothesis $\theta = 0$. However, (4.48) can still be used to construct a confidence interval for θ when it is assumed in advance that $\theta \neq 0$, or if the null hypothesis $\theta = 0$ is rejected.

If X_1 is normally distributed, or more generally, X_1 satisfies condition (4.35), then the condition (4.40) in Theorem 4.1.6 and Theorem 4.1.7 can be weakened to $(\log n)^2/\ell \to 0$.

4.1.6 Simulations

To intuitively illustrate the method in this section, we conduct simulations for interval estimation under various parameter settings, specifically focusing on θ (the jump at the change point) and σ^2 (the variance of $e(t)$). The effectiveness of the method is assessed by the proportion of estimated intervals that successfully capture the true location and magnitude of the change point.

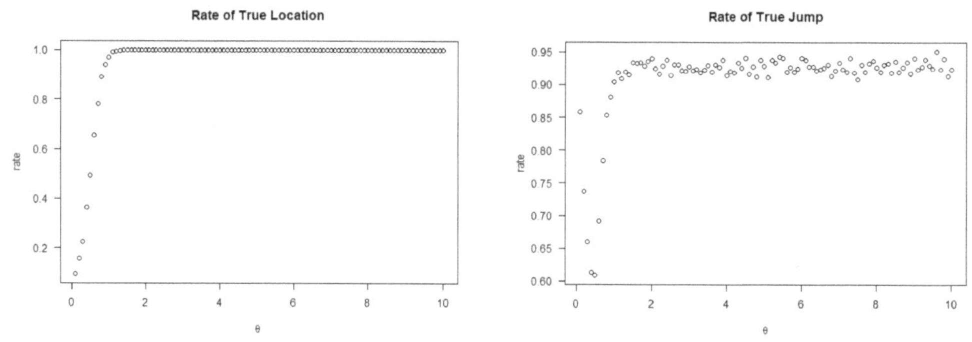

Figure 4.1: Coverage rates for jump location and jump size for different θ.

First, we examine different values of θ using 1000 simulations, with $\ell = 40$ and the error distribution following $\mathcal{N}(0, 1)$. Theoretically, a larger θ should lead to an improved coverage rate, as clearly demonstrated in Figure 4.1.

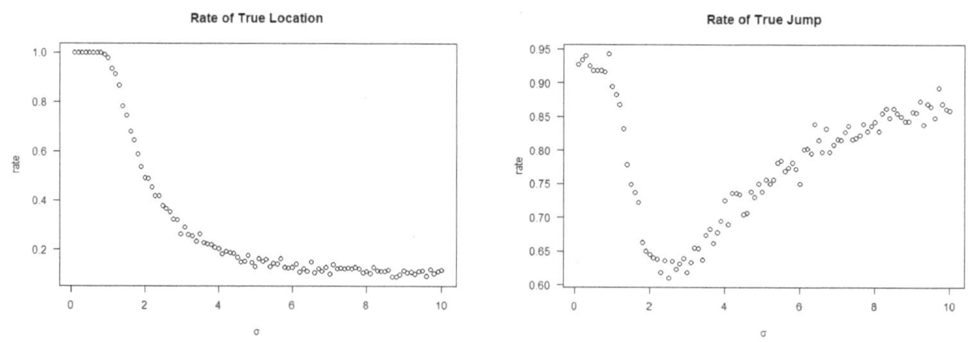

Figure 4.2: The two figures come from 1000 simulations with different σ settings. The ordinate is the rate of true value in confidence interval about location and jump of change point.

Next, we consider different error variances σ^2, keeping $\theta = 1$ and using the same settings as before. Figure 4.2 shows that smaller values of σ lead to better performance. However, an interesting result appears in the rate of capturing the true jump size in the right panel of Figure 4.2. When σ becomes sufficiently large, the rate of capturing the true value actually increases with σ. We believe this occurs because, when σ is much larger relative to θ, the confidence interval becomes excessively wide, making it more likely to contain the true value simply due to its increased length.

Overall, based on our simulation results, the interval estimation method proves to be highly effective in addressing the single change point problem and producing a reliable confidence interval.

4.2 Interval estimation of a single change point in linear regression model

4.2.1 Model and assumptions

The construction of confidence interval for change point in multiple linear regression model was first considered by [16], which formulated the following linear regression with a single change point at k_0:

$$\begin{aligned} y_t &= \boldsymbol{w}_t'\boldsymbol{\alpha} + \boldsymbol{z}_t'\boldsymbol{\delta}_1 + \epsilon_t, \quad t = 1, 2, \ldots, k_0, \\ y_t &= \boldsymbol{w}_t'\boldsymbol{\alpha} + \boldsymbol{z}_t'\boldsymbol{\delta}_2 + \epsilon_t, \quad t = k_0 + 1, \ldots, n, \end{aligned} \tag{4.49}$$

where notations are defined in the same way as in section 2.4.1. Model (4.49) can be rewritten as

$$\begin{aligned} y_t &= \boldsymbol{x}_t'\boldsymbol{\beta} + \epsilon_t, \quad t = 1, 2, \ldots, k_0, \\ y_t &= \boldsymbol{x}_t'\boldsymbol{\beta} + \boldsymbol{z}_t'\boldsymbol{\delta} + \epsilon_t, \quad t = k_0 + 1, \ldots, n. \end{aligned} \tag{4.50}$$

Following the same construction as in section 2.4.1, the sup-Wald-type test statistic is $\sup_{k\in[\pi n,(1-\pi)n]} W_n(k)$, where

$$W_n(k) = \frac{\hat{\boldsymbol{\delta}}_k' \left(\boldsymbol{Z}_2'\boldsymbol{M}\boldsymbol{Z}_2\right) \hat{\boldsymbol{\delta}}_k}{\hat{\sigma}^2(k)}$$

with $\boldsymbol{M} = \boldsymbol{I} - \boldsymbol{X}(\boldsymbol{X}'\boldsymbol{X})^{-1}\boldsymbol{X}$, $\hat{\sigma}^2(k) = S_n(k)/(n-p-q)$, $\pi \in (0, \frac{1}{2})$ a small number. From the identity $\overline{S} - S(k) = \hat{\boldsymbol{\delta}}_k' \left(\boldsymbol{Z}_2'\boldsymbol{M}\boldsymbol{Z}_2\right) \hat{\boldsymbol{\delta}}_k$ with $\overline{S}$ is the sum of squared residuals by regressing $\boldsymbol{Y}$ on $\boldsymbol{X}$ alone. We derive that Wald test statistics is a monotonic transformation of $S_n(k)$. It follows that

$$\begin{aligned} \hat{k} &= \arg\min_k S_n(k) = \arg\max_k \hat{\boldsymbol{\delta}}_k' \left(\boldsymbol{Z}_2'\boldsymbol{M}\boldsymbol{Z}_2\right) \hat{\boldsymbol{\delta}}_k \\ &= \arg\max_k W_n(k). \end{aligned} \tag{4.51}$$

The assumptions needed for the theoretical results are stated below. The notation $\|\cdot\|$ denotes the Euclidean norm, that is, $\|\boldsymbol{x}\| = \left(\sum_{i=1}^p x_i^2\right)^{1/2}$ for $\boldsymbol{x} \in \mathbb{R}^p$. For a matrix $\boldsymbol{A}$, $\|\boldsymbol{A}\|$ represents the vector-induced norm.

Assumption 4.1. *$k_0 = [\tau n]$, where $\tau \in (0,1)$ and $[\cdot]$ is the greatest integer function.*

Assumption 4.2. *The data $\{(\boldsymbol{y}_t, \boldsymbol{x}_t, \boldsymbol{z}_t), \quad 1 \le t \le n, \quad n \ge 1\}$ form a triangular array. For notation simplicity, the subscription n will be suppressed. In addition, $\boldsymbol{z}_t = \boldsymbol{R}'\boldsymbol{x}_t$, where $\boldsymbol{R}$ is $p \times q$,* rank$(\boldsymbol{R}) = q$, $\boldsymbol{z}_t \in \mathbb{R}^q$, $x_t \in \mathbb{R}^p$, $q \le p$.

Assumption 4.3. *The matrices $(1/j)\sum_{t=1}^j \boldsymbol{x}_t\boldsymbol{x}_t'$, $(1/j)\Sigma_{t=n-j+1}^n \boldsymbol{x}_t\boldsymbol{x}_t'$, $(1/j)\sum_{l=k_0-j+1}^{k_0} \boldsymbol{x}_t\boldsymbol{x}_t'$, and $(1/j)\sum_{t=k_0+1}^{k_l+j} \boldsymbol{x}_t\boldsymbol{x}_t'$ have minimum eigenvalues bounded away from zero in probability for all large j. That is, there exists $\lambda > 0$ such that for every $\epsilon > 0$, there exists j_0 such that $P(\lambda_j > \lambda) > 1 - \epsilon$ for all $j > j_0$, where λ_j denotes the minimum eigenvalue for each of the above matrices.*

For simplicity, we assume these matrices are invertible when $j \geq p$. In addition, these four matrices have stochastically bounded norms uniformly in j. That is, for example, $\sup_{j\geq 1}\left\|(1/j)\sum_{t=1}^{i}\boldsymbol{x}_t\boldsymbol{x}_t'\right\|$ is stochastically bounded.

Assumption 4.4. *$(\boldsymbol{X}'\boldsymbol{X}/n)$ converges in probability to a non-random and positive definite matrix.*

Assumption 4.5. *For random regressors, $\sup_t E\left\|\boldsymbol{x}_t\right\|^{4+\delta} \leq K$ for some $\delta > 0$ and $K < \infty$.*

Assumption 4.6. *The disturbances $\{\epsilon_t\}$ satisfy one of the following two alternative conditions:*

(a) $\{\epsilon_t, \mathscr{F}_t\}$ form a sequence of martingale differences for σ-field $\mathscr{F}_t = \{\epsilon_s, \boldsymbol{x}_{s+1}, \quad s \leq t\}$. Moreover, for all t, $E\left|\epsilon_t\right|^{4+\delta} < K$ for some $K < \infty$ and $\delta > 0$.

(b) The disturbance ϵ_t is independent of the regressor $\boldsymbol{x}_s$ for all t and s. For an increasing sequence of σ-field $\mathscr{F}_t$, $\{\epsilon_t, \mathscr{F}_t\}$ form a sequence of L^r, $r = 4 + \delta$, mixingales with mixingale numbers $\{\psi_t\}$ and mixingale norms $\{c_t\}$. See [120] for the definition of L^2 mixingales and [7] for general L^r mixingales. In addition, $\delta\Sigma_j j^{1+\delta}\psi_j < \infty$ for some $\delta > 0$ and $\max_j |c_j| < K < \infty$.

Assumption 4.7. *$\|\boldsymbol{\delta}_n\| \to 0$, and $\|n^{\frac{1}{2}-\alpha}\boldsymbol{\delta}_n\| \to \infty$ for some $\alpha \in \left(0, \frac{1}{2}\right)$.*

Assumption 4.8. *The process $\{\boldsymbol{z}_t, \epsilon_t\}_{t=-\infty}^{\infty}$ is strictly stationary.*

Assumption 4.1 suggests that the true change location is bounded away from the two end points. Assumption 4.2 allows for trending regressors written in the form $(t/n)^l, l > 0$, or, more generally, written as any function of the time trend $g(t/n)$. Expressing them in this format avoids a scaling matrix that would otherwise be required when deriving limiting distributions. This format is also required by Assumption 4.3 and Assumption 4.4 because no scaling matrix is used in these two assumptions. Assumption 4.3 requires that there be enough data around the change point, and at the beginning and the end of the sample, so that the change point can be identified. The latter half of Assumption 4.3 is typically implied by the strong law of large numbers for the sequence of $\{\boldsymbol{x}_t\boldsymbol{x}_t'\}$. Assumption 4.4 and Assumption 4.5 are standard for linear regressions, and they are needed for the central limit theorem. Assumption 4.6(a) allows for lagged dependent regressors (e.g., autoregressive models); Assumption 4.6(b) allows for general serial correlation in the disturbances. These conditions are sufficient to obtain the rate of convergence for the estimated change point.

4.2.2 Theoretical results

Let $\boldsymbol{X}_\Delta = \boldsymbol{X}_2 - \boldsymbol{X}_0 = (\boldsymbol{0}, \ldots, \boldsymbol{0}, \boldsymbol{x}_{k+1}, \ldots, \boldsymbol{x}_{k_0}, \boldsymbol{0}, \ldots, \boldsymbol{0})'$ for $k < k_0$, $\boldsymbol{X}_\Delta = -\left(\boldsymbol{X}_2 - \boldsymbol{X}_0\right) = (\boldsymbol{0}, \ldots, \boldsymbol{0}, \boldsymbol{x}_{k_0+1}, \ldots, \boldsymbol{x}_k, \boldsymbol{0}, \ldots, \boldsymbol{0})'$ for $k > k_0$, and $\boldsymbol{X}_\Delta = \boldsymbol{0}$ for $k = k_0$. It follows that $\boldsymbol{X}_2 = \boldsymbol{X}_0 + \boldsymbol{X}_\Delta \operatorname{sgn}(k_0 - k)$. Denote $\boldsymbol{Z}_\Delta = \boldsymbol{X}_\Delta\boldsymbol{R}$.

Lemma 4.1. *The following two inequalities hold:*

$$
\begin{aligned}
&(\boldsymbol{Z}_0'\boldsymbol{M}\boldsymbol{Z}_0) - (\boldsymbol{Z}_0'\boldsymbol{M}\boldsymbol{Z}_2)\left(\boldsymbol{Z}_2'\boldsymbol{M}\boldsymbol{Z}_2\right)^{-1}(\boldsymbol{Z}_2'\boldsymbol{M}\boldsymbol{Z}_0)\\
&\quad\geq \boldsymbol{R}'\left(\boldsymbol{X}_\Delta'\boldsymbol{X}_\Delta\right)\left(\boldsymbol{X}_2'\boldsymbol{Z}_2\right)^{-1}\left(\boldsymbol{X}_0'\boldsymbol{X}_0\right)\boldsymbol{R}, \quad k < k_0,\\
&(\boldsymbol{Z}_0'\boldsymbol{M}\boldsymbol{Z}_0) - (\boldsymbol{Z}_0'\boldsymbol{M}\boldsymbol{Z}_2)\left(\boldsymbol{Z}_2'\boldsymbol{M}\boldsymbol{Z}_2\right)^{-1}(\boldsymbol{Z}_2'\boldsymbol{M}\boldsymbol{Z}_0)\\
&\quad\geq \boldsymbol{R}'\boldsymbol{X}_\Delta'\boldsymbol{X}_\Delta\left(\boldsymbol{X}'\boldsymbol{X} - \boldsymbol{X}_2'\boldsymbol{X}_2\right)^{-1}\left(\boldsymbol{X}'\boldsymbol{X} - \boldsymbol{X}_0'\boldsymbol{X}_0\right)\boldsymbol{R}, \quad k \geq k_0.
\end{aligned}
$$

Lemma 4.2. *Assume the conditions of Assumptions 4.2–4.4. There exists $\lambda > 0$ such that for every $\epsilon > 0$, there exists $C < \infty$ such that $\inf_{|k-k_0|>C\|\boldsymbol{\delta}_n\|^{-2}} G_n(k) \geq \lambda\left\|\boldsymbol{\delta}_n\right\|^2$ with probability at least $1-\epsilon$.*

Lemma 4.3. *Under Assumptions 4.1–4.6 there exists a $B < \infty$ such that for every $c > 0$ and $m > 0$,*

$$
P\left(\sup_{n\geq k\geq m}\frac{1}{k}\left\|\sum_{t=1}^{k}\boldsymbol{z}_t\epsilon_t\right\| > c\right) \leq \frac{B}{c^2 m}.
$$

Lemma 4.4. *Under Assumptions 4.1–4.6, for every $\epsilon > 0$ and $\eta > 0$, there exists $n_0 > 0$, so that when $n > n_0$, $P\left(\left|\hat{k} - k_0\right| > n\eta\right) < \epsilon$.*

Lemma 4.5. *Under Assumptions 4.1–4.6, for $\boldsymbol{\delta}_n = \boldsymbol{\delta}$ fixed or for $\boldsymbol{\delta}_n$ satisfying Assumption 4.7,*

$$V_n(k) - V_n(k_0) = -\boldsymbol{\delta}_n' \boldsymbol{Z}_\Delta' \boldsymbol{Z}_\Delta \boldsymbol{\delta}_n + 2\boldsymbol{\delta}_n' \boldsymbol{Z}_\Delta' \boldsymbol{\epsilon} \operatorname{sgn}(k_0 - k) + o_p(1),$$

where $o_p(1)$ is uniform on $D(C)$.

All these lemmas were available in ([16]).

Proposition 4.1. *Assume the conditions of Assumptions 4.1–4.6. If $\boldsymbol{\delta}_n$ is fixed or $\|\boldsymbol{\delta}_n\| \to 0$ but satisfying Assumption 4.7, then $\hat{k} = k_0 + O_p\left(\|\delta_n\|^{-2}\right)$.*

Proof of Proposition 4.1. Define $V_n(k) = \hat{\boldsymbol{\delta}}_k' (\boldsymbol{Z}_2' \boldsymbol{M} \boldsymbol{Z}_2) \hat{\boldsymbol{\delta}}_k$. By (4.51), $\hat{k} = \arg\max_{\hat{k}} V_n(k)$. Notice that $\hat{\boldsymbol{\delta}}_k = (\boldsymbol{Z}_2'\boldsymbol{M}\boldsymbol{Z}_2)^{-1} (\boldsymbol{Z}_2'\boldsymbol{M}\boldsymbol{Z}_0) \boldsymbol{\delta}_n + (\boldsymbol{Z}_2'\boldsymbol{M}\boldsymbol{Z}_2)^{-1} \boldsymbol{Z}_2'\boldsymbol{M}\boldsymbol{\epsilon}$ and $\hat{\boldsymbol{\delta}}_{k_0} = \boldsymbol{\delta}_n + (\boldsymbol{Z}_0'\boldsymbol{M}\boldsymbol{Z}_0)^{-1} \boldsymbol{Z}_0'\boldsymbol{M}\boldsymbol{\epsilon}$. It follows that

$$V_n(k) - V_n(k_0) = \boldsymbol{\delta}_n' \{(\boldsymbol{Z}_0'\boldsymbol{M}\boldsymbol{Z}_2)(\boldsymbol{Z}_2'\boldsymbol{M}\boldsymbol{Z}_2)^{-1}(\boldsymbol{Z}_2'\boldsymbol{M}\boldsymbol{Z}_0) - (\boldsymbol{Z}_0'\boldsymbol{M}\boldsymbol{Z}_0)\}\boldsymbol{\delta}_n + H_n(k) \quad (4.52)$$

where

$$\begin{aligned} H_n(k) =& 2\boldsymbol{\delta}_n' (\boldsymbol{Z}_0'\boldsymbol{M}\boldsymbol{Z}_2)(\boldsymbol{Z}_2'\boldsymbol{M}\boldsymbol{Z}_2)^{-1} \boldsymbol{Z}_2'\boldsymbol{M}\boldsymbol{\epsilon} - 2\boldsymbol{\delta}_n' \boldsymbol{Z}_0'\boldsymbol{M}\boldsymbol{\epsilon} \\ &+ \boldsymbol{\epsilon}'\boldsymbol{M}\boldsymbol{Z}_2 (\boldsymbol{Z}_2'\boldsymbol{M}\boldsymbol{Z}_2)^{-1} \boldsymbol{Z}_2'\boldsymbol{M}\boldsymbol{\epsilon} - \boldsymbol{\epsilon}'\boldsymbol{M}\boldsymbol{Z}_0 (\boldsymbol{Z}_0'\boldsymbol{M}\boldsymbol{Z}_0)^{-1} \boldsymbol{Z}_0'\boldsymbol{M}\boldsymbol{\epsilon}. \end{aligned} \quad (4.53)$$

Define for $k \neq k_0$,

$$G_n(k) = \frac{\boldsymbol{\delta}_n' \left\{(\boldsymbol{Z}_0'\boldsymbol{M}\boldsymbol{Z}_0) - (\boldsymbol{Z}_0'\boldsymbol{M}\boldsymbol{Z}_2)(\boldsymbol{Z}_2'\boldsymbol{M}\boldsymbol{Z}_2)^{-1}(\boldsymbol{Z}_2'\boldsymbol{M}\boldsymbol{Z}_0)\right\} \boldsymbol{\delta}_n}{|k_0 - k|}.$$

When $k = k_0$, define $G_n(k_0) = \boldsymbol{\delta}_n' \boldsymbol{\delta}_n$. We have the following identities:

$$V_n(k) - V_n(k_0) = -|k_0 - k| G_n(k) + H_n(k), \quad \text{for all } k. \quad (4.54)$$

Because $V_n(\hat{k}) \geq V_n(k_0)$ by definition, it suffices to show that for each $\epsilon > 0$, there exists $C > 0$ such that

$$P\left(\sup_{|k-k_0|>C\|\delta_n\|^{-2}} V_n(k) \geq V_n(k_0)\right) < \epsilon.$$

By Lemma 4.4, it is sufficient to show that

$$P\left(\sup_{k \in K(C)} V_n(k) \geq V_n(k_0)\right) < \epsilon,$$

where $K(C) = \left\{k : |k - k_0| > C\|\delta_n\|^{-2} \text{ and } n\eta \leq k \leq (1-\eta)n\right\}$ for a small number $\eta > 0$. By (4.54), $V_n(k) \geq V_n(k_0)$ is equivalent to $H_n(k)/|k_0 - k| \geq G_n(k)$. By Lemma 4.2, it is sufficient to prove that

$$P\left(\sup_{k \in K'(C)} \left|\frac{H_n(k)}{k_0 - k}\right| > \lambda \|\delta_n\|^2\right) < \epsilon. \quad (4.55)$$

Note that $H_n(k)$ is defined in (4.53) . Consider (4.53), Use $\boldsymbol{Z}_0 = \boldsymbol{Z}_2 - \boldsymbol{Z}_\Delta \operatorname{sgn}(k_0 - k)$ twice, and we can rewrite the first line of (4.53) as

$$\begin{aligned}
&2\boldsymbol{\delta}_n' \left(\boldsymbol{Z}_0'\boldsymbol{M}\boldsymbol{Z}_2\right)\left(\boldsymbol{Z}_2'\boldsymbol{M}\boldsymbol{Z}_2\right)^{-1}\boldsymbol{Z}_2'\boldsymbol{M}\boldsymbol{\epsilon} - 2\boldsymbol{\delta}_n'\boldsymbol{Z}_0'\boldsymbol{M}\boldsymbol{\epsilon} \\
=&2\boldsymbol{\delta}_n'\boldsymbol{Z}_2'\boldsymbol{M}\boldsymbol{\epsilon} - 2\boldsymbol{\delta}_n'\left(\boldsymbol{Z}_\Delta'\boldsymbol{M}\boldsymbol{Z}_2\right)\left(\boldsymbol{Z}_2'\boldsymbol{M}\boldsymbol{Z}_2\right)^{-1} \\
&\times \boldsymbol{Z}_2'\boldsymbol{M}\boldsymbol{\epsilon}\operatorname{sgn}(k_0 - k) - 2\boldsymbol{\delta}_n'\boldsymbol{Z}_0'\boldsymbol{M}\boldsymbol{\epsilon} \\
=&\left[2\boldsymbol{\delta}_n'\boldsymbol{Z}_\Delta'\boldsymbol{M}\boldsymbol{\epsilon} - 2\boldsymbol{\delta}_n'\left(\boldsymbol{Z}_\Delta'\boldsymbol{M}\boldsymbol{Z}_2\right)\left(\boldsymbol{Z}_2'\boldsymbol{M}\boldsymbol{Z}_2\right)^{-1}\boldsymbol{Z}_2'\boldsymbol{M}\boldsymbol{\epsilon}\right]\operatorname{sgn}(k_0 - k).
\end{aligned} \tag{4.56}$$

Now $\boldsymbol{Z}_\Delta'\boldsymbol{M}\boldsymbol{\epsilon} = \boldsymbol{Z}_\Delta\boldsymbol{\epsilon} - \boldsymbol{Z}_\Delta'\boldsymbol{X}\left(\boldsymbol{X}'\boldsymbol{X}\right)^{-1}\boldsymbol{X}'\boldsymbol{\epsilon}$. Note that $\boldsymbol{Z}_\Delta'\boldsymbol{X} = |k_0 - k|\, O_p(1)$ and $\left(\boldsymbol{X}'\boldsymbol{X}\right)^{-1}\boldsymbol{X}'\boldsymbol{\epsilon} = n^{-1/2}O_p(1)$. For example, for $k_0 - k > 0$, $\boldsymbol{Z}_\Delta'\boldsymbol{X} = \sum_{t=k+1}^{k_0} z_t\boldsymbol{x}_t' = |k_0 - k|\, O_p(1)$. Thus, $\boldsymbol{\delta}_n'\boldsymbol{Z}_\Delta'\boldsymbol{M}\boldsymbol{\epsilon} = \boldsymbol{\delta}_n'\boldsymbol{Z}_\Delta\boldsymbol{\epsilon} - |k_0 - k|\, n^{-1/2}\left\|\boldsymbol{\delta}_n\right\| O_p(1)$. Similarly, $\left\|\boldsymbol{Z}_\Delta'\boldsymbol{M}\boldsymbol{Z}_2\right\| = |k_0 - k|\, O_p(1)$, and $\left(\boldsymbol{Z}_2'\boldsymbol{M}\boldsymbol{Z}_2\right)^{-1}\boldsymbol{Z}_2'\boldsymbol{M}\boldsymbol{\epsilon} = n^{-1/2}O_p(1)$ uniformly on $K(C)$. This implies that (4.56) is

$$2\boldsymbol{\delta}_n'\boldsymbol{Z}_\Delta'\boldsymbol{\epsilon}\operatorname{sgn}(k_0 - k) + |k_0 - k|\, n^{-1/2}\left\|\boldsymbol{\delta}_n\right\| O_p(1).$$

Furthermore, it is easy to argue that the second line of (4.53) is $O_p(1)$ uniformly on $K(C)$. In summary,

$$\begin{aligned}
H_n(k) =& 2\boldsymbol{\delta}_n'\boldsymbol{Z}_\Delta'\boldsymbol{\epsilon}\operatorname{sgn}(k_0 - k) \\
&+ |k_0 - k|\, n^{-1/2}\left\|\boldsymbol{\delta}_n\right\| O_p(1) + O_p(1).
\end{aligned}$$

Thus,

$$\frac{H_n(k)}{|k_0 - k|} = \boldsymbol{\delta}_n'\frac{2}{|k_0 - k|}\boldsymbol{Z}_\Delta'\boldsymbol{\epsilon}\operatorname{sgn}(k_0 - k) + n^{-1/2}\left\|\boldsymbol{\delta}_n\right\| O_p(1) + \frac{O_p(1)}{|k_0 - k|}. \tag{4.57}$$

We can now prove (4.55) using (4.57). Consider $k < k_0$. First,

$$\begin{aligned}
&P\left(\sup_{k\in K'(C)}\left\|\boldsymbol{\delta}_n'\frac{2}{k_0 - k}\sum_{t=k+1}^{k_0} z_t\epsilon_t\right\| > \frac{\lambda\left\|\boldsymbol{\delta}_n\right\|^2}{3}\right) \\
\le& P\left(\sup_{k\le k_0 - C\|\boldsymbol{\delta}_n\|^{-2}}\left|\frac{1}{k_0 - k}\sum_{t=k+1}^{k_0} z_t\epsilon_t\right| > \frac{\lambda\left\|\boldsymbol{\delta}_n\right\|}{6}\right).
\end{aligned}$$

By Lemma 4.3 (applied with reversed data order), the right-hand side above is bounded by

$$B\frac{36}{\lambda^2\left\|\boldsymbol{\delta}_n\right\|^2}\frac{1}{C\left\|\boldsymbol{\delta}_n\right\|^{-2}} = B\frac{36}{\lambda^2 C} < \frac{\epsilon}{3}$$

for large C. Next,

$$P\left(n^{-1/2}\left\|\boldsymbol{\delta}_n\right\| O_p(1) > \lambda\left\|\boldsymbol{\delta}_n\right\|^2/3\right) = P\left(O_p(1)/\left(n^{1/2}\left\|\boldsymbol{\delta}_n\right\|\right) > \lambda/3\right) < \epsilon/3,$$

because $\left(n^{1/2}\left\|\boldsymbol{\delta}_n\right\|\right)^{-1} \to 0$. Finally, for $k \le k_0 - C\left\|\boldsymbol{\delta}_n\right\|^{-2}$, we have

$$O_p(1)/|k_0 - k| \le \left\|\boldsymbol{\delta}_n\right\|^2 O_p(1)/C.$$

Thus,

$$P\left(\sup_{k\in K(C)} O_p(1)/|k_0 - k| > \lambda\left\|\boldsymbol{\delta}_n\right\|^2/3\right) \le P\left(O_p(1)/C > \lambda/3\right) < \epsilon/3$$

for large C. Summarizing these results yields (4.55). The proof of Proposition 4.1 is complete. □

We define a stochastic process $W^*(m)$ on the set of integers as follows: $W^*(0) = 0$, $W^*(m) = W_1(m)$ for $m < 0$ and $W^*(m) = W_2(m)$ for $m > 0$, with

$$\begin{aligned} W_1(m) &= -\boldsymbol{\delta}' \sum_{i=m+1}^{0} \boldsymbol{z}_i \boldsymbol{z}_i' \boldsymbol{\delta} + 2\boldsymbol{\delta}' \sum_{i=m+1}^{0} \boldsymbol{z}_i \epsilon_i, \quad m = -1, -2, \ldots, \\ W_2(m) &= -\boldsymbol{\delta}' \sum_{i=1}^{m} \boldsymbol{z}_i \boldsymbol{z}_i' \boldsymbol{\delta} - 2\boldsymbol{\delta}' \sum_{i=1}^{m} \boldsymbol{z}_i \epsilon_i, \quad m = 1, 2, \ldots. \end{aligned} \tag{4.58}$$

In the case of independence for $(\boldsymbol{z}_t, \epsilon_t)$ over t, the process W^* is a two-sided random walk with (stochastic) drift.

Proposition 4.2. *Under Assumptions 4.1–4.6, and 4.8, and assuming that* $(\boldsymbol{\delta}' \boldsymbol{z}_t)^2 \pm (\boldsymbol{\delta}' \boldsymbol{z}_t') \epsilon_t$ *has a continuous distribution, then*

$$\hat{k} - k_0 \xrightarrow{d} \arg\max_m W^*(m).$$

Proof of Proposition 4.2. For $k < k_0$,

$$-\boldsymbol{\delta}' \boldsymbol{Z}_\Delta' \boldsymbol{Z}_\Delta \boldsymbol{\delta} + 2\boldsymbol{\delta}' \boldsymbol{Z}_\Delta' \boldsymbol{\epsilon} \operatorname{sgn}(k_0 - k) = -\boldsymbol{\delta}' \left(\sum_{t=k+1}^{k_0} \boldsymbol{z}_t \boldsymbol{z}_t' \right) \boldsymbol{\delta} + 2\boldsymbol{\delta}' \sum_{t=k+1}^{k_0} \boldsymbol{z}_t \epsilon_t.$$

Under strict stationarity, the above has the same distribution as $W_1(k - k_0)$ defined in (4.58). Similarly, for $k > k_0$, the left-hand side above is

$$-\boldsymbol{\delta}' \left(\sum_{t=k_0+1}^{k} \boldsymbol{z}_t \boldsymbol{z}_t \right) \boldsymbol{\delta} - 2\boldsymbol{\delta}' \sum_{t=k_0+1}^{k} \boldsymbol{z}_t \epsilon_t,$$

which has the same distribution as $W_2(k - k_0)$.

Lemma 4.5 implies that $V_n(k) - V_n(k_0)$ converges in distribution to $W^*(k - k_0)$ and the convergence is uniform on any given bounded set of integers around k_0. Now the process $W^*(m)$ has a unique maximum with probability 1 when $(\boldsymbol{\delta}' \boldsymbol{z}_t)^2 + 2\boldsymbol{\delta}' \boldsymbol{z}_t \epsilon_t$ has a continuous distribution, because $P(W^*(m) = W^*(m')) = 0$ for $m \neq m'$.

Let $m^* = \arg\max_m W^*(m)$, then m^* is $O_p(1)$ because $W^*(m) \to -\infty$ with probability tending to 1 as $|m| \to \infty$. We thus have that for every $\epsilon > 0$, there exists $K < \infty$ such that $P(|m^*| > K) < \epsilon$.

Similarly by Proposition 4.1, $P(\left|\hat{k} - k_0\right| > K) < \epsilon$. Let $\hat{k}_K = \arg\max_{|k - k_0| \le K} V_n(k)$ and $m_K^* = \arg\max_{|m| \le K} W^*(m)$. The uniform convergence of $V_n(k) - V_n(k_0)$ to $W^*(k - k_0)$ on any bounded set of integers (the difference $|k - k_0|$ is bounded) implies that $\hat{k}_K - k_0 \xrightarrow{d} m_K^*$. That is, for large n, $\left| P\left(\hat{k}_K - k_0 = j\right) - P(m_K^* = j) \right| < \epsilon$ for all $|j| \le K$.

However, if $\left|\hat{k} - k_0\right| \le K$, then $\hat{k} = \hat{k}_K$ and if $|m^*| \le K$, then $m^* = m_K^*$. Thus, the difference $\left| P\left(\hat{k} - k_0 = j\right) - P(m^* = j) \right|$ is bounded by

$$\left| P\left(\hat{k}_K - k_0 = j\right) - P(m_K^* = j) \right| + P\left(\left|\hat{k} - k_0\right| > K\right) + P(|m^*| > K) < 3\epsilon.$$

Since ϵ can be arbitrarily small and K can be arbitrarily large, the proposition is proved. □

We further extend the result to general regression models. Moreover, we allow nonstationary regressors and disturbances. In this case, the limiting distribution turns out to be nonsymmetric. We derive the analytical density function and the cumulative distribution function (cdf) for this general limiting distribution to construct confidence intervals. To this end, we amend Assumption 4.7 to the following assumption:

Assumption 4.9. *$\boldsymbol{\delta}_n = \boldsymbol{\delta}_0 v_n$, where v_n is a positive number such that $v_n \to 0$ and $n^{\frac{1}{2}-\alpha} v_n \to \infty$ for some $\alpha \in \left(0, \frac{1}{2}\right)$ and $\boldsymbol{\delta}_0 \neq \mathbf{0}$.*

This amendment makes the asymptotic argument easier. Further, assume

Assumption 4.10. *(a) $(\boldsymbol{z}_t, \epsilon_t)$ is second-order stationary within each regime such that $E\boldsymbol{z}_t\boldsymbol{z}_t' = \boldsymbol{Q}_1$ and $E\epsilon_t^2 = \sigma_1^2$ for $t \leq k_0$ and $E\boldsymbol{z}_t\boldsymbol{z}_t' = \boldsymbol{Q}_2$ and $E\epsilon_t^2 = \sigma_2^2$ for $t > k_0$.*

(b) A functional central limit theorem holds for $\{\boldsymbol{z}_t\epsilon_t\}$ within each regime. That is,

$$k_0^{-1/2} \sum_{t=1}^{[rk_0]} \boldsymbol{z}_t\epsilon_t \Rightarrow \boldsymbol{B}_1(r),$$

$$(n-k_0)^{-1/2} \sum_{t=k_0+1}^{k_0+[r(n-k_0)]} \boldsymbol{z}_t\epsilon_t \Rightarrow \boldsymbol{B}_2(r),$$

where $\boldsymbol{B}_i(r)$ is a multivariate Gaussian process on $[0,1]$ with mean zero and covariance $E\boldsymbol{B}_i(u)\boldsymbol{B}_i(v) = \min\{u,v\}\boldsymbol{\Omega}_i$ for $i = 1, 2$, and $\boldsymbol{\Omega}_1 = \lim E\left(k_0^{-1/2}\sum_{t=1}^{k_0} \boldsymbol{z}_t\epsilon_t\right)^2$ and $\boldsymbol{\Omega}_2 = \lim E\left((n-k_0)^{-1/2}\sum_{t=k_0+1}^{n} \boldsymbol{z}_t\epsilon_t\right)^2$.

Define

$$\xi = \frac{\boldsymbol{\delta}_0'\boldsymbol{Q}_2\boldsymbol{\delta}_0}{\boldsymbol{\delta}_0'\boldsymbol{Q}_1\boldsymbol{\delta}_0} \quad \text{and} \quad \phi = \frac{\boldsymbol{\delta}_0'\boldsymbol{\Omega}_2\boldsymbol{\delta}_0}{\boldsymbol{\delta}_0'\boldsymbol{\Omega}_1\boldsymbol{\delta}_0}. \tag{4.59}$$

Let $W_i(s), i = 1, 2$, be two independent standard Wiener processes defined on $[0, \infty)$, starting at the origin when $s = 0$. Let

$$Z(s) = \begin{cases} W_1(-s) - |s|/2, & \text{if } s \leq 0, \\ \sqrt{\phi}W_2(s) - \xi|s|/2, & \text{if } s > 0. \end{cases} \tag{4.60}$$

Proposition 4.3. *Under Assumptions 4.1–4.6, 4.9, and 4.10,*

$$\frac{(\boldsymbol{\delta}_n'\boldsymbol{Q}_1\boldsymbol{\delta}_n)^2}{\boldsymbol{\delta}_n'\boldsymbol{\Omega}_1\boldsymbol{\delta}_n}\left(\hat{k} - k_0\right) \xrightarrow{d} \arg\max_s Z(s). \tag{4.61}$$

The distribution of $\hat{k}$ is not symmetric when $\xi \neq 1$ or $\phi \neq 1$. When $\boldsymbol{Q}_i$, $\boldsymbol{\Omega}_i$, and σ_i^2 are the same for each regime ($\xi = 1$ and $\phi = 1$), the limiting distribution reduces to (write $\boldsymbol{Q}_1 = \boldsymbol{Q}_2 = \boldsymbol{Q}$ and $\boldsymbol{\Omega}_1 = \boldsymbol{\Omega}_2 = \boldsymbol{\Omega}$),

$$\frac{(\boldsymbol{\delta}_n'\boldsymbol{Q}\boldsymbol{\delta}_n)^2}{\boldsymbol{\delta}_n'\boldsymbol{\Omega}\boldsymbol{\delta}_n}\left(\hat{k} - k_0\right) \xrightarrow{d} \arg\max_s \{W(s) - |s|/2\} \tag{4.62}$$

where $W(s) = W_1(-s)$ for $s \leq 0$ and $W(s) = W_2(s)$ for $s > 0$. Finally, when the errors are uncorrelated, we have $\boldsymbol{\Omega} = \boldsymbol{\sigma}^2 Q$, and the limiting distribution further reduces to

$$\frac{\boldsymbol{\delta}_n'\boldsymbol{Q}\boldsymbol{\delta}_n}{\sigma^2}\left(\hat{k} - k_0\right) \xrightarrow{d} \arg\max_s \{W(s) - |s|/2\}. \tag{4.63}$$

In this case, the distribution is symmetric about the origin.

Proof of Proposition 4.3. Because $\boldsymbol{\delta}_n = \boldsymbol{\delta}_0 v_n$, Proposition 4.1 implies that $\hat{k} = k_0 + O_p\left(v_n^{-2}\right)$. For any given $C > 0$, we shall derive the limiting process of $V_n(k) - V_n(k_0)$ for $k = k_0 + \left[sv_n^{-2}\right]$ and $s \in [-C, C]$. Consider $s \leq 0$ (i.e., $k \leq k_0$). By Lemma 4.5,

$$V_n(k) - V_n(k_0) = -\boldsymbol{\delta}_0'\left(v_n^2 \sum_{t=k+1}^{k_0} \boldsymbol{z}_t \boldsymbol{z}_t\right)\boldsymbol{\delta}_0 + 2\boldsymbol{\delta}_0'\left(v_n \sum_{t=k+1}^{k_0} \boldsymbol{z}_t \epsilon_t\right) + o_p(1).$$

By Assumptions 4.4 and 4.10, for $k = k_0 + \left[sv_n^{-2}\right]$, $v_n^2 \Sigma_{t=k+1}^{k_0} \boldsymbol{z}_t \boldsymbol{z}_t' \to |s|\boldsymbol{Q}_1$. In addition, $v_n \sum_{t=k+1}^{k_0} \boldsymbol{z}_t \epsilon_t \Rightarrow \boldsymbol{B}_1(-s)$, where $\boldsymbol{B}_1(x)$ is a Brownian motion process on $[0, \infty)$ with variance $x\boldsymbol{\Omega}_1$. Because $\boldsymbol{\delta}_0'\boldsymbol{B}_1(-s)$ has the same distribution as $\left(\boldsymbol{\delta}_0'\boldsymbol{\Omega}_1\boldsymbol{\delta}_0\right)^{1/2} W_1(-s)$, where W_1 is a standard Wiener process on $[0, \infty)$, it follows that, for $s \leq 0$, $V_n\left(k_0 + \left[sv_n^{-2}\right]\right) - V_n(k_0) \Rightarrow -|s|\boldsymbol{\delta}_0'\boldsymbol{Q}_1\boldsymbol{\delta}_0 + 2\left(\boldsymbol{\delta}_0'\boldsymbol{\Omega}_1\boldsymbol{\delta}_0\right)^{1/2} W_1(-s)$. Similarly, for $s > 0$, $V_n\left(k_0 + \left[sv_n^{-2}\right]\right) - V_n(k_0) \Rightarrow -s\boldsymbol{\delta}_0'\boldsymbol{Q}_2\boldsymbol{\delta}_0 + 2\left(\boldsymbol{\delta}_0'\boldsymbol{\Omega}_2\boldsymbol{\delta}_0\right)^{1/2} W_2(s)$, where W_2 is another Wiener process on $[0, \infty)$ independent of W_1. In summary,

$$\begin{aligned} V_n\left(k_0 + \left[sv_n^{-2}\right]\right) - V_n(k_0) &\Rightarrow G(s) \\ &= \begin{cases} -|s|\boldsymbol{\delta}_0'\boldsymbol{Q}_1\boldsymbol{\delta}_0 + 2\left(\boldsymbol{\delta}_0'\boldsymbol{\Omega}_1\boldsymbol{\delta}_0\right)^{1/2} W_1(-s), & \text{if } s < 0, \\ -s\boldsymbol{\delta}_0'\boldsymbol{Q}_2\boldsymbol{\delta}_0 + 2\left(\boldsymbol{\delta}_0'\boldsymbol{\Omega}_2\boldsymbol{\delta}_0\right)^{1/2} W_2(s), & \text{if } s \geq 0. \end{cases} \end{aligned}$$

By the continuous mapping theorem, $v_n^2\left(\hat{k} - k_0\right) \xrightarrow{d} \arg\max_s G(s)$. However, by a change in variable $s = bv$ with $b = \left(\boldsymbol{\delta}_0'\boldsymbol{\Omega}_1\boldsymbol{\delta}_0\right) / \left(\boldsymbol{\delta}_0'\boldsymbol{Q}_1\boldsymbol{\delta}_0\right)^2$, it can be shown that $\arg\max_s G(s) = b \arg\max_v Z(v)$, where $Z(v)$ is defined in (4.60). This implies that $b^{-1}v_n^2\left(\hat{k} - k_0\right) \xrightarrow{d} \arg\max_v Z(v)$. But $b^{-1}v_n^2 = \left(\boldsymbol{\delta}_n'\boldsymbol{Q}_1\boldsymbol{\delta}_n\right)^2 / \left(\boldsymbol{\delta}_n'\boldsymbol{\Omega}_1\boldsymbol{\delta}_n\right)$ in view of $\boldsymbol{\delta}_n = \boldsymbol{\delta}_0 v_n$. □

Next, we consider $\boldsymbol{z}_t$ being a vector of trending regressors such that $\boldsymbol{z}_t = \boldsymbol{g}(t/n) = [g_1(t/n), \ldots, g_q(t/n)]'$. Assumption 4.10 will not be satisfied, we need to derive the limiting distribution separately.

Proposition 4.4. *Let* $\boldsymbol{z}_t = \boldsymbol{g}(t/n) = [g_1(t/n), \ldots, g_q(t/n)]'$, *with* $g_i(x)$ *having a bounded derivative on* $[0, 1]$. *Assume that the functional central limit theorem holds for* ϵ_t *within each regime. Under Assumptions 4.1–4.6,*

$$\frac{\boldsymbol{\delta}_n'\boldsymbol{g}(\tau)\boldsymbol{g}(\tau)'\boldsymbol{\delta}_n}{\psi_1^2}\left(\hat{k} - k_0\right) \xrightarrow{d} \arg\max_s Z^{\dagger}(s),$$

where

$$Z^{\dagger}(s) = \begin{cases} W_1(-s) - |s|/2, & \textit{if } s \leq 0, \\ \sqrt{\psi_2/\psi_1}\, W_2(s) - |s|/2, & \textit{if } s > 0, \end{cases}$$

with $\psi_1 = \lim E(k_0^{-1/2}\Sigma_{t=1}^{k_0}\epsilon_t)^2$ *and* $\psi_2 = \lim E[(n - k_0)^{-1/2}\Sigma_{t=k_0+1}^{n}\epsilon_t]^2$.

The proof of Proposition 4.4 is similar to that of [15].

When ϵ_t is second-order stationary for the whole sample, $\psi_1 = \psi_2 = \psi = \sigma^2 + 2\sum_{j=2}^{\infty} E\left(\epsilon_1\epsilon_j\right)$, the limiting distribution reduces to $\boldsymbol{\delta}_n\boldsymbol{g}(\tau)\boldsymbol{g}(\tau)'\boldsymbol{\delta}_n\psi^{-2}\left(k - k_0\right) \to \arg\max\{W(s) - |s|/2\}$. Furthermore, when the ϵ_t are serially uncorrelated, $\psi^2 = \sigma^2 = E\epsilon_t^2$.

4.2.3 Construction of confidence intervals

The results derived above facilitate the construction of confidence intervals for the change point. This only requires consistent estimates for $\boldsymbol{Q}_i$, $\boldsymbol{\Omega}_i$, σ_i^2 for $i = 1, 2$, and $\boldsymbol{\delta}_n$. Various special cases are discussed below.

Case 1. The series (z_t, ϵ_t) is second-order stationary for the whole sample. In this case, the limiting distribution is characterized by (4.62) or (4.63). For serially uncorrelated disturbances, we only need estimates for δ_n, Q, and σ^2. An estimate for δ_n is already available. Q and σ^2 are estimated by $n^{-1}\Sigma_{t=1}^{n} z_t z_t'$ and $n^{-1}\Sigma_{t=1}^{n}\hat{\epsilon}_t^2$, respectively. Define $\hat{L} = \hat{\delta}_n'\hat{Q}\hat{\delta}_n/\hat{\sigma}^2$. It can be shown that $(\hat{L}-L)\left(\hat{k}-k_0\right) \xrightarrow{p} 0$. Thus, (4.63) holds when δ_n, Q, and σ^2 are replaced by their estimators. A $100(1-\alpha)\%$ confidence interval is given by

$$[\hat{k} - [c/\hat{L}] - 1, \hat{k} + [c/\hat{L}] + 1] \tag{4.64}$$

where c is the $(1-\alpha/2)$th quantile of the random variable $\arg\max\{W(s) - |s|/2\}$, and $[c/\hat{L}]$ is the integer part of $c/\hat{L}$. For serially correlated disturbances we need an estimate for Ω. Define $\hat{L} = \left(\hat{\delta}_n'\hat{Q}\hat{\delta}_n\right)^2 / \left(\hat{\delta}_n'\hat{\Omega}\hat{\delta}_n\right)$. The confidence interval is again given by (4.64) with this new $\hat{L}$.

Case 2. The series (z_t, ϵ_t) is stationary within each regime segmented by the change points. In this case, the limiting distribution is described by (4.61). The quantities to be estimated are Q_i, Ω_i, $i = 1, 2$, and δ_n. We estimate Q_1 by $\hat{k}^{-1}\sum_{t=1}^{\hat{k}} z_t z_t'$, and Q_2 by $(n-\hat{k})^{-1}\Sigma_{t=\hat{k}+1}^{n} z_t z_t'$. The Ω_i are estimated by the Newey and West method, but Ω_1 is computed using the subsample $[1, \hat{k}]$ and Ω_2 using the subsample $[\hat{k}+1, n]$. For serially uncorrelated ϵ_t, $\Omega_i = \sigma_i^2 Q_i$, where σ_1^2 is estimated by $\hat{k}^{-1}\sum_{t=1}^{\hat{k}}\hat{\epsilon}_t^2$ and σ_2^2 by $(n-\hat{k})^{-1}\Sigma_{t=\hat{k}+1}^{n}\hat{\epsilon}_t^2$. Once these quantities are obtained, we can estimate ξ and ϕ in (4.59) by the plug-in method. Note that δ_0 can be replaced by $\hat{\delta}_n$ in estimating ξ and ϕ. (Multiply both the numerator and the denominator by v_n and use $\delta_n = \delta_0 v_n$.) When ξ and ϕ are known, the quantiles for $\arg\max_z Z(s)$ can be computed. Let c_1 denote the $(\alpha/2)$th quantile, and c_2 the $(1-\alpha/2)$th quantile. Define $\hat{L} = \left(\hat{\delta}_n'\hat{Q}_1\hat{\delta}_n\right)^2 / \left(\hat{\delta}_n'\hat{\Omega}_1\hat{\delta}_n\right)$. Because $P\left[c_1 < \hat{L}\left(\hat{k}-k_0\right) < c_2\right] \to 1-\alpha$, a $100(1-\alpha)\%$ confidence interval is given by

$$\left[\hat{k} - \left[c_2/\hat{L}\right] - 1, \hat{k} - \left[c_1/\hat{L}\right] + 1\right]. \tag{4.65}$$

Case 3. Trending regressors. The limiting distribution is given in Proposition 4.4. First, consider the simple case that ϵ_t is stationary. Let $\hat{f}(0)$ denote an estimate of the spectral density of ϵ_t at zero (without 2π). For serially uncorrelated ϵ_t, $\hat{f}(0) = \hat{\sigma}^2 = n^{-1}\Sigma_{t=1}^{n}\hat{\epsilon}_t^2$. Define $\hat{L} = \hat{\delta}_n' g(\hat{k}/n) g(\hat{k}/n)'\hat{\delta}_n/\hat{f}(0)$. In this case, $\psi_1 = \psi_2$ and $Z^{\dagger}(s) = W(s) - |s|/2$. The quantiles are computed from (4.69). The confidence interval is given by (4.64) with this choice of $\hat{L}$.

When the errors ϵ_t are only stationary within each regime, let $\hat{f}_1(0)$ be an estimate of spectral density of ϵ_t for the first regime (using the subsample $[1, \hat{k}]$) and $\hat{f}_2(0)$ for the second regime. We estimate ψ_2/ψ_1 by $\hat{f}_2(0)/\hat{f}_1(0)$. The quantiles of $\arg\max Z^{\dagger}(s)$ are computed from (4.67) and (4.68) with $\phi = \psi_2/\psi_1$ and $\xi = 1$. Define $\hat{L} = \hat{\delta}_n g(\hat{k}/n) g(\hat{k}/n)'\hat{\delta}_n/\hat{f}_1(0)$. Then, the confidence interval is computed using (4.65).

4.2.4 Density function of $\arg\max_v Z(v)$

Recall that the process $Z(v)$ involves two independent Brownian motions with different scales ($\phi \neq 1$) and two different drifts $\xi \neq 1$. To obtain the density function of $l^* = \arg\max_v Z(v)$, we use the argument of [28], who considered the case in which the two Brownian motions have the same scale. In addition to the density function, we also derive the CDF of l^*, which is useful for obtaining distribution quantiles when constructing confidence intervals. Let m be the maximum and l the location of the maximum of $X(t) = \lambda B(t) - \theta t/2$, $t \geq 0$, where $B(t)$ is a standard Brownian motion process on $[0, \infty)$. That is, $l = \arg\max_t X(t)$ and $m = \max_t X(t)$. Using the result of

[28] with a scale transformation, we obtain the joint density of (l, m) as

$$\begin{aligned} f_{l,m}(x, y) =& \frac{\theta}{\lambda}(2\pi)^{-1} \exp\left(-\theta^2 \frac{x}{8}\right) \\ & \times \exp\left[-\frac{1}{2}\left(\frac{y^2}{\lambda^2}x^{-1} + \frac{\theta}{\lambda}y\right)\right], \quad x > 0, \quad y > 0. \end{aligned} \tag{4.66}$$

The marginal distribution of m is exponential with mean λ^2/θ. Now let (l_1, m_1) be the location and the maximum of $\{W_1(-v) - |v|/2\}$, $v < 0$. The joint density of (l_1, m_1) is $f_1(|x|, y)$, where f_1 is the same as (4.66) with $\theta = \lambda = 1$. Let (l_2, m_2) be the location and the maximum of $\{\sqrt{\phi}W_2(v) - \xi v/2\}$, $v > 0$, with joint density $f_2(x, y)$ (given by (4.66) with $\lambda = \sqrt{\phi}$ and $\theta = \xi$). Because (l_1, m_1) is independent of (l_2, m_2), the joint density of (l_1, m_1, l_2, m_2) is $f_1(|x|, y) f_2(x', y')$. Let $l^* = \arg\max_v Z(v)$. Then $l^* = l_1$ if $m_1 > m_2$ and $l^* = l_2$ if $m_1 < m_2$. Thus the density function $g(x)$ of l^* for $x < 0$ is

$$\begin{aligned} g(x)dx &= P(l_1 \in (x, x + dx), m_1 > m_2) \\ &= dx \int_0^\infty \left[f_1(|x|, y) \int_0^\infty \int_0^y f_2(x', y')\, dx'dy'\right] dy \\ &= dx \int_0^\infty f_1(|x|, y) P(m_2 < y)\, dy \\ &= dx \int_0^\infty f_1(|x|, y) \left[1 - \exp\left(-\frac{\theta}{\lambda^2}y\right)\right] dy. \end{aligned}$$

Carrying out the integration, we obtain

$$\begin{aligned} g(x) =& 2^{-1}\Phi\left(-2^{-1}\sqrt{|x|}\right) + 2^{-1}(1 + 2\alpha)\exp\left[2^{-1}\alpha(1 + \alpha)|x|\right] \\ & \times \Phi\left[-2^{-1}(1 + 2\alpha)\sqrt{|x|}\right], \quad x < 0, \end{aligned}$$

where $\alpha = \xi\phi^{-1}$ and $\Phi(\cdot)$ is the cdf of a standard normal random variable. Similarly, for $x > 0$, $g(x) = \int_0^\infty f_2(x, y) P(m_1 < y)\, dy = \int_0^\infty f_2(x, y)(1 - e^{-y})dy$. Carrying out the integration, we obtain

$$\begin{aligned} g(x) =& -2^{-1}\beta^2\Phi\left(-2^{-1}\beta\sqrt{x}\right) + (\xi + 2^{-1}\beta^2)\exp\left[2^{-1}(\phi + \xi)x\right] \\ & \times \Phi\left[-\left(\phi^{1/2} + 2^{-1}\beta\right)\sqrt{x}\right], \quad x > 0, \end{aligned}$$

where $\beta = \xi/\phi^{1/2}$. The CDF $G(x)$ for $x < 0$ is then obtained as

$$\begin{aligned} G(x) =& -(2\pi)^{-1/2}|x|^{1/2}\exp\left(-8^{-1}|x|\right) \\ & - c\exp(a|x|)\Phi\left(-b|x|^{1/2}\right) \\ & + \left(d - 2 + 2^{-1}|x|\right)\Phi\left(-2^{-1}|x|^{1/2}\right), \quad x < 0, \end{aligned} \tag{4.67}$$

where

$$a = 2^{-1}\frac{\xi}{\phi}\left(1 + \frac{\xi}{\phi}\right), \quad b = \frac{1}{2} + \frac{\xi}{\phi}, \quad c = \frac{\phi(\phi + 2\xi)}{\xi(\phi + \xi)},$$

$$d = \frac{(\phi + 2\xi)^2}{(\phi + \xi)\xi}.$$

Similarly, for $x > 0$,

$$\begin{aligned} G(x) =& 1 + \xi\phi^{-1/2}(2\pi)^{-1/2}x^{1/2}\exp\left(-8^{-1}\xi^2\phi^{-1}x\right) \\ & + c\exp(ax)\Phi\left(-bx^{1/2}\right) + \left(-d + 2 - 2^{-1}\xi^2\phi^{-1}x\right) \\ & \times \Phi\left(-2^{-1}\xi\phi^{-1/2}x^{1/2}\right), \quad x > 0 \end{aligned} \tag{4.68}$$

where

$$a = \frac{\phi + \xi}{2}, \quad b = \frac{2\phi + \xi}{2\sqrt{\phi}}, \quad c = \frac{\xi(2\phi + \xi)}{(\phi + \xi)\phi},$$
$$d = \frac{(2\phi + \xi)^2}{(\phi + \xi)\phi}.$$

The density function and the CDF specialize for the symmetric case when $\phi = 1$ and $\xi = 1$, and we have

$$\begin{aligned} G(x) =& 1 + (2\pi)^{-1/2}\sqrt{x}e^{-x/8} - \frac{1}{2}(x+5) \\ &\times \Phi\left(-\frac{\sqrt{x}}{2}\right) + \frac{3}{2}e^{x}\Phi\left(-\frac{3\sqrt{x}}{2}\right), \end{aligned} \tag{4.69}$$

for $x > 0$ and $G(x) = 1 - G(-x)$.

4.3 Interval estimates of two change points

4.3.1 Model and assumptions

In this section we consider the following epidemic alternative model:

$$y_i = \begin{cases} \mu_1 + \epsilon_i, & i = 1, \ldots, k_1^*, \\ \mu_2 + \epsilon_i, & i = k_1^* + 1, \ldots, k_2^*, \\ \mu_1 + \epsilon_i, & i = k_2^* + 1, \ldots, T. \end{cases} \tag{4.70}$$

Here y_i is the one-dimensional observed value at time i, μ_1 and μ_2 are the constant means which satisfy $\mu_1 \neq \mu_2$, ϵ_i is the random error term. The two change points have unknown locations at k_1^* and k_2^*, respectively.

The definition of the epidemic alternative model can be found in Section 9.3 of [51]. [272] published a survey of the available test procedures for this model together with their comparison.

The main model assumptions are listed as follows:

A1. $\{\epsilon_t, t = 1, \ldots, T\}$ is a martingale difference sequence with $\sigma^2 \equiv E\epsilon_t^2 < \infty, t = 1, \ldots, T$.

A2. $\min\left(\frac{k_1^*}{T}, \frac{k_2^* - k_1^*}{T}, \frac{T - k_2^*}{T}\right) \geq c_0$, for some constant $c_0 \in (0, 1/3]$ and $\lim_{T\to\infty} \frac{k_2^* - k_1^*}{T} = \eta \in (c_0, 1 - c_0)$.

A3. $\rho_T = \mu_1 - \mu_2 \to 0$ and $\sqrt{T}|\rho_T| \to \infty$ as $T \to \infty$.

A4. $E\epsilon_t^r < \infty$ for some constant $r > 2$.

Remark 4.4. *Assumption A1 assumes that the random error sequence is a martingale difference sequence with the same variance. Assumption A2 assumes that the two change points are sufficiently separated from each other and that there is enough data around each change point. Assumption A3 assumes that if $\rho_T = \mu_1 - \mu_2 \to 0$, then it cannot go to zero too fast.*

For any two candidate change points, whose locations satisfy $1 < k_1 < k_2 < T$, we can naturally derive the following estimates for μ_1 and μ_2:

$$\hat{\mu}_1 = \frac{1}{T - k_2 + k_1}\left(\sum_{t=1}^{k_1} y_t + \sum_{t=k_2+1}^{T} y_t\right), \quad \hat{\mu}_2 = \frac{1}{k_2 - k_1}\left(\sum_{t=k_1+1}^{k_2} y_t\right).$$

In terms of the relative positions of k_1, k_2, k_1^*, and k_2^*, we have the following six cases and we discuss the expectations of $\hat{\mu}_1$ and $\hat{\mu}_2$ in each case.
(i) $k_1 < k_2 \leq k_1^* < k_2^*$. In this case, we have

$$\begin{aligned} E\hat{\mu}_1 =& \frac{1}{T-k_2+k_1} E\left(\sum_{t=1}^{k_1} y_t + \sum_{t=k_2+1}^{k_1^*} y_t + \sum_{t=k_1^*+1}^{k_2^*} y_t + \sum_{t=k_2^*+1}^{T} y_t\right) \\ =& \mu_1 + \frac{k_2^* - k_1^*}{T-k_2+k_1}(\mu_2 - \mu_1), \\ E\hat{\mu}_2 =& \frac{1}{k_2-k_1}\sum_{t=k_1+1}^{k_2} Ey_t = \mu_1. \end{aligned} \tag{4.71}$$

(ii) $k_1 \leq k_1^* < k_2 \leq k_2^*$. In this case, we have

$$\begin{aligned} E\hat{\mu}_1 =& \mu_1 + \frac{k_2^* - k_2}{T-k_2+k_1}(\mu_2 - \mu_1), \\ E\hat{\mu}_2 =& \mu_2 + \frac{k_1^* - k_1}{k_2-k_1}(\mu_1 - \mu_2). \end{aligned} \tag{4.72}$$

(iii) $k_1 \leq k_1^* < k_2^* \leq k_2$. In this case, we have

$$\begin{aligned} E\hat{\mu}_1 =& \mu_1, \\ E\hat{\mu}_2 =& \mu_2 + \frac{(k_2 - k_2^*) + (k_1^* - k_1)}{k_2-k_1}(\mu_1 - \mu_2). \end{aligned} \tag{4.73}$$

(iv) $k_1^* \leq k_1 < k_2 \leq k_2^*$. In this case, we have

$$\begin{aligned} E\hat{\mu}_1 =& \mu_1 + \frac{(k_2^* - k_2) + (k_1 - k_1^*)}{T-k_2+k_1}(\mu_2 - \mu_1), \\ E\hat{\mu}_2 =& \mu_2. \end{aligned} \tag{4.74}$$

(v) $k_1^* \leq k_1 < k_2^* \leq k_2$. In this case, we have

$$\begin{aligned} E\hat{\mu}_1 =& \mu_1 + \frac{k_1 - k_1^*}{T-k_2+k_1}(\mu_2 - \mu_1), \\ E\hat{\mu}_2 =& \mu_2 + \frac{k_2^* - k_2}{k_2-k_1}(\mu_2 - \mu_1). \end{aligned} \tag{4.75}$$

(vi) $k_1^* < k_2^* \leq k_1 < k_2$. In this case, we have

$$\begin{aligned} E\hat{\mu}_1 =& \mu_1 + \frac{k_2^* - k_1^*}{T-k_2+k_1}(\mu_2 - \mu_1), \\ E\hat{\mu}_2 =& \mu_2. \end{aligned} \tag{4.76}$$

Denote $S_T(k_1, k_2)$ as the sum of squared residuals, that is,

$$S_T(k_1,k_2) = \sum_{t=1}^{k_1}(y_t - \hat{\mu}_1)^2 + \sum_{t=k_1+1}^{k_2}(y_t - \hat{\mu}_2)^2 + \sum_{t=k_2+1}^{T}(y_t - \hat{\mu}_1)^2. \tag{4.77}$$

We may estimate k_1 and k_2 of model (4.70) by minimizing $S_T(k_1, k_2)$. Let $\bar{y} = \frac{1}{T}\sum_{t=1}^{T} y_t$ and $S = \sum_{t=1}^{T}(y_t - \bar{y})^2$. Thus, minimizing $S_T(k_1, k_2)$ is equivalent to minimizing the statistic

$(S - S_T(k_1, k_2))/T$. Since $(T - k_2 + k_1)\hat{\mu}_1 + (k_2 - k_1)\hat{\mu}_2 = T\bar{y}$, we have

$$\begin{aligned}\frac{1}{T}(S - S_T(k_1, k_2)) &= \frac{1}{T}\left[(T - k_2 + k_1)\hat{\mu}_1^2 + (k_2 - k_1)\hat{\mu}_2^2 - T\bar{y}^2\right] \\ &= \frac{(T - k_2 + k_1)(k_2 - k_1)}{T^2}(\hat{\mu}_2 - \hat{\mu}_1)^2.\end{aligned} \tag{4.78}$$

Denote

$$W_{k_1,k_2} = \frac{1}{T}\sqrt{(T - k_2 + k_1)(k_2 - k_1)}(\hat{\mu}_2 - \hat{\mu}_1). \tag{4.79}$$

We propose to estimate the two change points k_1 and k_2 by

$$\begin{aligned}(\hat{k}_1, \hat{k}_2) &= \arg \max_{k_1 > cT, k_2 < (1-c)T, k_2 - k_1 > cT} \frac{1}{T}(S - S_T(k_1, k_2)) \\ &= \arg \max_{k_1 > cT, k_2 < (1-c)T, k_2 - k_1 > cT} |W_{k_1,k_2}|,\end{aligned}$$

where the constant c is defined in Assumption A2. Denote $\hat{\delta}_1 = k_1^* - \hat{k}_1$, $\hat{\delta}_2 = k_2^* - \hat{k}_2$,

$$\hat{\tau}_1 = \frac{|\delta_1|}{T} \text{ and } \hat{\tau}_2 = \frac{|\delta_2|}{T}. \tag{4.80}$$

We have the following theorem.

Theorem 4.3.1. *Under Assumptions A1–A2, we have*

$$\hat{\tau}_1 + \hat{\tau}_2 \to_p 0. \tag{4.81}$$

Under Assumptions A1–A4, we have

$$|\hat{\delta}_1| + |\hat{\delta}_2| = O_p(\rho_T^{-2}).$$

and

$$\rho_T^2(|\hat{\delta}_1| + |\hat{\delta}_2|) \to_d \arg\max_{s>0}\left\{\sigma W(s) - \frac{s}{2}\right\},$$

where $W(s)$ is a standard Wiener process defined on $[0, \infty]$.

Remark 4.5. *We estimate ρ_T by $\hat{\rho}_T = \hat{\mu}_1 - \hat{\mu}_2$ and σ^2 by $\hat{\sigma}^2 = \frac{1}{T}\left(\sum_{t=1}^{\hat{k}_1}(y_t - \hat{\mu}_1)^2 + \sum_{t=\hat{k}_2+1}^{T}(y_t - \hat{\mu}_1)^2 + \sum_{t=\hat{k}_1+1}^{\hat{k}_2}(y_t - \hat{\mu}_2)^2\right)$. Then, we have*

$$\frac{\hat{\rho}_T^2}{\hat{\sigma}^2}(|\hat{\delta}_1| + |\hat{\delta}_2|) \to_d \arg\max_{s>0}\left\{W(s) - \frac{s}{2}\right\},$$

where the CDF of $\arg\max_{s>0}\left\{W(s) - \frac{s}{2}\right\}$ is $2G(s) - 1$ and $G(s)$ is defined by [16]:

$$G(s) = 1 + (2\pi)^{-1/2}\sqrt{s}e^{-s/8} - \frac{1}{2}(s + 5) \times \Phi\left(-\frac{\sqrt{s}}{2}\right) + \frac{3}{2}e^s\Phi\left(-\frac{3\sqrt{s}}{2}\right),$$

for $s > 0$.

Denote $h(k_1, k_2) = \frac{1}{T}\sqrt{(T - k_2 + k_1)(k_2 - k_1)}(\mu_2 - \mu_1)$. Before we prove Theorem 4.3.1, we have the following two lemmas.

Lemma 4.6. *(Rényi-Hájek-Chow inequality). Suppose that* $\{X_n, n \geq 1\}$ *be a martingale difference sequence with* $\sigma_n^2 \equiv EX_n^2 < \infty, n = 1, 2, \ldots$. *Set constants* $c_1 \geq c_2 \geq \cdots \geq c_n > 0, 1 \leq m \leq n$, *and let* $S_n = \sum_{j=1}^{n} X_j$. *Then, for any* $x > 0$,

$$\mathbb{P}\left\{\max_{m\leq j\leq n} c_j |S_j| \geq x\right\} \leq \frac{1}{x^2}\left(c_m^2 \sum_{j=1}^{m} \sigma_j^2 + \sum_{j=m+1}^{n} c_j^2\sigma_j^2\right).$$

Lemma 4.7. *Under Assumption A1, for any* $\varepsilon > 0$,

$$\mathbb{P}\left(\max_{k_1>cT,k_2<(1-c)T,k_2-k_1>cT} |W_{k_1,k_2} - h(k_1,k_2)| \geq \varepsilon\right) \leq \frac{27\sigma^2(1-2c)}{8c^2\varepsilon^2 T}, \tag{4.82}$$

where c is a constant and $0 < c < 1/3$.

Proof. By the definition of W_{k_1,k_2} in (4.79),

$$W_{k_1,k_2} - h(k_1,k_2) = \frac{1}{T}\sqrt{(T-k_2+k_1)(k_2-k_1)}\left[(\hat{\mu}_2 - \hat{\mu}_1) - E(\hat{\mu}_2 - \hat{\mu}_1)\right].$$

For $\forall\varepsilon > 0$, we have

$$\begin{aligned}
&\mathbb{P}\left(\max_{k_1>cT,k_2<(1-c)T,k_2-k_1>cT} |W_{k_1,k_2} - h(k_1,k_2)| \geq \varepsilon\right)\\
=&\mathbb{P}\left(\max_{k_1>cT,k_2<(1-c)T,k_2-k_1>cT} \left|\frac{1}{T}\sqrt{(T-k_2+k_1)(k_2-k_1)}\left[(E\hat{\mu}_1 - \hat{\mu}_1) + (\hat{\mu}_2 - E\hat{\mu}_2)\right]\right| \geq \varepsilon\right)\\
\leq&\mathbb{P}\left(\max_{k_1>cT,k_2<(1-c)T,k_2-k_1>cT} \left|\frac{1}{2}\left[(E\hat{\mu}_1 - \hat{\mu}_1) + (\hat{\mu}_2 - E\hat{\mu}_2)\right]\right| \geq \varepsilon\right)\\
\leq&\mathbb{P}\left(\max_{(1-2c)T>k_1>cT,(1-2c)T>k_2-k_1>cT} \left|\frac{1}{2}\frac{1}{T-k_2+k_1}\sum_{t=1}^{k_1}\varepsilon_t\right| \geq \frac{1}{3}\varepsilon\right)\\
&+\mathbb{P}\left(\max_{(1-c)T>k_2>2cT,(1-2c)T>k_2-k_1>cT} \left|\frac{1}{2}\frac{1}{T-k_2+k_1}\sum_{t=k_2+1}^{T}\varepsilon_t\right| \geq \frac{1}{3}\varepsilon\right)\\
&+\mathbb{P}\left(\max_{(1-2c)T>k_2-k_1>cT} \left|\frac{1}{2}\frac{1}{k_2-k_1}\sum_{t=k_1+1}^{k_2}\varepsilon_t\right| \geq \frac{1}{3}\varepsilon\right)\\
\leq&\mathbb{P}\left(\max_{(1-2c)T>k_1>cT} \left|\sum_{t=1}^{k_1}\varepsilon_t\right| \geq \frac{4cT}{3}\varepsilon\right) + \mathbb{P}\left(\max_{(1-c)T>k_2>2cT} \left|\sum_{t=k_2+1}^{T}\varepsilon_t\right| \geq \frac{4cT}{3}\varepsilon\right)\\
&+\mathbb{P}\left(\max_{(1-2c)T>k_2-k_1>cT} \left|\sum_{t=k_1+1}^{k_2}\varepsilon_t\right| \geq \frac{2cT}{3}\varepsilon\right).
\end{aligned}$$

From Lemma 4.6, we have

$$\begin{aligned}
&\mathbb{P}\left(\max_{k_1>cT,k_2<(1-c)T,k_2-k_1>cT} |W_{k_1,k_2} - h(k_1,k_2)| \geq \varepsilon\right)\\
&\leq \frac{9\sigma^2(1-2c)T}{(4cT\varepsilon)^2} + \frac{9\sigma^2(1-2c)T}{(4cT\varepsilon)^2} + \frac{9\sigma^2(1-2c)T}{(2cT\varepsilon)^2} = \frac{27\sigma^2(1-2c)}{8c^2\varepsilon^2 T}.
\end{aligned}$$

□

Proof of Theorem 4.3.1. First, it is easy to see that

$$|W_{\hat{k}_1,\hat{k}_2}| \le |W_{\hat{k}_1,\hat{k}_2} - h(\hat{k}_1,\hat{k}_2)| + |h(\hat{k}_1,\hat{k}_2)|, \quad |h(k_1^*,k_2^*)| \le |W_{k_1^*,k_2^*} - h(k_1^*,k_2^*)| + |W_{k_1^*,k_2^*}|.$$

Note that $|W_{\hat{k}_1,\hat{k}_2}| \ge |W_{k_1^*,k_2^*}|$. By the triangle inequality, we have

$$|h(k_1^*,k_2^*)| - |h(\hat{k}_1,\hat{k}_2)| \le 2 \max_{k_1>cT, k_2<(1-c)T, k_2-k_1>cT} |W_{k_1,k_2} - h(k_1,k_2)|. \tag{4.83}$$

We only give the proof for Case (ii), in which $k_1 \le k_1^* < k_2 \le k_2^*$, since the proofs for the Cases (iii)–(v) are similar. By (4.72), we have

$$\begin{aligned} h(k_1,k_2) =& \frac{1}{T}\sqrt{(T-k_2+k_1)(k_2-k_1)}E(\hat{\mu}_2-\hat{\mu}_1) \\ =& \frac{1}{T}\sqrt{(T-k_2+k_1)(k_2-k_1)}\left(1-\frac{k_1^*-k_1}{k_2-k_1}-\frac{k_2^*-k_2}{T-k_2+k_1}\right)(\mu_2-\mu_1). \end{aligned}$$

It is easy to see that

$$h(k_1^*,k_2^*) = \frac{1}{T}\sqrt{(T-k_2^*+k_1^*)(k_2^*-k_1^*)}(\mu_2-\mu_1).$$

Since $k_1 \le k_1^* < k_2 \le k_2^*$ in Case (ii), we write $a_0 = k_2^* - k_1^*$, $a_1 = k_2 - k_1$, and $a_2 = k_2 - k_1^*$. Then, we have

$$\begin{aligned} & |h(k_1^*,k_2^*)| - |h(k_1,k_2)| \\ =& \frac{|\mu_2-\mu_1|}{T}\left[\sqrt{(T-a_0)a_0} - \sqrt{(T-a_1)a_1} \times \left|1-\frac{a_1-a_2}{a_1}-\frac{a_0-a_2}{T-a_1}\right|\right] \\ =& \frac{|\mu_2-\mu_1|}{T}\left[\sqrt{(T-a_0)a_0} - \frac{|a_0a_1-Ta_2|}{\sqrt{(T-a_1)a_1}}\right] \\ =& \frac{|\mu_2-\mu_1|}{T}\frac{(T-a_0)(T-a_1)a_0a_1-(Ta_2-a_0a_1)^2}{\sqrt{(T-a_1)a_1}\left(\sqrt{(T-a_0)(T-a_1)a_0a_1}+|Ta_2-a_0a_1|\right)} \\ =& \frac{|\mu_2-\mu_1|\left(a_0a_1(T+2a_2-(a_0+a_1))-Ta_2^2\right)}{\sqrt{(T-a_1)a_1}\left(\sqrt{(T-a_0)(T-a_1)a_0a_1}+|Ta_2-a_0a_1|\right)}, \end{aligned} \tag{4.84}$$

and

$$(h(k_1^*,k_2^*))^2 - (h(k_1,k_2))^2 = \frac{(\mu_2-\mu_1)^2\left[a_0a_1(T+2a_2-(a_0+a_1)]-Ta_2^2\right)}{T(T-a_1)a_1}. \tag{4.85}$$

Define $\delta_1 = k_1^* - k_1$ and $\delta_2 = k_2^* - k_2$. Consider the following two cases:

1. $a_0a_1 \ge Ta_2$.

   By the fact that $0 < a_0 + a_1 - a_2 = k_2^* - k_1 < T$, we have

   $$a_0a_1[T+2a_2-(a_0+a_1)] - Ta_2^2 > Ta_2[T+a_2-(a_0+a_1)] > 0.$$

2. $a_0a_1 < Ta_2$.

   Since $a_0 + a_1 - 2a_2 = \delta_1 + \delta_2$ and $a_0a_1 - a_2^2 = (\delta_1+a_2)(\delta_2+a_2) - a_2^2 = a_2(\delta_1+\delta_2) + \delta_1\delta_2$, we have

   $$a_0a_1[T+2a_2-(a_0+a_1)] - Ta_2^2 = (Ta_2 - a_0a_1)(\delta_1+\delta_2) + \delta_1\delta_2 T > 0. \tag{4.86}$$

Therefore, by (4.84), we obtain that

$$|h(k_1^*, k_2^*)| - |h(k_1, k_2)| > 0. \tag{4.87}$$

By Lemma 4.7, (4.83), and (4.87), we have

$$0 \leq |h(k_1^*, k_2^*)| - |h(\hat{k}_1, \hat{k}_2)| \to_p 0,$$

which implies that

$$\frac{k_1^* - \hat{k}_1}{T} \to_p 0 \text{ and } \frac{k_2^* - \hat{k}_2}{T} \to_p 0.$$

Assume that $\hat{k}_1 \leq k_1^* < \hat{k}_2 \leq k_2^*$. Define $\hat{a}_1 = \hat{k}_2 - \hat{k}_1$, $\hat{a}_2 = \hat{k}_2 - k_1^*$, $\hat{\delta}_1 = k_1^* - \hat{k}_1$, and $\hat{\delta}_2 = k_2^* - \hat{k}_2$. Then, by $\hat{a}_1 = a_0 + \hat{\delta}_1 - \hat{\delta}_2$, $\hat{a}_2 = a_0 - \hat{\delta}_2$, and $a_0/T \to \eta \in (0,1)$, we have

$$\frac{T\hat{a}_2 - a_0\hat{a}_1}{T^2} = \frac{(T-a_0)a_0}{T^2} - \frac{(T-a_0)\hat{\delta}_2 - a_0\hat{\delta}_1}{T^2} \to_p \eta(1-\eta) > 0. \tag{4.88}$$

In view of (4.85)–(4.86), it can be shown that

$$\begin{aligned}(h(k_1^*, k_2^*))^2 - (h(\hat{k}_1, \hat{k}_2))^2 &= \frac{|\mu_2 - \mu_1|^2 \left[(T\hat{a}_2 - a_0\hat{a}_1)(\hat{\delta}_1 + \hat{\delta}_2) + \hat{\delta}_1\hat{\delta}_2 T\right]}{T(T-\hat{a}_1)\hat{a}_1} \\ &= \frac{|\mu_2 - \mu_1|^2 (\hat{\delta}_1 + \hat{\delta}_2)}{T} + o_p\left(\frac{|\mu_2 - \mu_1|^2 (\hat{\delta}_1 + \hat{\delta}_2)}{T}\right).\end{aligned} \tag{4.89}$$

Since $W^2_{\hat{k}_1,\hat{k}_2} \geq W^2_{k_1^*,k_2^*}$ and

$$\begin{aligned}W^2_{\hat{k}_1,\hat{k}_2} - W^2_{k_1^*,k_2^*} =& (W_{\hat{k}_1,\hat{k}_2} - h(\hat{k}_1,\hat{k}_2))^2 - (W_{k_1^*,k_2^*} - h(k_1^*,k_2^*))^2 + [h(\hat{k}_1,\hat{k}_2)]^2 - [h(k_1^*,k_2^*)]^2 \\ &+ 2\left(W_{\hat{k}_1,\hat{k}_2} - h(\hat{k}_1,\hat{k}_2)\right)\left(h(\hat{k}_1,\hat{k}_2) - h(k_1^*,k_2^*)\right) \\ &+ 2\left(W_{\hat{k}_1,\hat{k}_2} - h(\hat{k}_1,\hat{k}_2) - W_{k_1^*,k_2^*} + h(k_1^*,k_2^*)\right) h(k_1^*,k_2^*),\end{aligned} \tag{4.90}$$

we have

$$\begin{aligned}[h(k_1^*,k_2^*)]^2 - [h(\hat{k}_1,\hat{k}_2)]^2 \leq& (W_{\hat{k}_1,\hat{k}_2} - h(\hat{k}_1,\hat{k}_2))^2 - (W_{k_1^*,k_2^*} - h(k_1^*,k_2^*))^2 \\ &+ 2\left(W_{\hat{k}_1,\hat{k}_2} - h(\hat{k}_1,\hat{k}_2)\right)\left(h(\hat{k}_1,\hat{k}_2) - h(k_1^*,k_2^*)\right) \qquad (4.91) \\ &+ 2\left(W_{\hat{k}_1,\hat{k}_2} - h(\hat{k}_1,\hat{k}_2) - W_{k_1^*,k_2^*} + h(k_1^*,k_2^*)\right) h(k_1^*,k_2^*) \\ =& I + II + III + IV.\end{aligned} \tag{4.92}$$

Note that by (4.84) and (4.88), we have

$$\begin{aligned}&|h(k_1^*,k_2^*) - h(k_1,k_2)| \\ =& \frac{|\mu_2 - \mu_1|}{T}\left|\sqrt{(T-\hat{a}_0)\hat{a}_0} - \sqrt{(T-\hat{a}_1)\hat{a}_1} \times \left(1 - \frac{\hat{a}_1 - \hat{a}_2}{\hat{a}_1} - \frac{\hat{a}_0 - \hat{a}_2}{T - \hat{a}_1}\right)\right| \\ =& \frac{|\mu_2 - \mu_1|\left(\hat{a}_0\hat{a}_1(T + 2\hat{a}_2 - (\hat{a}_0 + \hat{a}_1)) - T\hat{a}_2^2\right)}{\sqrt{(T-\hat{a}_1)\hat{a}_1}\left|\sqrt{(T-\hat{a}_0)(T-\hat{a}_1)\hat{a}_0\hat{a}_1} + (T\hat{a}_2 - \hat{a}_0\hat{a}_1)\right|} \\ =& \frac{|\mu_1 - \mu_2|(\hat{\delta}_1 + \hat{\delta}_2)}{2\sqrt{\eta(1-\eta)}T} + o_p\left(\frac{|\mu_1 - \mu_2|(\hat{\delta}_1 + \hat{\delta}_2)}{T}\right).\end{aligned}$$

From the above result and Lemma 4.7, we obtain that

$$III = O_p\left(\frac{|\mu_1 - \mu_2|(\hat{\delta}_1 + \hat{\delta}_2)}{T^{3/2}}\right).$$

Since

$$\begin{aligned}
& W_{\hat{k}_1,\hat{k}_2} - h(\hat{k}_1,\hat{k}_2) - W_{k_1^*,k_2^*} + h(k_1^*,k_2^*) \\
&= \left[\frac{1}{T}\left(\sqrt{\frac{T-\hat{a}_1}{\hat{a}_1}} - \sqrt{\frac{T-a_0}{a_0}}\right)\sum_{t=k_1^*+1}^{\hat{k}_2}(y_t - \mu_2)\right. \\
&\left. -\frac{1}{T}\left(\sqrt{\frac{\hat{a}_1}{T-\hat{a}_1}} - \sqrt{\frac{a_0}{T-a_0}}\right)\left(\sum_{t=1}^{\hat{k}_1}(y_t-\mu_1) + \sum_{t=k_2^*+1}^{T}(y_t-\mu_1)\right)\right] \\
&+ \left[\frac{1}{T}\left(\sqrt{\frac{\hat{a}_1}{T-\hat{a}_1}} + \sqrt{\frac{T-a_0}{a_0}}\right)\sum_{t=\hat{k}_2+1}^{k_2^*}(y_t-\mu_2) - \frac{1}{T}\left(\sqrt{\frac{T-\hat{a}_1}{\hat{a}_1}} + \sqrt{\frac{a_0}{T-a_0}}\right)\sum_{t=\hat{k}_1+1}^{k_1^*}(y_t-\mu_1)\right] \\
&= O_p\left(\frac{\hat{\delta}_1+\hat{\delta}_2}{T^{3/2}}\right) + O_p\left(\frac{\sqrt{\hat{\delta}_1+\hat{\delta}_2}}{T}\right) \\
&= O_p\left(\frac{\sqrt{\hat{\delta}_1+\hat{\delta}_2}}{T}\right),
\end{aligned} \tag{4.93}$$

where the last equation follows from the fact that $(\hat{\delta}_1+\hat{\delta}_2)/T \to_p 0$, by Lemma 4.7, it can be shown that

$$\begin{aligned}
|I + II| &= \left|(W_{\hat{k}_1,\hat{k}_2} - h(\hat{k}_1,\hat{k}_2))^2 - (W_{k_1^*,k_2^*} - h(k_1^*,k_2^*))^2\right| \\
&= \left|(W_{\hat{k}_1,\hat{k}_2} - h(\hat{k}_1,\hat{k}_2) - (W_{k_1^*,k_2^*} - h(k_1^*,k_2^*))\right| \\
&\times \left|(W_{\hat{k}_1,\hat{k}_2} - h(\hat{k}_1,\hat{k}_2) + (W_{k_1^*,k_2^*} - h(k_1^*,k_2^*))\right| \\
&= O_p\left(\frac{\sqrt{\hat{\delta}_1+\hat{\delta}_2}}{T}\right) \times O_p\left(\frac{1}{\sqrt{T}}\right) = O_p\left(\frac{\sqrt{\hat{\delta}_1+\hat{\delta}_2}}{T^{3/2}}\right).
\end{aligned}$$

In view that $|h(k_1^*,k_2^*)| = O(|\mu_2-\mu_1|)$, by (4.93), we have

$$IV = O_p\left(\frac{|\mu_2-\mu_1|\sqrt{\hat{\delta}_1+\hat{\delta}_2}}{T}\right).$$

Combining (4.89) with (4.92) yields

$$\begin{aligned}
[h(k_1^*,k_2^*)]^2 - [h(\hat{k}_1,\hat{k}_2)]^2 &= \frac{|\mu_2-\mu_1|^2(\hat{\delta}_1+\hat{\delta}_2)}{T} + o_p\left(\frac{|\mu_2-\mu_1|^2(\hat{\delta}_1+\hat{\delta}_2)}{T}\right) \\
&\leq I + II + III + IV = O_p\left(\frac{\sqrt{\hat{\delta}_1+\hat{\delta}_2}|\mu_1-\mu_2|}{T}\right),
\end{aligned}$$

Therefore, we obtain that

$$|\hat{\delta}_1| + |\hat{\delta}_2| = O_p\left(\frac{1}{|\mu_2-\mu_1|^2}\right).$$

Denote $|\delta_1| + |\delta_2| = s\rho_T^{-2}$, $s > 0$, where ρ_T is defined in Assumption A3. By (4.90), Assumption A3, and the Central Limit Theorem for a martingale difference sequence, we have

$$\begin{aligned}
T(W^2_{\hat{k}_1,\hat{k}_2} - W^2_{k_1^*,k_2^*}) &= T(I + II + III + [h(\hat{k}_1,\hat{k}_2)]^2 - [h(k_1^*,k_2^*)]^2 + IV) \\
&= o_p(1) - s + 2\rho_T \left(\sum_{t=\hat{k}_2+1}^{k_2^*} (y_t - \mu_2) - \sum_{t=\hat{k}_1+1}^{k_1^*} (y_t - \mu_1) \right) \\
&\to_d 2\sigma W(s) - s.
\end{aligned}$$

Since $(\hat{k}_1, \hat{k}_2) = \arg\max W^2_{k_1,k_2}$, we obtain that

$$\hat{s} = \rho_T^2(|\hat{\delta}_1| + |\hat{\delta}_2|) \to_d \arg\max_{s>0} \{\sigma W(s) - s/2\}.$$

The proof is completed.

4.3.2 Numerical studies

In this section, we conduct simulation studies to evaluate the performance of the proposed epidemic change-point detection method. For comparative analysis, we use three multiple change-point detection procedures: the WBS method combined with the sSIC stopping criterion in [102], the traditional BS method, and the SMUCE method in [102]. We refer to these procedures as WBS, BS, and SMUCE, respectively. WBS and BS are implemented in the R package `wbs`, and SMUCE is implemented in the R package `stepR`. All results shown in this section are based on 1000 replicates for the three-state model (4.70).

Scenario I. Set $T = 1000$, $\mu_1 = 1$ and μ_2=3, and then randomly generate y_t with ϵ_t (for $t = 1, \ldots, T$) being independently drawn from the standard normal distribution. Then estimate $\hat{k}_1$ and $\hat{k}_2$ according to the method introduced in Section 2. Repeat 1000 times to obtain the cumulative histogram of values of $\hat{\rho}_T^2/\hat{\sigma}^2(|\hat{\delta}_1| + |\hat{\delta}_2|)$ to obtain the empirical distribution. Then we obtain the curve of $2G(s)-1$, the CDF of $\arg\max_{s>0} \{W(s) - \frac{s}{2}\}$, through the definition of $G(s)$ in Remark 4.5. The blue areas in Figure 4.3 and 4.4 are the cumulative histograms, and the black curve is the curve of $2G(s) - 1$. From Figure 6.9, we see that the empirical distribution of $\hat{\rho}_T^2/\hat{\sigma}^2(|\hat{\delta}_1| + |\hat{\delta}_2|)$ is basically consistent with the curve of $\hat{\rho}_T^2/\hat{\sigma}^2(|\hat{\delta}_1| + |\hat{\delta}_2|)$, verifying the result of Theorem 4.3.1.

In addition, we randomly generate y_t, where ϵ_t for $t = 1, \ldots, T$ are independently drawn from the t_{10} distribution. Similarly, we put the $2G(s) - 1$ curve together with the cumulative histogram of $\hat{\rho}_T^2/\hat{\sigma}^2(|\hat{\delta}_1| + |\hat{\delta}_2|)$ values, as shown in Figures 4.3 and 4.4. From this figure, we can also see that the empirical distribution is close to the $2G(s)-1$ curve, which verifies the result of Theorem 4.3.1.

Scenario II. Set $\mu_1 = 1$ and $\mu_2 = 1.8$, and then randomly generate y_t, where ϵ_t (for $t = 1, \ldots, T$) are independently drawn from the standard normal distribution. The change-points k_1^* and k_2^* are randomly sampled from the range 1 to T but satisfy Assumption A2, where η in Assumption A2 is drawn from the uniform distribution $U(0.05, 0.2)$.

We consider five cases of T: $T = 500, 600, 700, 800$, and 900, and record the simulation results of the four different methods, i.e., the "Proposed method" introduced in Section 2, WBS, BS, and SMUCE. Let $\hat{N}$ denote the number of estimated change points, and let $|\hat{N} \neq 2|$ denote the total number of experiments in which the estimated number of change points differs from the true number of change points 2. Table 4.1 reports the simulation results of $\hat{\tau}_1 + \hat{\tau}_2$ and $|\hat{N} \neq 2|$ for each method, where $\hat{\tau}_1$ and $\hat{\tau}_2$ are defined in (4.80) to measure the difference between the estimated change points and the true change points.

As can be seen from Table 4.1, the proposed method can give fairly accurate change point estimates. It performs best when $T > 500$, followed by WBS, SMUCE, and BS. In addition, the convergence rate of the estimated change point locations is consistent with our theoretical analysis

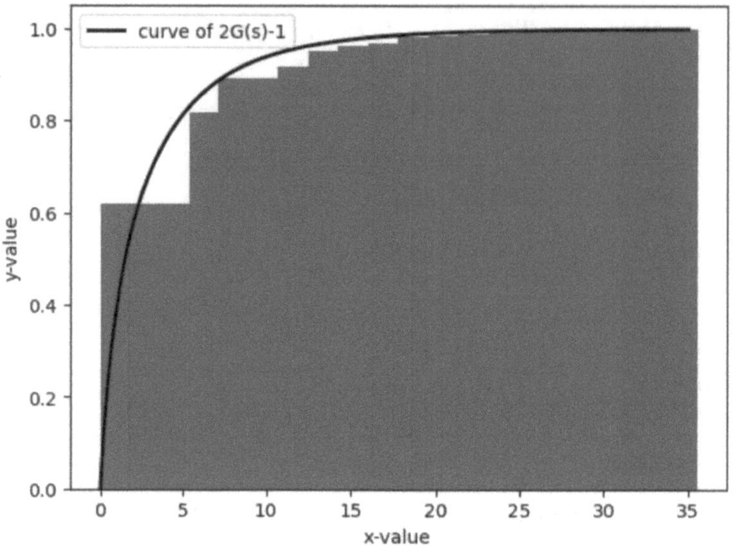

Figure 4.3: Cumulative histogram of the values of $\hat{\rho}_T^2/\hat{\sigma}^2(|\hat{\delta}_1|+|\hat{\delta}_2|)$ and curve of $2G(s)-1$ where y_i is generated with ϵ_i following the standard normal distribution, for $i=1,\ldots,T$.

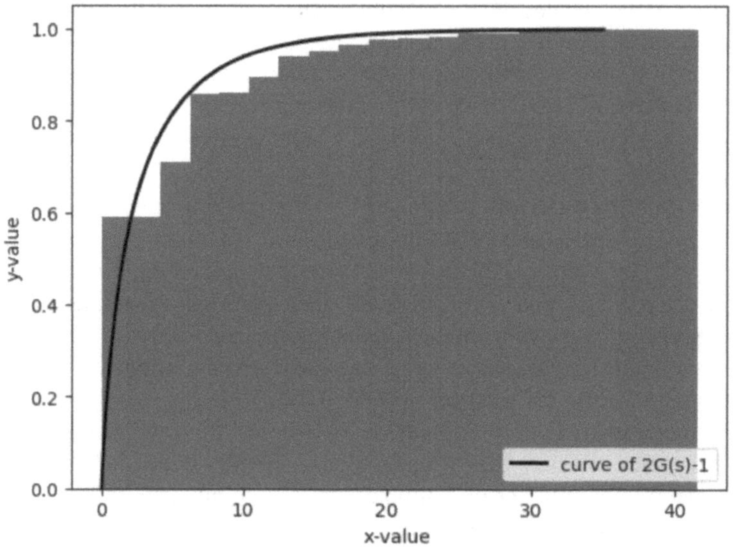

Figure 4.4: Cumulative histogram of the values of $\hat{\rho}_T^2/\hat{\sigma}^2(|\hat{\delta}_1|+|\hat{\delta}_2|)$ and curve of $2G(s)-1$ where y_i is generated with ϵ_i following the t_{10} distribution, for $i=1,\ldots,T$.

in Section 3. The other three methods also perform well in terms of the consistency between the estimated change point locations and the true locations, but they tend to be less precise in terms of the estimated number of change points. For example, when $T = 500$, the BS method overestimates or underestimates the number of change points in about 40% of the experiments. As T increases, the frequency with which the other three methods correctly estimate the number of change points also increases.

Table 4.1: The average of $\hat{\tau}_1 + \hat{\tau}_2$ (%) and the number of experiments where the estimated number of change points differs from the true number of change points in Scenario I.

	Proposed method		WBS		BS		SMUCE	
T	$\hat{\tau}_1 + \hat{\tau}_2$	$\lvert\hat{N} \neq 2\rvert$	$\hat{\tau}_1 + \hat{\tau}_2$	$\lvert\hat{N} \neq 2\rvert$	$\hat{\tau}_1 + \hat{\tau}_2$	$\lvert\hat{N} \neq 2\rvert$	$\hat{\tau}_1 + \hat{\tau}_2$	$\lvert\hat{N} \neq 2\rvert$
500	2.41	0	2.18	76	2.72	414	2.3	160
600	1.83	0	1.83	26	2.15	359	1.86	103
700	1.46	0	1.52	35	1.9	267	1.56	68
800	1.27	0	1.32	25	1.7	212	1.34	53
900	1.05	0	1.11	22	1.58	165	1.08	29

Scenario III. Let $y_t, t = 1, \ldots, T$, be a sequence of independent exponential random variables with density function

$$f(y_t, \mu_t) = \frac{1}{\mu_t} e^{-y_t/\mu_t}.$$

The positive real numbers μ_t follow the epidemic change theme, that is,

$$\mu_t = \begin{cases} \mu_1, & t = 1, \ldots, k_1^*, \\ \mu_2, & t = k_1^* + 1, \ldots, k_2^*, \\ \mu_1, & t = k_2^* + 1, \ldots, T. \end{cases}$$

Let $\mu_1 = 1$, $\mu_2 = 2.5$. The other settings are the same as in Scenario I. Table 2 reports the comparison results of the four methods. These results show that the proposed method can still achieve satisfactory accuracy in non-Gaussian scenarios. However, WBS, BS and SMUCE, as popular procedures mainly used for multiple change-point detection, often have difficulty in accurately estimating the number of change points in the epidemic change model. Furthermore, this problem does not improve with increasing sample size, especially for models containing random variables with heavier tails than sub-Gaussian random variables.

Table 4.2: The average of $\hat{\tau}_1 + \hat{\tau}_2$ (%) and the number of experiments where the estimated number of change points differs from the true number of change points in Scenario II.

	Proposed method		WBS		BS		SMUCE	
T	$\hat{\tau}_1 + \hat{\tau}_2$	$\lvert\hat{N} \neq 2\rvert$	$\hat{\tau}_1 + \hat{\tau}_2$	$\lvert\hat{N} \neq 2\rvert$	$\hat{\tau}_1 + \hat{\tau}_2$	$\lvert\hat{N} \neq 2\rvert$	$\hat{\tau}_1 + \hat{\tau}_2$	$\lvert\hat{N} \neq 2\rvert$
500	7.64	0	2.5	992	1.86	774	2.11	826
600	5.68	0	1.73	992	1.71	764	2.08	828
700	4.29	0	2.71	995	1.32	767	1.46	865
800	2.92	0	1.56	996	1.03	777	1.21	879
900	2.76	0	0.89	996	0.93	767	1.05	904

4.3.3 Real-data analysis

This study analyzes the daily peaks of the Nikkei 225 Index, from February 17, 2006, to December 10, 2013 (1915 consecutive trading days). The Nikkei 225 Index is a price-weighted equity index consisting of 225 stocks in the Prime Market of the Tokyo Stock Exchange. Since the data spans the 2008–2009 global financial crisis, it is possible to formulate an epidemic change model in which the two change points in the model correspond precisely to the beginning and end of the financial crisis. Note that the data can be downloaded from Econstats at `http://www.econstats.com/index.htm`, which is publicly available.

We apply the proposed method as well as WBS, BS, and SMUCE to this financial dataset. Table 4.3 reports the number of change points $\hat{N}$ identified by the four methods. It is well known that financial returns and macroeconomic variables exhibit heavy tails for rare events [92], so this study draws similar conclusions to the analysis of Scenario II in the previous section. It is noted that except for the proposed method, the other three methods detect too many change points. In many applications of multi-change point analysis, if we have some prior information about the number of change points, this knowledge can often be used to simplify our analysis instead of overcomplicating it.

Figure 4.5 shows the time series of the data. It can be seen from the figure that during the time period we observed (2006–2013), the Japanese stock market experienced two major changes: a rapid decline in the second half of 2008 and a recovery in 2012–2013. The two change points identified by our method are September 8, 2008 and April 5, 2013. The former coincided with the outbreak of the financial crisis. In addition, it can be observed that by around April 2013, the Japanese stock market had basically recovered from the impact of the financial crisis.

Table 4.3: Estimated number of change points for Nikkei 225 data by the four methods.

	Proposed method	WBS	BS	SMUCE
$\hat{N}$	2	50	166	135

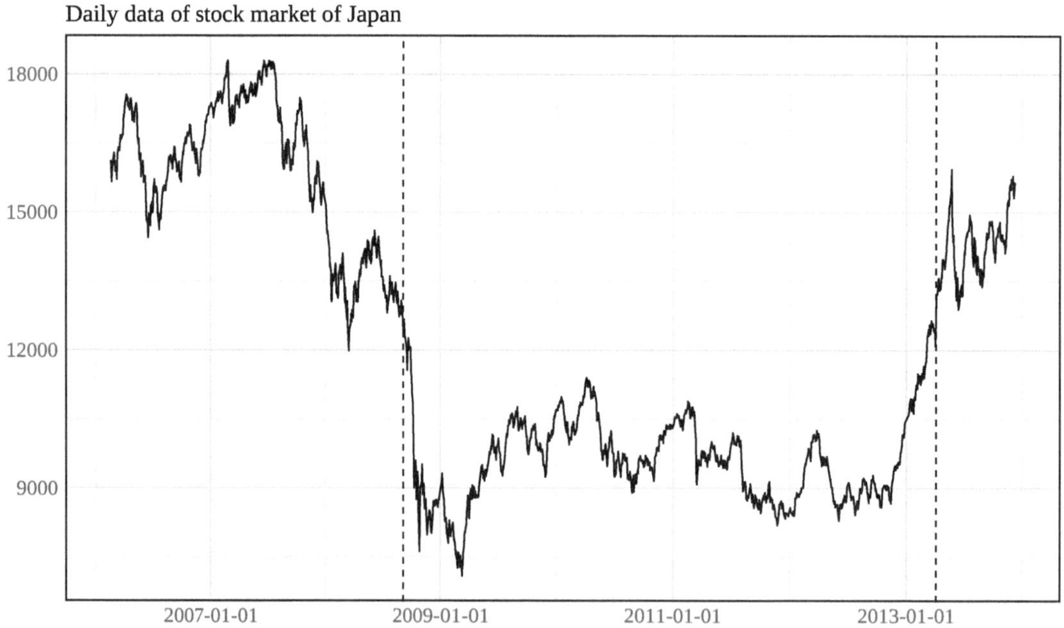

Figure 4.5: The impact of the 2008–2009 global financial crisis on the Japanese stock market. The vertical dashed lines indicate the change points estimated by the proposed method.

4.4 Interval estimates of multiple change points

Assume $(\boldsymbol{x}_i, y_i)$, $i = 1, \ldots, n$, satisfy the following linear regression model with s change points located at $a_1 < \ldots < a_s$ as:

$$
\begin{aligned}
y_i &= \boldsymbol{x}_i^\top \left[\boldsymbol{\beta}_1 + \sum_{l=1}^{s} \boldsymbol{\delta}_l I\left(a_l < i \leq n\right) \right] + \varepsilon_i, \quad i = 1, \ldots, n, \\
&= \begin{cases} \boldsymbol{x}_i^\top \boldsymbol{\beta}_1 + \varepsilon_i, & \text{if } 1 < i < a_1, \\ \boldsymbol{x}_i^\top \left(\boldsymbol{\beta}_1 + \boldsymbol{\delta}_1\right) + \varepsilon_i & \text{if } a_1 < i < a_2 \\ \ldots, & \ldots, \\ \boldsymbol{x}_i^\top \left(\boldsymbol{\beta}_1 + \sum_{l=1}^{s} \boldsymbol{\delta}_l\right) + \varepsilon_i & \text{if } a_s < i < n \end{cases}
\end{aligned} \tag{4.94}
$$

where n is the sample size, $\boldsymbol{x}_i = \left(x_{i,1}, \ldots, x_{i,q}\right)^\top$ is a sequence of q-dimensional predictors, $\boldsymbol{\beta}_1 = \left(\beta_{1,1}, \ldots, \beta_{q,1}\right)^\top \neq 0$ is an unknown q-dimensional vector of regression coefficients, s is an unknown number of change points, $1 < a_1 < \ldots < a_s < n$ are unknown change points, $\boldsymbol{\delta}_l = \left(\delta_{1,l}, \ldots, \delta_{q,l}\right)^\top, l = 1, \ldots, s$ are unknown changes in regression coefficient vectors at change point, and ε_i's are random errors.

The estimation of regression coefficients $\boldsymbol{\beta}_1$ and $\left(\boldsymbol{\beta}_1 + \sum_{l=1}^{j} \boldsymbol{\delta}_l\right), j = 1, \ldots s$, is hindered by the unknown parameters $s, a_1, \ldots, a_s$. We intend to substitute s in (4.94) with a predetermined number of segments denoted as p_n to facilitate the estimation of regression coefficients. This substitution enables us to identify the total number of change points in return. Specifically, we partition the data sequence into p_n segments, where $p_n \to \infty$ as $n \to \infty$. This ensures that all segments, except for the initial one, have a length of m, while the first segment's length is $n - (p_n - 1)\, m$. Let $\mathcal{Q}_1 = \{1, 2, \ldots, n - (p_n - 1)\, m\}$ and $\mathcal{Q}_i = \{n - (p_n - i + 1)\, m + 1, \ldots, n - (p_n - i)\, m\}$, $i = 2, \ldots, p_n$. For each $j = 1, \ldots, s$, we assume each change point a_j within one of the segments, including those at the partition boundary. We denote this segment as k_j, i.e., $a_j \in \mathcal{Q}_{k_j}$. This partitioning results in the following model:

$$
y_i = \boldsymbol{x}_i^\top \left[\boldsymbol{\beta}_1 + \sum_{l=2}^{p_n} \boldsymbol{d}_l I\left(n - (p_n - l + 1)\, m < i \leq n\right) \right] - \omega_i + \varepsilon_i, \tag{4.95}
$$

where

$$
\begin{cases} \boldsymbol{d}_{k_j} = \boldsymbol{\delta}_j \neq \mathbf{0}, & \text{for } j = 1, \ldots, s, \\ \boldsymbol{d}_l = \mathbf{0}, & \text{for any } l \notin \{k_1, \ldots, k_s\}, \end{cases}
$$

and $\omega_i = \boldsymbol{x}_i^\top \sum_{j=1}^{s} \boldsymbol{d}_{k_j} I\left(i \in \mathcal{T}_j\right)$ with $\mathcal{T}_j = \{n - (p_n - k_j + 1)\, m + 1, \ldots, a_{j,n}\}$. $\boldsymbol{\omega} = \left(\omega_1, \ldots, \omega_n\right)^\top$ represents the artificial n-dimensional error resulting from the segmentation, with $\omega_i = 0$ for all $i \notin \cup_{j=1}^{s} \mathcal{Q}_{k_j}$.

Therefore, (4.95) can be rewritten in matrix form as follows:

$$
\boldsymbol{y}_n = \mathbf{X}_n \boldsymbol{\theta} + \boldsymbol{e}_n, \tag{4.96}
$$

where $\boldsymbol{y}_n = \left(y_1, y_2, \ldots, y_n\right)^\top$,

$$
\mathbf{X}_n = \left(\mathbf{X}^{(1)}, \ldots, \mathbf{X}^{(p_n)}\right) = \begin{pmatrix} \mathbf{X}_{(1)} & \mathbf{0} & \cdots & \mathbf{0} \\ \mathbf{X}_{(2)} & \mathbf{X}_{(2)} & \cdots & \mathbf{0} \\ \vdots & \vdots & \ddots & \vdots \\ \mathbf{X}_{(p_n)} & \mathbf{X}_{(p_n)} & \cdots & \mathbf{X}_{(p_n)} \end{pmatrix},
$$

$\boldsymbol{X}_{(1)} = \left(\boldsymbol{x}_1, \cdots, \boldsymbol{x}_{n-(p_n-1)m}\right)^\top$ and $\boldsymbol{X}_{(j)} = \left(\boldsymbol{x}_{n-(p_n-j+1)m+1}, \dots, \boldsymbol{x}_{n-(p_n-j)m}\right)^\top$. In addition, the definitions of $\boldsymbol{\theta} = \left(\boldsymbol{\theta}_1^\top, \dots, \boldsymbol{\theta}_{p_n}^\top\right)^\top = \left(\boldsymbol{\beta}_1^\top, \boldsymbol{d}_2^\top, \dots, \boldsymbol{d}_{p_n}^\top\right)^\top$, $\boldsymbol{e}_n = -\boldsymbol{\omega} + \boldsymbol{\varepsilon}_n$ with $\boldsymbol{\varepsilon}_n = (\varepsilon_1, \dots, \varepsilon_n)^\top$.

We observe that $\boldsymbol{d}_l = \boldsymbol{0}$ if $l \neq 1, k_{j,n}$, or $k_{j,n} + 1$ for $j = 1, \dots, s$. Let $\mathcal{A}_n = \{1, k_{1,n}, k_{1,n} + 1, \dots, k_{s,n}, k_{s,n} + 1\}$. It is important to note that $\boldsymbol{\theta}_j = \boldsymbol{0}_{q\times 1}$ obviously for all $j \notin \mathcal{A}_n$. Consequently, the estimation of $\mathcal{A}_n$ boils down to identifying the non-zero elements from $\boldsymbol{\theta}$, which is the central focus of the subsequent subsection.

4.4.1 Selection of segments containing change points

In this subsection, our goal is to select the segments $\mathcal{Q}_{k_j}$ for $j = 1, \cdots, s$, each containing a change point. Based on the previous statements, the problem of detecting multiple change points is reduced to a variable selection issue commonly encountered in high-dimensional data analysis.

We adopt the methodology from combining orthogonal greedy algorithm (OGA), high-dimensional information criterion (HDIC), and Trim (OGA+HDIC+Trim) for segment selection. OGA is a straightforward stepwise forward regression method that uses sequential orthogonalization of predictor variables to enable componentwise linear least squares estimation. The OGA+HDIC method aims to identify the smallest set of relevant variables along the OGA path, with HDIC serving as a extended information criterion for high-dimensional data. Lastly, the Trim method defines a subset to further exclude certain irrelevant variables from OGA+HDIC. An `R` package `Ohit` is available at CRAN to implement this kind of procedure. In particular, following the preceding section and using scalar representations, (4.96) can be rewritten as follows:

$$\boldsymbol{y}_n = \sum_{j=1}^{r_n} \boldsymbol{z}_j \theta_j + \boldsymbol{e}_n, \tag{4.97}$$

where $\boldsymbol{X}_n = (z_1, \dots, z_{r_n}), \boldsymbol{\theta} = (\theta_1, \cdots, \theta_{r_n})^\top$ and $r_n = p_n \cdot q$. Without loss of generality, replacing $\boldsymbol{y}_n$ by $\boldsymbol{y}_n - \boldsymbol{y}_n^\top \boldsymbol{1}_n$ and $\boldsymbol{z}_j$ by $\boldsymbol{z}_j - \boldsymbol{z}_j^\top \boldsymbol{1}_n$. Define $\hat{\sigma}^2_{J_d} = n^{-1}\boldsymbol{y}_n^\top (\boldsymbol{I}_n - \boldsymbol{H}_{J_d})\boldsymbol{y}_n$, where $\boldsymbol{H}_{J_d}$ is the orthogonal projection matrix onto the linear span of $\{z_j, j \in J_d\}$ $(\boldsymbol{H}_\emptyset = 0)$. The algorithm can be outlined as follows:

Algorithm 8 OGA+HDIC+Trim

Require: response vector $\boldsymbol{y}_n \in \mathbb{R}^n$, regressor matrix $\boldsymbol{X}_n \in \mathbb{R}^{n\times r_n}$.
Initialzation: set $d = 0$, $\boldsymbol{r}^{(0)} = \boldsymbol{y}_n$ and $\hat{J}_0 = \emptyset$.
while $d = 0, \dots, D_n$ **do**
  Compute $\hat{j}_{d+1} = \arg\max_{1\le j\le r_n, j\notin \hat{J}_{d-1}} |\boldsymbol{z}_j^\top \boldsymbol{r}^{(d)}|/(n^{1/2}\|\boldsymbol{z}_j\|)$ and update $\hat{J}_{d+1} = \hat{J}_d \bigcup \hat{j}_{d+1}$;
  Compute $\boldsymbol{z}^\perp_{\hat{j}_{d+1}}$ via $\boldsymbol{z}^\perp_{\hat{j}_{d+1}} = (\boldsymbol{I}_n - \sum_{\ell=1}^d \boldsymbol{H}^\perp_{\hat{j}_\ell})\boldsymbol{z}_{\hat{j}_{d+1}}$, where $\boldsymbol{H}^\perp_{\hat{j}_\ell} = \boldsymbol{z}_{\hat{j}_\ell}\boldsymbol{z}^\top_{\hat{j}_\ell}/\|\boldsymbol{z}_{\hat{j}_\ell}\|^2$;
  Compute $\boldsymbol{r}^{(d+1)}$ via $\boldsymbol{r}^{(d+1)} = (\boldsymbol{I}_n - \boldsymbol{H}^\perp_{\hat{j}_{d+1}})\boldsymbol{r}^{(d)}$.
end while
Compute the minimum of HDIC via

$$\begin{aligned} \text{HDIC}(\hat{J}_d) &= \log \hat{\sigma}^2_{\hat{J}_d} + \sharp(\hat{J}_d) c_n \log(r_n)/n, \\ \hat{d}_n &= \arg\min_{1\le d\le D_n} \text{HDIC}(\hat{J}_d). \end{aligned} \tag{4.98}$$

return $\hat{\mathcal{J}}_n$ via

$$\hat{\mathcal{J}}_n = \begin{cases} \{\hat{j}_\ell : \text{HDIC}\left(\hat{J}_d - \{\hat{j}_\ell\}\right) > \text{HDIC}(\hat{J}_d),\ 1 \le \ell \le \hat{d}_n\}, & \text{if } \hat{d}_n > 1; \\ \{\hat{j}_1\}, & \text{if } \hat{d}_n = 1. \end{cases}$$

In this context, the value of D_n represents the maximum number of iterations, and the theoretical result on the convergence rate of OGA provided in [131] suggests $D_n = O\left(n^{1/2}(\log r_n)^{-1/2}\right)$. The parameter c_n satisfies the conditions $c_n \to \infty$ and $c_n \log r_n = o\left(n^{1-2\gamma}\right)$ with $\gamma \in [0,1)$. Hence, we obtain $\widehat{\mathcal{A}}_n$ from $\hat{\mathcal{J}}_n$. Let $\widehat{\boldsymbol{\theta}} = \left(\widehat{\boldsymbol{\theta}}_1^\top, \ldots, \widehat{\boldsymbol{\theta}}_{p_n}^\top\right)^\top$ be the OGA estimator. We denote the estimate of $\mathcal{A}_n$ as

$$\widehat{\mathcal{A}_n} = \left\{j : \widehat{\boldsymbol{\theta}}_j \neq \mathbf{0}, j = 1, \ldots, p_n\right\} = \left\{\widehat{k}_1, \ldots, \widehat{k}_{\hat{s}}\right\}, \tag{4.99}$$

which represents a set that contains change points. By following the steps outlined above and accounting for the change point at the partition boundary, we obtain the number of change points $\hat{s}$ and the corresponding segments

$$\hat{\mathcal{Q}}_{(j)} = \hat{\mathcal{Q}}_{k_{j-1}} \bigcup \hat{\mathcal{Q}}_{k_j} = \left\{n - \left(p_n - \hat{k}_j + 2\right)m + 1, n - \left(p_n - \hat{k}_j\right)m + 1\right\}. \tag{4.100}$$

4.4.2 Refining change points

In the refining stage, we determine the specific locations of the change points by conducting the sup-Wald-type test ([16]) on the estimated segments in Section 4.4.1. For each $j = 1, \ldots, \hat{s}$, we consider the segment with a potential change point located at a_j, which can be expressed as follows:

$$y_i = \boldsymbol{x}_i^\top \left[\boldsymbol{\beta}_j + \boldsymbol{\delta}_j I\left(a_j < i \leq n_j^{(r)}\right)\right] + \varepsilon_i, \quad n_j^{(l)} \leq i \leq n_j^{(r)}, \tag{4.101}$$

where $\boldsymbol{\beta}_j$ and $\boldsymbol{\delta}_j$ are q-dimensional vectors of unknown parameters and $\boldsymbol{\beta}_j = \boldsymbol{\beta}_1 + \sum_{l=1}^{j-1} \boldsymbol{\delta}_l$, $n_j^{(l)} = n - \left(p_n - \hat{k}_j + 2\right)m + 1$ and $n_j^{(r)} = n - \left(p_n - \hat{k}_j\right)m$, $j = 2, \ldots, s+1$ are the location of observations. Denote $N_j = n_j^{(r)} - n_j^{(l)}$. To test the hypothesis $H_0 : \boldsymbol{\beta}_j = \boldsymbol{\beta}_{j+1}$ in (4.101), we compute the sup-Wald-type test statistic to detect the change point

$$\widehat{\zeta}_{j,h_1} = \arg \max_{h_1 \in (0, N_j)} \hat{\boldsymbol{\delta}}_{j;h_1}^\top \left(\boldsymbol{Z}_{j,h_1}^\top \boldsymbol{M}_j \boldsymbol{Z}_{j,h_1}\right) \hat{\boldsymbol{\delta}}_{j;h_1} \tag{4.102}$$

with $\boldsymbol{Z}_{j,h_1} = \left(\mathbf{0}, \cdots, \mathbf{0}, \boldsymbol{x}_{h_1+1}, \boldsymbol{x}_{h_1+2}, \cdots, \boldsymbol{x}_{n_j^{(r)}}\right)^\top$, $\boldsymbol{X}_j = \left(\boldsymbol{x}_{n_j^{(l)}}, \boldsymbol{x}_{n_j^{(l)}+1} \cdots, \boldsymbol{x}_{n_j^{(r)}}\right)^\top$, $\boldsymbol{M}_j = \boldsymbol{I}_n - \boldsymbol{X}_j\left(\boldsymbol{X}_j^\top \boldsymbol{X}_j\right)^{-1} \boldsymbol{X}_j^\top$, and $\left(\hat{\boldsymbol{\beta}}_{j;h_1}, \hat{\boldsymbol{\delta}}_{j;h_1}\right)$ being the least-squares estimator of $(\boldsymbol{\beta}_j, \boldsymbol{\delta}_j)$ by regressing $\boldsymbol{y}_{(j)} = \left(y_{n_j^{(l)}}, \cdots, y_{n_j^{(r)}}\right)^\top$ on $\boldsymbol{X}_j$ and $\boldsymbol{Z}_{j,h_1}$. $\hat{\zeta}_{j,h_1}$, $j = 1, \ldots, \hat{s}$ are the estimates of multiple change points, the consistency and limit distribution of which are provided in Section 4.4.4.

Currently, the partition remains unknown. In light of (4.100), we set $m < \min\left(a_{j+1} - a_j\right)/2$, $j = 1, \ldots, s-1$, to ensure that each segment $Q_{(j)}$ contains at most one change point. We also avoid choosing excessively small values for m. When m is too small, it leads to inconsistencies in estimating regression parameters and increased computational time.

Hence, by the theorem, we denote $m = \lfloor c_0 \sqrt{n} \rfloor$, where c_0 is an adjusting parameter. Here, we explore a range of values for c_0 within $[0.1, 1.2]$. The best value of m is determined using the Bayesian Information Criterion (BIC)

$$\hat{m} = \arg\min_m \log\left\{\sum_{i=1}^n \left(y_i - \boldsymbol{x}_i^\top \left[\hat{\boldsymbol{\beta}}_{1,h1} + \sum_{j=1}^{\hat{s}} \hat{\boldsymbol{\delta}}_{j,h_1} I\left(\hat{\zeta}_{j,h1} < i \leq n\right)\right]\right)^2\right\} + \hat{s} \cdot q \log n.$$

4.4.3 Bootstrap confident interval for change points

In this subsection, we outline the construction of bootstrap confidence intervals(CI) for multiple change points estimates derived in the previous section. The two-stage procedure inherently encompasses uncertainty quantification for both the number of change points and their respective locations.

We define the estimated residuals $\hat{\varepsilon}_i$ and the centered residuals $\tilde{\varepsilon}_i$ as follows:

$$\hat{\varepsilon}_i = y_i - \boldsymbol{x}_i^\top \left[\hat{\boldsymbol{\beta}}_{1,h_1} + \sum_{l=1}^{\hat{s}} \hat{\boldsymbol{\delta}}_{l,h_1} I\left(\hat{\zeta}_{l,h_1} < i \leq n\right)\right],$$
$$\tilde{\varepsilon}_i = \hat{\varepsilon}_i - \frac{1}{\hat{\zeta}_{j,h_1} - \hat{\zeta}_{j-1,h_1}} \sum_{l=\hat{\zeta}_{j-1,h_1+1}}^{\hat{\zeta}_{j,h_1}} \hat{\varepsilon}_l, \; \hat{\zeta}_{j-1,h_1} < i \leq \hat{\zeta}_{j,h_1}, \tag{4.103}$$

where $\hat{\zeta}_{0,h_1} = 1$ and $\hat{\zeta}_{\hat{s}+1,h_1} = n$ represent the start and end points of the data. Let $\varepsilon^*_{\hat{\zeta}_{j-1,h_1}+1}, \ldots, \varepsilon^*_{\hat{\zeta}_{j,h_1}}$ be independently and identically distributed (i.i.d.) random variables sampled from the empirical distribution function of the set $\left\{\tilde{\varepsilon}_{\hat{\zeta}_{j-1,h_1}+1}, \ldots, \tilde{\varepsilon}_{\hat{\zeta}_{j,h_1}}\right\}$. We then consider the bootstrap observations defined as:

$$y_i^* = \boldsymbol{x}_i^\top \left[\hat{\boldsymbol{\beta}}_{1,h_1} + \sum_{l=1}^{\hat{s}} \hat{\boldsymbol{\delta}}_{l,h_1} I\left(\hat{\zeta}_{l,h_1} < i \leq n\right)\right] + \varepsilon_i^*, \; i = 1, \ldots, n, \tag{4.104}$$

where ε_i^* represents the bootstrap version of the residuals. To obtain an approximation to the distribution of $\hat{\zeta}_{j;h_1}$ in (4.102), a bootstrap statistic of the estimator defined as

$$\widehat{\zeta}^*_{j^*;h_2} = \arg\max_{h_2} \hat{\boldsymbol{\delta}}^{*\top}_{j^*;h_2} \left(\boldsymbol{Z}^\top_{j^*,h_2} \boldsymbol{M}_j \boldsymbol{Z}_{j^*,h_2}\right) \hat{\boldsymbol{\delta}}^*_{j^*;h_2}, \; j^* = 1, \ldots, \hat{s}^* \tag{4.105}$$

where the number of change points $\hat{s}^*$ and the $\widehat{\mathcal{A}}_n^*$ are obtained by performing OGA on $\boldsymbol{y}_n^* = (y_1^*, \ldots, y_n^*)^\top$ as in (4.99). The data sequence segmentation remains unchanged. In this case, $\boldsymbol{Z}_{j^*,h_2} = \left(\boldsymbol{0}, \ldots, \boldsymbol{0}, \boldsymbol{x}_{h_2+1}, \ldots, \boldsymbol{x}_{n^{(r)}_{j^*}}\right)^\top$ and $\left(\hat{\boldsymbol{\beta}}^*_{j^*;h_2}, \hat{\boldsymbol{\delta}}^*_{j^*;h_2}\right)$ represents the least-squares estimator obtained by regressing $\boldsymbol{y}^*_{(j^*)} = \left(y^*_{n_{j^*}(l)}, \ldots, y^*_{n^{(r)}_{j^*}}\right)^\top$ on $\boldsymbol{X}_{j^*}$ and $\boldsymbol{Z}_{j^*,h_2}$. In what follows, we describe the construction of bootstrap CI for multiple change points.

1. Generate a bootstrap sample $(y_1^*, \ldots, y_n^*)^\top$ by randomly sampling residuals $(\varepsilon_1^*, \ldots, \varepsilon_n^*)^\top$ from the set $\{\tilde{\varepsilon}_1, \ldots, \tilde{\varepsilon}_n\}$ as in (4.104);

2. Apply the two-stage procedure and compute the local maximizers obtained as in (4.105) for each estimated segment;

3. For a given bootstrap sample size B, repeat Steps $1-2$ B times and record $\widehat{\zeta}^{*(b)}_{j^*;h_2}, j = 1, \cdots, \hat{s}$ for $b = 1, \ldots, B$.

Following the above procedure, the bootstrap-based approximation for the change point estimates can be constructed. Generally, for any $\alpha \in (0,1)$, the $100(1-\alpha)\%$-quantile bootstrap confidence interval for a single change point is given by:

$$\mathrm{CI}_j(\alpha) = \left[\hat{\zeta}_{j,h_1} + q_U(\alpha/2), \hat{\zeta}_{j,h_1} + q_L(\alpha/2)\right], \tag{4.106}$$

where

$$q_U(\alpha/2) = \sup\left\{x; \frac{1}{B}\sum_{b=1}^{B} I\left(\hat{\zeta}^{*(b)}_{j^*,h_2} - \widehat{\zeta}_{j;h_1} \leq x\right) \leq \alpha/2\right\},$$

and

$$q_L(\alpha/2) = \inf\left\{x; \frac{1}{B}\sum_{b=1}^{B} I\left(\hat{\zeta}^{*(b)}_{j^*,h_2} - \widehat{\zeta}_{j;h_1} \geq x\right) \geq 1-\alpha/2\right\}.$$

If $\widehat{\mathcal{A}}_n \subseteq \widehat{\mathcal{A}}^*_n$, $\widehat{\zeta}^*_{j;h_2}$ represents an estimate of $\hat{\zeta}_{j;h_1}$ for $j = 1,\ldots,\hat{s}$. If some elements in $\widehat{\mathcal{A}}_n$ are not in the set $\widehat{\mathcal{A}}^*_n$, $\hat{\zeta}_{j;h_1}$ does not have a corresponding bootstrap estimate for some j. Therefore, the bootstrap CI of $\widehat{\zeta}_{j;h_1}$ is constructed by using $\left\{\widehat{\zeta}^{*(b)}_{j;h_2} : \widehat{\zeta}^{*(b)}_{j;h_2} \notin \emptyset, b = 1,\ldots,B^*\right\}$ instead of $\left\{\widehat{\zeta}^{*(1)}_{j^*;h_2},\ldots,\widehat{\zeta}^{*(B)}_{j^*;h_2}\right\}$ in (4.106), which yields

$$\mathrm{CI}^*_j(\alpha) = \left[\hat{\zeta}_{j,h_1} + q^*_U(\alpha/2), \hat{\zeta}_{j,h_1} + q^*_L(\alpha/2)\right], \tag{4.107}$$

where

$$q^*_U(\alpha/2) = \sup\left\{x; \frac{1}{B^*}\sum_{b=1}^{B^*} I\left(\hat{\zeta}^{*(b)}_{j,h_2} - \widehat{\zeta}_{j;h_1} \leq x\right) \leq \alpha/2\right\}$$

and

$$q^*_L(\alpha/2) = \inf\left\{x; \frac{1}{B^*}\sum_{b=1}^{B^*} I\left(\hat{\zeta}^{*(b)}_{j,h_2} - \widehat{\zeta}_{j;h_1} \geq x\right) \geq 1-\alpha/2\right\}.$$

4.4.4 Theoretical validity of bootstrap confidence intervals

We impose the following assumptions to investigate the performance of bootstrap confidence intervals for multiple change points.

Assumption 4.11. *If* $s \geqslant 1$, $a_j/n \to \tau_j > 0$, *for* $1 \leq j \leq s$. *Further if* $s \geqslant 2$, $\min_{1\leq j\leq s-1}(\tau_{j+1} - \tau_j) > 0$.

Assumption 4.12. $\sum_{i=t_1}^{t_2} \boldsymbol{x}_i\boldsymbol{x}_i^\top / (t_1 - t_2) \to V$ *as* $t_2 - t_1 \to \infty$, *where* V *is a strictly positive definite matrix.*

Assumption 4.13. $\{\varepsilon_i, i = 1,2,\ldots\}$ *is a sequence of independently and identically distributed (i.i.d.) sub-Gaussian random variables with* $E(\varepsilon_i) = 0$ *and* $E(\varepsilon_i^2) = \sigma^2$.

Assumption 4.14. $a_j - n_j^{(l)} = [2\tau m]$, *where* $\tau \in (0,1)$ *and* $[\cdot]$ *is the greatest integer function.*

Assumption 4.15. $\sup_{(t_2-t_1)\geq 1}\left\|\sum_{i=t_1}^{t_2} \boldsymbol{x}_i\boldsymbol{x}_i^\top / (t_2 - t_1)\right\|$ *is stochastically bounded.*

Assumption 4.16. *Assume* $\boldsymbol{\delta}_j \to \mathbf{0}$, *and* $\boldsymbol{\delta}_j^{-1}N_j^{-1/2+\alpha} = o(1)$ *for some* $\alpha \in (0,1/2)$ *as* $n \to \infty$.

By Assumption 4.11, it follows that $m/(a_{j+1} - a_j) \to 0$, i.e., there is at most one change point in each segment for large n. Assumption 4.12 is similar to Condition (b) of [295]. Assumption 4.13 introduces condition on the error terms. Assumption 4.14 assumes that the shift point is bounded away from the end points, which is required for the asymptotic derivation. Assumption 4.15 requires that there be enough data around the change point and at the beginning and at the end of the sample, so that the change point can be identified. Next, we establish the consistency of the number of change points $\hat{s}$ and the change point estimator $\hat{\zeta}_{j,h_1}$. The convergence speed of the change point estimator not only describes the speed at which the estimator converges to the true value, but is also necessary for deriving the limit distribution. Thus we will need the rate of convergence for shrinking shifts. Apparently, the magnitude of the shift cannot be too small, otherwise, it will be impossible to distinguish the noise and identify the change point. Based on this, we specify the necessary Assumption 4.16.

The following theorem provides the consistency of the estimated number of change points.

Theorem 4.4.1. *Suppose that $m \to \infty$, $p_n \to \infty$, $D_n = O\left((n/\log(r_n))^{1/2}\right)$, and $\log(r_n)/n \to \infty$ as $n \to \infty$. If Assumptions 4.11–4.13 hold, $\boldsymbol{\delta}_j$ is fixed or $\boldsymbol{\delta}_j \to 0$ but satisfying Assumption 4.16, then we have*

$$\lim_{n\to\infty} P(\hat{s} = s) = 1$$
$$\lim_{n\to\infty} P\left(a_{j,n} \in \hat{\mathcal{I}}_{(j)}, j = 1, \ldots, s \mid \hat{s} = s\right) = 1, \tag{4.108}$$

where $\hat{s}$ and $\hat{k}_j, j = 1, \ldots, \hat{s}$ are given in (4.99).

The proof of Theorem 4.4.1 extends Theorem 4 in [131] to multiple change points detection cases. The consistency and asymptotic distribution of change points estimators are given below.

Theorem 4.4.2. *Under Assumptions 4.11–4.15, if $\boldsymbol{\delta}_j$ is fixed or $\boldsymbol{\delta}_j \to 0$ but satisfying Assumption 4.16. Then, as $n \to \infty$, $\hat{\zeta}_{j,h_1} - a_j = O_p\left(\|\boldsymbol{\delta}_j\|^{-2}\right)$ and*

$$\frac{\boldsymbol{\delta}_j^\top V \boldsymbol{\delta}_j}{\sigma^2}\left(\hat{\zeta}_{j,h_1} - a_j\right) \to_d \arg\max\{W(c) - |c|/2 : c \in \mathcal{R}\},$$

where $\{W(c) : c \in \mathcal{R}\}$ is a two-sided Wiener process.

Proof of Theorem 4.4.2. To prove this theorem, we define $V(h_1) = \hat{\boldsymbol{\delta}}_{j;h_1}^\top (\boldsymbol{Z}_{j,h_1}^\top \boldsymbol{M}_j \boldsymbol{Z}_{j,h_1}) \hat{\boldsymbol{\delta}}_{j;h_1}$. For the sake of simplicity, we denote $h_0 = a_j$, which means h_0 is the change point in $\mathcal{I}_{k_j}$. If $h_1 = h_0$, we can obtain $\boldsymbol{Z}_{j,h_1} = \boldsymbol{Z}_{j,h_0}$, where $\boldsymbol{Z}_{j,h_0} = (\mathbf{0}, \ldots, \mathbf{0}, \boldsymbol{x}_{h_0}, \ldots, \boldsymbol{x}_{n_j^{(r)}})^\top$. By (4.102), $\hat{\zeta}_{j,h_1} = \arg\max_{h_1} V(h_1)$. Notice that

$$\hat{\boldsymbol{\delta}}_{j;h_1} = \left(\boldsymbol{Z}_{j,h_1}^\top \boldsymbol{M}_j \boldsymbol{Z}_{j,h_1}\right)^{-1} \left(\boldsymbol{Z}_{j,h_1}^\top \boldsymbol{M}_j \boldsymbol{Z}_{j,h_0} \boldsymbol{\delta}_j + \boldsymbol{Z}_{j,h_1}^\top \boldsymbol{M}_j \boldsymbol{\varepsilon}\right),$$
$$\hat{\boldsymbol{\delta}}_{j,h_0} = \boldsymbol{\delta}_j + \left(\boldsymbol{Z}_{j,h_0}^\top \boldsymbol{M}_j \boldsymbol{Z}_{j,h_0}\right)^{-1} \boldsymbol{Z}_{j,h_0}^\top \boldsymbol{M}_j \boldsymbol{\varepsilon}.$$

It follows that

$$\begin{aligned}
&V(h_1) - V(h_0) \\
&= \boldsymbol{\delta}_j^\top \{(\boldsymbol{Z}_{j,h_0}^\top \boldsymbol{M}_j \boldsymbol{Z}_{j,h_1}) \left(\boldsymbol{Z}_{j,h_1}^\top \boldsymbol{M}_j \boldsymbol{Z}_{j,h_1}\right)^{-1} (\boldsymbol{Z}_{j,h_1}^\top \boldsymbol{M}_j \boldsymbol{Z}_{j,h_0}) - (\boldsymbol{Z}_{j,h_0}^\top \boldsymbol{M}_j \boldsymbol{Z}_{j,h_0})\} \boldsymbol{\delta}_j \\
&\quad + v(h_1),
\end{aligned}$$

where

$$\begin{aligned}
v(h_1) &= 2\boldsymbol{\delta}_j^\top (\boldsymbol{Z}_{j,h_0}^\top \boldsymbol{M}_j \boldsymbol{Z}_{j,h_1})(\boldsymbol{Z}_{j,h_1}^\top \boldsymbol{M}_j \boldsymbol{Z}_{j,h_1})^{-1} \boldsymbol{Z}_{j,h_0}^\top \boldsymbol{M}_j \boldsymbol{\varepsilon} - 2\boldsymbol{\delta}_j^\top \boldsymbol{Z}_{j,h_0}^\top \boldsymbol{M}_j \boldsymbol{\varepsilon} \\
&+ \boldsymbol{\varepsilon}^\top \boldsymbol{M}_j \boldsymbol{Z}_{j,h_1} (\boldsymbol{Z}_{j,h_1}^\top \boldsymbol{M}_j \boldsymbol{Z}_{j,h_1})^{-1} \boldsymbol{Z}_{j,h_1}^\top \boldsymbol{M}_j \boldsymbol{\varepsilon} - \boldsymbol{\varepsilon}^\top \boldsymbol{M}_j \boldsymbol{Z}_{j,h_0} (\boldsymbol{Z}_{j,h_0}^\top \boldsymbol{M}_j \boldsymbol{Z}_{j,h_0})^{-1} \boldsymbol{Z}_{j,h_0}^\top \boldsymbol{M}_j \boldsymbol{\varepsilon}.
\end{aligned}$$

Define for $h_1 \neq k_0$,

$$\begin{aligned}
g_1(h_1) = \boldsymbol{\delta}_j^\top &\left|(\boldsymbol{Z}_{j,h_0}^\top \boldsymbol{M}_j \boldsymbol{Z}_{j,h_1})(\boldsymbol{Z}_{j,h_1}^\top \boldsymbol{M}_j \boldsymbol{Z}_{j,h_1})^{-1}(\boldsymbol{Z}_{j,h_1}^\top \boldsymbol{M}_j \boldsymbol{Z}_{j,h_0})\right. \\
&\left. - \boldsymbol{Z}_{j,h_0} \boldsymbol{M}_j \boldsymbol{Z}_{j,h_0}\right| \boldsymbol{\delta}_j / |k_0 - h_1|
\end{aligned}$$

When $h_1 = h_0$, define $g(h_1) = \boldsymbol{\delta}_j^\top \boldsymbol{\delta}_j$. We have the following identity:

$$V(h_1) - V(h_0) = -|h_0 - h_1| g(h_1) + v(h_1). \tag{4.109}$$

Denote

$$\boldsymbol{Z}_{\Delta_1} = \begin{cases} \boldsymbol{Z}_{j,h_1} - \boldsymbol{Z}_{j,h_0} = (\mathbf{0}, \cdots, \mathbf{0}, \boldsymbol{x}_{h_1+1}, \ldots, \boldsymbol{x}_{h_0}, \mathbf{0}, \ldots, \mathbf{0})^\top, & h_1 < h_0 \\ \boldsymbol{Z}_{j,h_0} - \boldsymbol{Z}_{j,h_1} = (\mathbf{0}, \cdots, \mathbf{0}, \boldsymbol{x}_{h_0+1}, \ldots, \boldsymbol{x}_{h_1}, \mathbf{0}, \ldots, \mathbf{0})^\top, & h_1 > h_0 \\ \mathbf{0}, & h_1 = h_0 \end{cases}.$$

To prove this theorem, we use $\boldsymbol{Z}_{j,h_0} = \boldsymbol{Z}_{j,h_1} - \boldsymbol{Z}_{\Delta_1}\,\mathrm{sgn}(h_0 - h_1)$ to deduce that

$$\begin{aligned}
&v(h_1) = v_1(h_1) + v_2(h_1) + v_3(h_1) + v_4(h_1) + v_5(h_1)\\
&= \left[2\boldsymbol{\delta}_j^\top \boldsymbol{Z}_{\Delta_1}^\top \boldsymbol{\varepsilon}\right] \mathrm{sgn}(h_0 - h_1)\\
&- \left[2\boldsymbol{\delta}_j^\top \boldsymbol{Z}_{\Delta_1}^\top \boldsymbol{X}_j(\boldsymbol{X}_j^\top \boldsymbol{X}_j)^{-1}\boldsymbol{X}_j^\top \boldsymbol{\varepsilon}\right] \mathrm{sgn}(h_0 - h_1)\\
&- \left[2\boldsymbol{\delta}_j^\top (\boldsymbol{Z}_{\Delta_1}^\top \boldsymbol{M}_j \boldsymbol{Z}_{j,h_1})(\boldsymbol{Z}_{j,h_1}^\top \boldsymbol{M}_j \boldsymbol{Z}_{j,h_1})^{-1}\boldsymbol{Z}_{j,h_1}^\top \boldsymbol{M}_j \boldsymbol{\varepsilon}\right] \mathrm{sgn}(h_0 - h_1)\\
&+ \boldsymbol{\varepsilon}^\top \boldsymbol{M}_j \boldsymbol{Z}_{j,h_1}(\boldsymbol{Z}_{j,h_1}^\top \boldsymbol{M}_j \boldsymbol{Z}_{j,h_1})^{-1}\boldsymbol{Z}_{j,h_1}^\top \boldsymbol{M}_j \boldsymbol{\varepsilon}\\
&- \boldsymbol{\varepsilon}^\top \boldsymbol{M}_j \boldsymbol{Z}_{j,h_0}(\boldsymbol{Z}_{j,h_0}^\top \boldsymbol{M}_j \boldsymbol{Z}_{j,h_0})^{-1}\boldsymbol{Z}_{j,h_0}^\top \boldsymbol{M}_j \boldsymbol{\varepsilon}.
\end{aligned} \tag{4.110}$$

First, it is sufficient to show the rate of consistency for the change point estimator. By Lemma 4.2, there exists $\lambda > 0$ such that for every $\epsilon > 0$, $\inf_{|h_1 - h_0| > C\|\boldsymbol{\delta}_j\|^{-2}} g(h_1) \geq \lambda \|\boldsymbol{\delta}_j\|^2$ with probability at least $1 - \epsilon$. Thus, we need to prove

$$\begin{aligned}
&P(|\hat{\zeta}_{j,h_1} - a_j| > C\|\boldsymbol{\delta}_j\|^{-2})\\
\leq &P\left(\sup_{|h_1 - k_0| > C\|\boldsymbol{\delta}_j\|^{-2}} V(h_1) \geq V(k_0)\right)\\
\leq &P\left(\sup_{h_1 \in K(C)} \left|\frac{v(h_1)}{(k_0 - h_1)\|\boldsymbol{\delta}_j\|^2}\right| > \lambda\right) \leq \epsilon.
\end{aligned} \tag{4.111}$$

where $K(C) = \left\{h_1 : |h_1 - k_0| > C\|\boldsymbol{\delta}_j\|^{-2} \text{ and } n_j^{(l)} + \eta N_j \leq h_1 \leq n_j^{(r)} - \eta N_j\right\}$ for a small number $\eta > 0$. It is easy to get

$$\begin{aligned}
\sup_{h_1 \in K(C)} \frac{|v_2(h_1)|}{|h_0 - h_1|\|\boldsymbol{\delta}_j\|^2} &\leq 2\|\boldsymbol{\delta}_j\|^{-1}\|\boldsymbol{Z}_{\Delta_1}^\top \boldsymbol{X}_j/(h_0 - h_1)\|\|(\boldsymbol{X}_j^\top \boldsymbol{X}_j)^{-1}\boldsymbol{X}_j^\top \boldsymbol{\varepsilon}\|\\
&= O_p(\|\boldsymbol{\delta}_j\|^{-1}N_j^{-1/2}) = o_p(1),\\
\sup_{h_1 \in K(C)} \frac{v_3(h_1)}{|h_0 - h_1|\|\boldsymbol{\delta}_j\|^2} &\leq \|\boldsymbol{\delta}_j\|^{-1}\|(\boldsymbol{Z}_{\Delta_1}^\top \boldsymbol{M}_j \boldsymbol{Z}_{j,h_1})/(h_0 - h_1)\|\|(\boldsymbol{Z}_{j,h_1}^\top \boldsymbol{M}_j \boldsymbol{Z}_{j,h_1})^{-1}\boldsymbol{Z}_{j,h_1}^\top \boldsymbol{M}_j \boldsymbol{\varepsilon}\|\\
&= O_p(\|\boldsymbol{\delta}_j\|^{-1}N_j^{-1/2}) = o_p(1)\\
\sup_{h_1 \in K(C)} \frac{|v_4(h_1)|}{|h_0 - h_1|\|\boldsymbol{\delta}_j\|^2} &\leq \|\boldsymbol{\delta}_j\|^{-2}\|(\boldsymbol{Z}_{j,h_1}^\top \boldsymbol{M}_j \boldsymbol{Z}_{j,h_1})^{-1/2}\boldsymbol{Z}_{j,h_1}^\top \boldsymbol{M}_j \boldsymbol{\varepsilon}\|^2/|h_0 - h_1|\\
&= O_p(\|\boldsymbol{\delta}_j\|^{-2}|h_0 - h_1|^{-1}) = O_p(1).\\
\sup_{h_1 \in K(C)} \frac{|v_5(h_1)|}{|h_0 - h_1|\|\boldsymbol{\delta}_j\|^2} &\leq \|\boldsymbol{\delta}_j\|^{-2}\|(\boldsymbol{Z}_{j,h_0}^\top \boldsymbol{M}_j \boldsymbol{Z}_{j,h_0})^{-1/2}\boldsymbol{Z}_{j,h_0}^\top \boldsymbol{M}_j \boldsymbol{\varepsilon}\|^2/|h_0 - h_1|\\
&= O_p(\|\boldsymbol{\delta}_j\|^{-2}|h_0 - h_1|^{-1}) = O_p(1).
\end{aligned}$$

By Lemma 4.3, there exists a large C such that

$$\begin{aligned}
&P\left(\sup_{h_1 \in K(C)} \frac{|v_1(h_1)|}{|h_0 - h_1|\|\boldsymbol{\delta}_j\|^2} > \frac{\lambda}{5}\right)\\
\leq &P\left(\sup_{h_1 \in K(C)} \|\boldsymbol{\delta}_j\|^{-1}\|\boldsymbol{Z}_{\Delta_1}^\top \boldsymbol{\varepsilon}/(h_0 - h_1)\| > \frac{\lambda}{10}\right) \leq \frac{100B}{\lambda^2 C} \leq \frac{\epsilon}{5}.
\end{aligned}$$

This completes the proof of (4.111).

To study the limiting distribution, we study the behavior of $V(h_1)$ on $D(C) = \{h_1 : |h_1 - h_0| \leq C\|\boldsymbol{\delta}_j\|^{-2}\}$.

$$\begin{aligned}
|h_0 - h_1|g(h_1) &= \boldsymbol{\delta}_j^\top\{\left(\boldsymbol{Z}_{j,h_0}^\top \boldsymbol{M}_j \boldsymbol{Z}_{j,h_1}\right)\left(\boldsymbol{Z}_{j,h_1}^\top \boldsymbol{M}_j \boldsymbol{Z}_{j,h_1}\right)^{-1}\left(\boldsymbol{Z}_{j,h_1}^\top \boldsymbol{M}_j \boldsymbol{Z}_{j,h_0}\right) - \left(\boldsymbol{Z}_{j,h_0}^\top \boldsymbol{M}_j \boldsymbol{Z}_{j,h_0}\right)\}\boldsymbol{\delta}_j\\
&= \boldsymbol{\delta}_j^\top\{\boldsymbol{Z}_{\Delta_1}^\top \boldsymbol{Z}_{\Delta_1} - \boldsymbol{Z}_{\Delta_1}^\top \boldsymbol{X}_j(\boldsymbol{X}_j^\top \boldsymbol{X}_j)^{-1}\boldsymbol{X}_j^\top \boldsymbol{Z}_{\Delta_1}\}\boldsymbol{\delta}_j\\
&- \boldsymbol{\delta}_j\{(\boldsymbol{Z}_{\Delta_1}^\top \boldsymbol{M}_j \boldsymbol{Z}_{j,h_1})(\boldsymbol{Z}_{j,h_1}^\top \boldsymbol{M}_j \boldsymbol{Z}_{j,h_1})^{-1}(\boldsymbol{Z}_{j,h_1}^\top \boldsymbol{M}_j \boldsymbol{Z}_{\Delta_1})\}\boldsymbol{\delta}_j.
\end{aligned}$$

Since $\|\boldsymbol{X}_j^\top \boldsymbol{Z}_{\Delta_1}\| = O_p(\|\boldsymbol{\delta}_j\|^{-2})$ and $\|\boldsymbol{Z}_{j,h_1}^\top \boldsymbol{M}_j \boldsymbol{Z}_{\Delta_1}\| = O_p(\|\boldsymbol{\delta}_j\|^{-2})$, we have

$$|h_0 - h_1| g(h_1) = \boldsymbol{\delta}_j^\top \boldsymbol{Z}_{\Delta_1}^\top \boldsymbol{Z}_{\Delta_1} \boldsymbol{\delta}_j + o_p(1).$$

Next, we consider about $v(h_1)$. Because $|h_1 - h_0| \leq C\|\boldsymbol{\delta}_j\|^{-2}$, $v_2(h_1) = O_p(\|\boldsymbol{\delta}_j\|^{-1} N_j^{-1/2}) = o_p(1)$, $v_3(h_1) = O_p(\|\boldsymbol{\delta}_j\|^{-1} N_j^{-1/2}) = o_p(1)$. We also have

$$\begin{aligned} v_4(h_1) + v_5(h_1) &= \boldsymbol{\varepsilon}^\top \boldsymbol{M}_j \boldsymbol{Z}_{j,h_0} \left[(\boldsymbol{Z}_{j,h_1}^\top \boldsymbol{M}_j \boldsymbol{Z}_{j,h_1})^{-1} - (\boldsymbol{Z}_{j,h_0}^\top \boldsymbol{M}_j \boldsymbol{Z}_{j,h_0})^{-1} \right] \boldsymbol{Z}_{j,h_0}^\top \boldsymbol{M}_j \boldsymbol{\varepsilon} \\ &+ \boldsymbol{\varepsilon}^\top \boldsymbol{M}_j \boldsymbol{Z}_{\Delta_1} (\boldsymbol{Z}_{j,h_1}^\top \boldsymbol{M}_j \boldsymbol{Z}_{j,h_1})^{-1} \boldsymbol{Z}_{j,h_1}^\top \boldsymbol{M}_j \boldsymbol{\varepsilon} \\ &+ \boldsymbol{\varepsilon}^\top \boldsymbol{M}_j \boldsymbol{Z}_{j,h_0} (\boldsymbol{Z}_{j,h_1}^\top \boldsymbol{M}_j \boldsymbol{Z}_{j,h_1})^{-1} \boldsymbol{Z}_{\Delta_1}^\top \boldsymbol{M}_j \boldsymbol{\varepsilon} = o_p(1) \end{aligned}$$

on $D(C)$. According to the above process, we can obtain

$$V(h_1) - V(h_0) = -\boldsymbol{\delta}_j^\top \mathbf{Z}_{\Delta_1}^\top \mathbf{Z}_{\Delta_1} \boldsymbol{\delta}_j + \left[2\boldsymbol{\delta}_j^\top \mathbf{Z}_{\Delta_1}^\top \boldsymbol{\varepsilon} \right] \operatorname{sgn}(h_0 - h_1) + o_p(1). \tag{4.112}$$

In summary, combined with the proof of Proposition 4.2, we arrive at the limiting distribution stated in Theorem 4.4.2.

□

We first present some lemmas that facilitate the proofs of bootstrap asymptotic validity. We first look at convergence properties of the variance of $\varepsilon^*_{n_j^{(l)}}$. Since $E^*(\varepsilon^*_{n_j^{(l)}}) = N_j^{-1} \sum_{i=n_j^{(l)}}^{n_j^{(r)}} \tilde{\varepsilon}_i = 0$ by (4.103), we have that $\operatorname{var}^*(\varepsilon^*_{n_j^{(l)}}) = N_j^{-1} \sum_{i=n_j^{(l)}}^{n_j^{(r)}} \tilde{\varepsilon}_i^2$.

Lemma 4.8. *If $\boldsymbol{\delta}_j$ is fixed or $\boldsymbol{\delta}_j \to \mathbf{0}$ but satisfying Assumption 4.16, then as $N_j \to \infty$,*

$$\operatorname{var}^*(\varepsilon^*_{n_j^{(l)}}) = N_j^{-1} \sum_{i=n_j^{(l)}}^{n_j^{(r)}} (\varepsilon_i - \bar{\varepsilon}_{N_j})^2 + O_p\left(N_j^{-1}\right).$$

Proof of Lemma 4.8. In view of (4.94) and (4.103), we have

$$\begin{aligned} \tilde{\varepsilon}_i = &\left[\boldsymbol{x}_i I(a_j < i \leq n_j^{(r)}) - \frac{1}{n_j^{(r)} - a_j} \sum_{l=a_j+1}^{n_j^{(r)}} \boldsymbol{x}_l \right]^\top \boldsymbol{\delta}_j + (\boldsymbol{x}_i - \overline{\boldsymbol{x}}_j)^\top (\boldsymbol{\beta}_j - \hat{\boldsymbol{\beta}}_{j,h_1}) \\ &- \left[\boldsymbol{x}_i I(\hat{\zeta}_{j,h_1} < i \leq n_j^{(r)}) - \frac{1}{N_j} \sum_{l=\hat{\zeta}_{j,h_1}+1}^{n_j^{(r)}} \boldsymbol{x}_l \right]^\top \hat{\boldsymbol{\delta}}_{j,h_1} + \varepsilon_i - \bar{\varepsilon}_{N_j}, \end{aligned} \tag{4.113}$$

where $\overline{\boldsymbol{x}}_j = \frac{1}{N_j} \sum_{l=n_j^{(l)}}^{n_j^{(r)}} \boldsymbol{x}_l$ and $\bar{\varepsilon}_{N_j} = \frac{1}{N_j} \sum_{l=n_j^{(l)}}^{n_j^{(r)}} \varepsilon_l$. Assume that $a_j < \hat{\zeta}_{j,h_1}$ (the other case can be dealt with similarly).

It follows from (4.113) that

$$\begin{aligned} \operatorname{var}^*(\varepsilon^*_{n_j^{(l)}}) = &\frac{1}{N_j} \sum_{i=n_j^{(l)}}^{n_j^{(r)}} (\varepsilon_i - \bar{\varepsilon}_{N_j})^2 + \frac{2}{N_j} \sum_{i=n_j^{(l)}}^{n_j^{(r)}} (\boldsymbol{x}_i - \overline{\boldsymbol{x}}_j)^\top (\boldsymbol{\beta}_j - \hat{\boldsymbol{\beta}}_{j,h_1})(\varepsilon_i - \bar{\varepsilon}_{N_j}) \\ &+ \frac{1}{N_j} \sum_{i=n_j^{(l)}}^{n_j^{(r)}} \left[(\boldsymbol{x}_i - \overline{\boldsymbol{x}}_j)^\top (\boldsymbol{\beta}_j - \hat{\boldsymbol{\beta}}_{j,h_1}) \right]^2 + \frac{1}{N_j^2} \left(\sum_{i=\hat{\zeta}_{j,h_1}+1}^{n_j^{(r)}} \boldsymbol{x}_i^\top (\boldsymbol{\delta}_j - \hat{\boldsymbol{\delta}}_{j,h_1}) \right)^2 \\ &+ \frac{1}{N_j^2} \left(\sum_{i=a_j+1}^{\hat{\zeta}_{j,h_1}} \boldsymbol{x}_i^\top \boldsymbol{\delta}_j \right)^2 + \frac{2}{N_j^2} \sum_{i=a_j+1}^{\hat{\zeta}_{j,h_1}} \boldsymbol{x}_i^\top \boldsymbol{\delta}_j \sum_{l=\hat{\zeta}_{j,h_1}+1}^{n_j^{(r)}} \boldsymbol{x}_l^\top (\boldsymbol{\delta}_j - \hat{\boldsymbol{\delta}}_{j,h_1}). \end{aligned} \tag{4.114}$$

According to Corollary 1 in [16], we have $\boldsymbol{\beta}_j - \hat{\boldsymbol{\beta}}_{j,h_1} = O_p(N_j^{-1/2})$ and $\boldsymbol{\delta}_j - \hat{\boldsymbol{\delta}}_{j,h_1} = O_p(N_j^{-1/2})$. Combined with $\hat{\zeta}_{j,h_1} = a_j + O_p(\|\boldsymbol{\delta}_j\|^{-2})$, it can be obtained that $N_j^{-1}\sum_{i=a_j+1}^{\hat{\zeta}_{j,h_1}} \boldsymbol{x}_i^\top \boldsymbol{\delta}_j = O_p(N_j^{-1}\|\boldsymbol{\delta}_j\|^{-1})$ and $N_j^{-1}\sum_{l=\hat{\zeta}_{j,h_1}+1}^{n_j^{(r)}} \boldsymbol{x}_l^\top(\boldsymbol{\delta}_j - \hat{\boldsymbol{\delta}}_{j,h_1}) = O_p(N_j^{-1/2})$. This completes the proof. □

Lemma 4.9. *For every $\epsilon > 0$,*

$$\limsup_{N_j\to\infty} P^*\left(\left|N_j^{-1}\sum_{i=n_j^{(l)}}^{n_j^{(r)}} (\varepsilon_i^*)^2 - \text{var}^*(\varepsilon^*_{n_j^{(l)}})\right| \geq \epsilon\right) \leq \epsilon\sigma^2.$$

Proof of Lemma 4.9. We have for every $\epsilon > 0$ and every $\eta > 0$,

$$\begin{aligned}
&P^*\left(\left|N_j^{-1}\sum_{i=n_j^{(l)}}^{n_j^{(r)}} \varepsilon_i^{*2} - \text{var}^*(\varepsilon^*_{n_j^{(l)}})\right| \geq \epsilon\right)\\
=&P^*\left(\left|N_j^{-1}\sum_{i=n_j^{(l)}}^{n_j^{(r)}} \varepsilon_i^{*2} - N_j^{-1}\sum_{i=n_j^{(l)}}^{n_j^{(r)}} \tilde{\varepsilon}_i^2\right| \geq \epsilon\right)\\
\leq&P^*\left(\left|N_j^{-1}\sum_{i=n_j^{(l)}}^{n_j^{(r)}} \varepsilon_i^{*2} I(|\varepsilon_i^*| \geq \eta\sqrt{N_j}) - N_j^{-1}\sum_{i=n_j^{(l)}}^{n_j^{(r)}} \tilde{\varepsilon}_i^2 I(|\tilde{\varepsilon}_i| \geq \eta\sqrt{N_j})\right| \geq \epsilon\right)\\
+&P^*\left(\left|N_j^{-1}\sum_{i=n_j^{(l)}}^{n_j^{(r)}} \varepsilon_i^{*2} I(|\varepsilon_i^*| < \eta\sqrt{N_j}) - N_j^{-1}\sum_{i=n_j^{(l)}}^{n_j^{(r)}} \tilde{\varepsilon}_i^2 I(|\tilde{\varepsilon}_i| < \eta\sqrt{N_j})\right| \geq \epsilon\right).
\end{aligned} \tag{4.115}$$

Clearly, for every $\epsilon > 0$ and every $\eta > 0$, as $N_j \to \infty$, we have

$$N_j^{-1}\sum_{i=n_j^{(l)}}^{n_j^{(r)}} \tilde{\varepsilon}_i^2 I(|\tilde{\varepsilon}_i| \geq \eta\sqrt{N_j}) \to 0 \quad \text{a.s.}$$

It follows that

$$P^*\left(N_j^{-1}\sum_{i=n_j^{(l)}}^{n_j^{(r)}} \varepsilon_i^{*2} I\left(|\varepsilon_i^*| \geq \eta\sqrt{N_j}\right) \geq \epsilon/2\right) \leq \frac{2}{\epsilon}N_j^{-1}\sum_{i=n_j^{(l)}}^{n_j^{(r)}} \tilde{\varepsilon}_i^2 I\left(|\tilde{\varepsilon}_i| \geq \eta\sqrt{N_j}\right) \to 0 \quad \text{a.s.}$$

The second part of (4.115)

$$\begin{aligned}
&\limsup_{N_j\to\infty} P^*\left(\left|N_j^{-1}\sum_{i=n_j^{(r)}}^{n_j^{(l)}} \varepsilon_i^{*2} I\left(|\varepsilon_i^*| < \eta\sqrt{N_j}\right) - N_j^{-1}\sum_{i=n_j^{(r)}}^{n_j^{(l)}} \tilde{\varepsilon}_i^2 I\left(|\tilde{\varepsilon}_i| < \eta\sqrt{N_j}\right)\right| \geq \epsilon/2\right)\\
\leq& \frac{4}{\epsilon^2}\limsup_{N_j\to\infty} N_j^{-2}\sum_{i=n_j^{(r)}}^{n_j^{(l)}} \tilde{\varepsilon}_i^4 I\left(|\tilde{\varepsilon}_i| < \eta\sqrt{N_j}\right)\\
\leq& \frac{4}{\epsilon^2}\eta^2\limsup_{n\to\infty} N_j^{-1}\sum_{i=n_j^{(r)}}^{n_j^{(l)}} \tilde{\varepsilon}_i^2 = \frac{4}{\epsilon^2}\eta^2\sigma^2 \quad \text{a.s.}
\end{aligned}$$

Choosing $\eta^2 = \epsilon^3/8$ gives the assertion in Lemma 4.9. □

Next, we derive the consistency of $\hat{\zeta}^*_{j,h_2}$ in Lemma 4.10, which facilitates the proofs of our main results.

Lemma 4.10. *Under Assumptions 4.11–4.15, if $\boldsymbol{\delta}_j$ is fixed or $\boldsymbol{\delta}_j \to \mathbf{0}$ but satisfying Assumption 4.16, then*

$$\hat{\zeta}^*_{j,h_2} - \hat{\zeta}_{j,h_1} = O_p(\|\hat{\boldsymbol{\delta}}_{j,h_1}\|^{-2}).$$

Proof of Lemma 4.10. To prove this lemma, we define $V^*(h_2) = (\hat{\boldsymbol{\delta}}^*_{j;h_2})^\top(\boldsymbol{Z}^\top_{j,h_2}\boldsymbol{M}_j\boldsymbol{Z}_{j,h_2})\hat{\boldsymbol{\delta}}^*_{j;h_2}$. By (4.105), $\hat{\zeta}^*_{j,h_2} = \arg\max_{h_2} V^*(h_2)$. Denote $h_1 = \hat{\zeta}_{j,h_1}$. If $h_2 = h_1$, we can obtain $\boldsymbol{Z}_{j,h_2} = \boldsymbol{Z}_{j,h_1}$. It can be noticed from (4.104) that

$$\begin{aligned}
\hat{\boldsymbol{\delta}}^*_{j;h_2} &= \left(\boldsymbol{Z}^\top_{j,h_2}\boldsymbol{M}_j\boldsymbol{Z}_{j,h_2}\right)^{-1}\left(\boldsymbol{Z}^\top_{j,h_2}\boldsymbol{M}_j\boldsymbol{Z}_{j,h_1}\hat{\boldsymbol{\delta}}_{j,h_1} + \boldsymbol{Z}^\top_{j,h_2}\boldsymbol{M}_j\boldsymbol{\varepsilon}^*\right),\\
\hat{\boldsymbol{\delta}}^*_{j;h_1} &= \hat{\boldsymbol{\delta}}_{j,h_1} + \left(\boldsymbol{Z}^\top_{j,h_1}\boldsymbol{M}_j\boldsymbol{Z}_{j,h_1}\right)^{-1}\boldsymbol{Z}^\top_{j,h_1}\boldsymbol{M}_j\boldsymbol{\varepsilon}^*.
\end{aligned}$$

It follows that

$$\begin{aligned}
&V^*(h_2) - V^*(h_1)\\
&= \hat{\boldsymbol{\delta}}^\top_{j,h_1}\{(\boldsymbol{Z}^\top_{j,h_1}\boldsymbol{M}_j\boldsymbol{Z}_{j,h_2})\left(\boldsymbol{Z}^\top_{j,h_2}\boldsymbol{M}_j\boldsymbol{Z}_{j,h_2}\right)^{-1}\left(\boldsymbol{Z}^\top_{j,h_2}\boldsymbol{M}_j\boldsymbol{Z}_{j,h_1}\right) - \left(\boldsymbol{Z}^\top_{j,h_1}\boldsymbol{M}_j\boldsymbol{Z}_{j,h_1}\right)\}\hat{\boldsymbol{\delta}}_{j,h_1}\\
&\quad + v^*(h_2),
\end{aligned}$$

where

$$\begin{aligned}
v^*(h_2) &= 2\hat{\boldsymbol{\delta}}^\top_{j,h_1}(\boldsymbol{Z}^\top_{j,h_1}\boldsymbol{M}_j\boldsymbol{Z}_{j,h_2})(\boldsymbol{Z}^\top_{j,h_2}\boldsymbol{M}_j\boldsymbol{Z}_{j,h_2})^{-1}\boldsymbol{Z}^\top_{j,h_2}\boldsymbol{M}_j\boldsymbol{\varepsilon}^* - 2\hat{\boldsymbol{\delta}}^\top_{j,h_1}\boldsymbol{Z}^\top_{j,h_1}\boldsymbol{M}_j\boldsymbol{\varepsilon}^*\\
&+ \boldsymbol{\varepsilon}^{*\top}\boldsymbol{M}_j\boldsymbol{Z}_{j,h_2}(\boldsymbol{Z}^\top_{j,h_2}\boldsymbol{M}_j\boldsymbol{Z}_{j,h_2})^{-1}\boldsymbol{Z}^\top_{j,h_2}\boldsymbol{M}_j\boldsymbol{\varepsilon}^* - \boldsymbol{\varepsilon}^{*\top}\boldsymbol{M}_j\boldsymbol{Z}_{j,h_1}(\boldsymbol{Z}^\top_{j,h_1}\boldsymbol{M}_j\boldsymbol{Z}_{j,h_1})^{-1}\boldsymbol{Z}^\top_{j,h_1}\boldsymbol{M}_j\boldsymbol{\varepsilon}^*.
\end{aligned}$$

Define for $h_2 \neq h_1$,

$$g_2^*(h_2) = \frac{\hat{\boldsymbol{\delta}}^\top_{j,h_1}\left|(\boldsymbol{Z}^\top_{j,h_1}\boldsymbol{M}_j\boldsymbol{Z}_{j,h_2})(\boldsymbol{Z}^\top_{j,h_2}\boldsymbol{M}_j\boldsymbol{Z}_{j,h_2})^{-1}(\boldsymbol{Z}^\top_{j,h_2}\boldsymbol{M}_j\boldsymbol{Z}_{j,h_1}) - \boldsymbol{Z}_{j,h_1}\boldsymbol{M}_j\boldsymbol{Z}_{j,h_1}\right|\hat{\boldsymbol{\delta}}_{j,h_1}}{|h_1 - h_2|}.$$

When $h_2 = h_1$, define $g_2^*(h_1) = \hat{\boldsymbol{\delta}}^\top_{j,h_1}\hat{\boldsymbol{\delta}}_{j,h_1}$. We have the following identities:

$$V^*(h_2) - V^*(h_1) = -|h_1 - h_2|g_2^*(h_2) + v^*(h_2) \tag{4.116}$$

Again because $V^*(h_2) \geq V^*(h_1)$ by definition, it suffices to show

$$P^*\left(\sup_{|h_2 - h_1| > C\|\hat{\boldsymbol{\delta}}_{j,h_1}\|^{-2}} V^*(h_2) \geq V^*(h_1)\right) < \epsilon.$$

By Lemmas 4.8 and 4.9, for any $\epsilon > 0$ and $\eta > 0$, we have $P(|h_2 - h_1| > N_j\eta) < \epsilon$ for large N_j. Thus, to prove the above equation, it is sufficient to show that

$$P^*\left(\sup_{h_2 \in K^*(C)} V(h_2) \geq V(h_1)\right) < \epsilon,$$

where $K^*(C) = \left\{h_2 : |h_2 - h_1| > C\|\hat{\boldsymbol{\delta}}_{j,h_1}\|^{-2} \textit{ and } n_j^{(l)} + N_j\eta \leq h_2 \leq n_j^{(r)} - \eta N_j\right\}$ for a small number $\eta > 0$. Finding the maximum value over the set $K^*(C)$ amounts to restricted searching, but this is legitimate only after establishing consistency. Thus, it suffices to show that

$$P^*\left(\sup_{h_2 \in K^*(C)} \frac{|v^*(h_2)|}{|h_1 - h_2|\|\hat{\boldsymbol{\delta}}_{j,h_1}\|^2} > \lambda\right) < \epsilon. \tag{4.117}$$

Next consider $v^*(h_2)$. Let

$$\boldsymbol{Z}_{\Delta_2} = \begin{cases} \boldsymbol{Z}_{j,h_2} - \boldsymbol{Z}_{j,h_1} = (\mathbf{0},\ldots,\mathbf{0},\boldsymbol{x}_{h_2+1},\ldots,\boldsymbol{x}_{h_1},\mathbf{0},\ldots,\mathbf{0})^\top, & h_2 < h_1 \\ \boldsymbol{Z}_{j,h_1} - \boldsymbol{Z}_{j,h_2} = (\mathbf{0},\ldots,\mathbf{0},\boldsymbol{x}_{h_1+1},\ldots,\boldsymbol{x}_{h_2},\mathbf{0},\ldots,\mathbf{0})^\top, & h_2 > h_1 \\ \mathbf{0}, & h_2 = h_1 \end{cases}.$$

It follows that $\boldsymbol{Z}_{j,h_2} = \boldsymbol{Z}_{j,h_1} + \boldsymbol{Z}_{\Delta_2}\,\mathrm{sgn}\,(h_1 - h_2)$. Then, we have

$$\begin{aligned} v^*(h_2) &= v_1^*(h_2) + v_2^*(h_2) + v_3^*(h_2) + v_4^*(h_2) + v_5^*(h_2) \\ &= \left[2\hat{\boldsymbol{\delta}}_{j,h_1}^\top \boldsymbol{Z}_{\Delta_2}^\top \boldsymbol{\varepsilon}^*\right] \mathrm{sgn}(h_1 - h_2) \\ &- \left[2\hat{\boldsymbol{\delta}}_{j,h_1}^\top \boldsymbol{Z}_{\Delta_2}^\top \boldsymbol{X}_j (\boldsymbol{X}_j^\top \boldsymbol{X}_j)^{-1} \boldsymbol{X}_j^\top \boldsymbol{\varepsilon}^*\right] \mathrm{sgn}(h_1 - h_2) \\ &- \left[2\hat{\boldsymbol{\delta}}_{j,h_1}^\top (\boldsymbol{Z}_{\Delta_2}^\top \boldsymbol{M}_j \boldsymbol{Z}_{j,h_2})(\boldsymbol{Z}_{j,h_2}^\top \boldsymbol{M}_j \boldsymbol{Z}_{j,h_2})^{-1} \boldsymbol{Z}_{j,h_2}^\top \boldsymbol{M}_j \boldsymbol{\varepsilon}^*\right] \mathrm{sgn}(h_1 - h_2) \\ &+ \boldsymbol{\varepsilon}^{*\top} \boldsymbol{M}_j \boldsymbol{Z}_{j,h_2} (\boldsymbol{Z}_{j,h_2}^\top \boldsymbol{M}_j \boldsymbol{Z}_{j,h_2})^{-1} \boldsymbol{Z}_{j,h_2}^\top \boldsymbol{M}_j \boldsymbol{\varepsilon}^* \\ &- \boldsymbol{\varepsilon}^{*\top} \boldsymbol{M}_j \boldsymbol{Z}_{j,h_1} (\boldsymbol{Z}_{j,h_1}^\top \boldsymbol{M}_j \boldsymbol{Z}_{j,h_1})^{-1} \boldsymbol{Z}_{j,h_1}^\top \boldsymbol{M}_j \boldsymbol{\varepsilon}^*. \end{aligned} \tag{4.118}$$

By Lemmas 4.8 and 4.9, we deduce that $\left(\boldsymbol{X}_j^\top \boldsymbol{X}_j\right)^{-1} \boldsymbol{X}_j^\top \boldsymbol{\varepsilon}^* = N_j^{-1/2} O_{p^*}(1)$. Since $N_j^{-1/2+\alpha} \boldsymbol{\delta}_j^{-1} \to 0$ and

$$N_j^{-1/2+\alpha} \boldsymbol{\delta}_j^{-1} - N_j^{-1/2+\alpha} \hat{\boldsymbol{\delta}}_{j,h_1}^{-1} = N_j^{-1/2+\alpha} \boldsymbol{\delta}_j^{-1} (\hat{\boldsymbol{\delta}}_{j,h_1} - \boldsymbol{\delta}_j) \hat{\boldsymbol{\delta}}_{j,h_1}^{-1} = O_{p^*}(N_j^{-1+\alpha} \boldsymbol{\delta}_j^{-2})$$

by corollary 1 in [16], $N_j^{-1/2+\alpha} \hat{\boldsymbol{\delta}}_{j,h_1}^{-1}$ tends to zero. It follows from $\boldsymbol{Z}_{\Delta_2}^\top \boldsymbol{X}_j = |h_1 - h_2| O_p(1)$ that

$$\begin{aligned} \sup_{h_2 \in K^*(C)} \frac{|v_2^*(h_2)|}{|h_1 - h_2| \|\hat{\boldsymbol{\delta}}_{j,h_1}\|^2} &\le 2 \|\hat{\boldsymbol{\delta}}_{j,h_1}\|^{-1} \|\boldsymbol{Z}_{\Delta_2}^\top \boldsymbol{X}_j / (h_1 - h_2)\| \|(\boldsymbol{X}_j^\top \boldsymbol{X}_j)^{-1} \boldsymbol{X}_j^\top \boldsymbol{\varepsilon}^*\| \\ &= O_{p^*}(\|\hat{\boldsymbol{\delta}}_{j,h_1}\|^{-1} N_j^{-1/2}) = o_{p^*}(1). \end{aligned}$$

Similarly,

$$\|(\boldsymbol{Z}_{\Delta_2}^\top \boldsymbol{M}_j \boldsymbol{Z}_{j,h_2})\| = \|h_1 - h_2\| O_{p^*}(1),$$

and

$$(\boldsymbol{Z}_{j,h_2}^\top \boldsymbol{M}_j \boldsymbol{Z}_{j,h_2})^{-1} \boldsymbol{Z}_{j,h_2}^\top \boldsymbol{M}_j \boldsymbol{\varepsilon}^* = N_j^{-1/2} O_{p^*}(1)$$

uniformly on $K^*(C)$. This implies

$$\begin{aligned} &\sup_{h_2 \in K^*(C)} \frac{v_3(h_2)}{|h_1 - h_2| \|\hat{\boldsymbol{\delta}}_{j,h_1}\|^2} \\ &\le \|\hat{\boldsymbol{\delta}}_{j,h_1}\|^{-1} \|(\boldsymbol{Z}_{\Delta_2}^\top \boldsymbol{M}_j \boldsymbol{Z}_{j,h_2}) / (h_1 - h_2)\| \|(\boldsymbol{Z}_{j,h_2}^\top \boldsymbol{M}_j \boldsymbol{Z}_{j,h_2})^{-1} \boldsymbol{Z}_{j,h_2}^\top \boldsymbol{M}_j \boldsymbol{\varepsilon}^*\| \\ &= O_{p^*}(\|\hat{\boldsymbol{\delta}}_{j,h_1}\|^{-1} N_j^{-1/2}) = o_{p^*}(1). \end{aligned}$$

Furthermore, it is easy to argue that expression

$$\sup_{h_2 \in K^*(C)} \frac{|v_4^*(h_2)|}{|h_1 - h_2| \|\hat{\boldsymbol{\delta}}_{j,h_1}\|^2}$$

and

$$\sup_{h_2 \in K^*(C)} \frac{|v_5^*(h_1)|}{|h_1 - h_2| \|\hat{\boldsymbol{\delta}}_{j,h_1}\|^2}$$

is $O_{p^*}(1)$ uniformly on $K^*(C)$. By Lemma 4.3,

$$P^*\left(\sup_{h_2\in K^*(C)} \frac{|v_1^*(h_2)|}{|h_1-h_2|\|\hat{\boldsymbol{\delta}}_{j,h_1}\|^2} > \frac{\lambda}{5}\right)$$
$$\leq P^*\left(\sup_{h_2\in K^*(C)} \|\hat{\boldsymbol{\delta}}_{j,h_1}\|^{-1}\|\boldsymbol{Z}_{\Delta_2}^\top\boldsymbol{\varepsilon}^*/(h_1-h_2)\| > \frac{\lambda}{10}\right) \leq \frac{100B}{\lambda^2 C} \leq \frac{\epsilon}{5}.$$

for large C. In summary, we have established the consistency of the bootstrap estimator. □

Subsequently, we establish the validity of the bootstrap CI in (4.107). To facilitate future reference, let us collect some notations here. Any symbol with a superscript $*$ denotes an object under the bootstrap probability measure instead of the original measure as used in some other section. For instance, $E^*(\cdot)$ represents the conditional expectation with respect to the bootstrap probability measure, conditioned on the original data. Similarly, $P^*(\cdot)$ represents the conditional probability under the bootstrap measure.

Theorem 4.4.3. *Under Assumptions 4.11–4.16,*

$$\sup_{x\in\mathcal{R}}\left|P^*\left(\hat{\zeta}^*_{j,h_2}-\hat{\zeta}_{j,h_1}\leq x\right)-P\left(\hat{\zeta}_{j,h_1}-a_j\leq x\right)\right| \to_p 0.$$

Proof of Theorem 4.4.3. To study the limiting distribution, we consider the expectation of $V^*(h_2)-V^*(h_1)$ on $D^*(C)=\{h_2:|h_2-h_1|\leq C\|\hat{\boldsymbol{\delta}}_{j,h_1}\|^{-2}\}$.

We use $\boldsymbol{Z}_{j,h_1}=\boldsymbol{Z}_{j,h_2}-\boldsymbol{Z}_{\Delta_2}\operatorname{sgn}(h_1-h_2)$ to obtain

$$\begin{aligned}|h_1-h_2|\,g_2^*(h_2) &= \hat{\boldsymbol{\delta}}_{j,h_1}^\top\{\boldsymbol{Z}_{j,h_1}\boldsymbol{M}_j\boldsymbol{Z}_{j,h_1}-\boldsymbol{Z}_{j,h_1}^\top\boldsymbol{M}_j\boldsymbol{Z}_{j,h_2}(\boldsymbol{Z}_{j,h_2}\boldsymbol{M}_j\boldsymbol{Z}_{j,h_2})^{-1}\boldsymbol{Z}_{j,h_2}^\top\boldsymbol{M}_j\boldsymbol{Z}_{j,h_1}\}\hat{\boldsymbol{\delta}}_{j,h_1}\\ &= \hat{\boldsymbol{\delta}}_{j,h_1}^\top\{\boldsymbol{Z}_{\Delta_2}^\top\boldsymbol{Z}_{\Delta_2}-\boldsymbol{Z}_{\Delta_2}^\top\boldsymbol{X}_j(\boldsymbol{X}_j^\top\boldsymbol{X}_j)^{-1}\boldsymbol{X}_j^\top\boldsymbol{Z}_{\Delta_2}\}\hat{\boldsymbol{\delta}}_{j,h_1}\\ &\quad-\hat{\boldsymbol{\delta}}_{j,h_1}^\top\boldsymbol{Z}_{\Delta_2}^\top\boldsymbol{M}_j\boldsymbol{Z}_{j,h_2}(\boldsymbol{Z}_{j,h_2}\boldsymbol{M}_j\boldsymbol{Z}_{j,h_2})^{-1}\boldsymbol{Z}_{j,h_2}^\top\boldsymbol{M}_j\boldsymbol{Z}_{\Delta_2}\hat{\boldsymbol{\delta}}_{j,h_1}\end{aligned} \tag{4.119}$$

Since $\|\boldsymbol{X}_j^\top\boldsymbol{Z}_{\Delta_2}\|=O_{p^*}(\|\hat{\boldsymbol{\delta}}_{j,h_1}\|^{-2})$ and $\|\boldsymbol{Z}_{j,h_2}^\top\boldsymbol{M}_j\boldsymbol{Z}_{\Delta_2}\|=O_{p^*}(\|\hat{\boldsymbol{\delta}}_{j,h_1}\|^{-2})$, we have

$$|h_1-h_2|g^*(h_2)=\hat{\boldsymbol{\delta}}_{j,h_1}^\top\boldsymbol{Z}_{\Delta_2}^\top\boldsymbol{Z}_{\Delta_2}\hat{\boldsymbol{\delta}}_{j,h_1}+o_{p^*}(1).$$

Next consider $v^*(h_2)$. Because $|h_1-h_2|\leq C\|\hat{\boldsymbol{\delta}}_{j,h_1}\|^{-2}$, $v_2^*(h_2)$ and $v_3^*(h_2)$ are bounded by $O_{p^*}(N_j^{-1/2}\|\hat{\boldsymbol{\delta}}_{j,h_1}\|^{-1})=o_{p^*}(1)$. $v_4^*(h_2)$ and $v_5^*(h_2)$ can be easily shown to be bounded by $o_{p^*}(1)$ uniformly on $D^*(C)$. Combining these results, we obtain

$$V^*(h_2)-V^*(h_1)=-\hat{\boldsymbol{\delta}}_{j,h_1}^\top\boldsymbol{Z}_{\Delta_2}^\top\boldsymbol{Z}_{\Delta_2}\hat{\boldsymbol{\delta}}_{j,h_1}+2\hat{\boldsymbol{\delta}}_{j,h_1}^\top\boldsymbol{Z}_{\Delta_2}^\top\boldsymbol{\varepsilon}^*\operatorname{sgn}(h_1-h_2)+o_{p^*}(1).$$

Combined with (4.112), the proof is completed. □

Combining Theorem 4.4.3 with Theorem 4.4.2 yields the validity of bootstrap for multiple change points.

Corollary 4.1. *Under Assumptions 4.11–4.16,*

$$\sup_{x\in\mathcal{R}}\left|P^*\left(\cap_{j=1}^s\left\{\hat{\zeta}^*_{j,h_2}-\hat{\zeta}_{j,h_1}\leq x\right\}\right)-P\left(\cap_{j=1}^s\left\{\hat{\zeta}_{j,h_1}-a_j\leq x\right\}\right)\right| \to_p 0.$$

The asymptotic validity of the proposed bootstrap CI in (4.107) follows from the fact that $P\left(a_j\in \mathrm{CI}_j^*(\alpha)\right)\to 1-\alpha$ for each $j=1,\ldots,s$, and $P\left(\cap_{j=1}^s\left\{a_j\in\mathrm{CI}_j^*(\alpha)\right\}\right)\to 1-\alpha$.

4.4.5 Simulation

In this subsection, we conduct two simulation studies to examine the finite sample performance of the two-stage procedure involving OGA and the sup-Wald-type test ($\text{TSP}_{\text{oga, wald}}$) under various parameter settings. We first illustrate the performance of change points detection in Section 4.4.2 and compare the results with the two-stage multiple change points detection procedure with LASSO ($\text{TSMCD}_{\text{lasso}}$) by [136]. We also construct confidence intervals using the bootstrap method proposed in Section 4.4.3. We denote the CIs in (4.107) by $\text{bootstrap}_{\text{oga,wald}}$. In light of [61], we can also construct bootstrap CIs for multiple change points without applying OGA on $\boldsymbol{y}_n^*$, which is a local bootstrap approximation. We denote the corresponding bootstrap CIs by bootstrap wald. For the sake of comparison, the corresponding theoretical confidence intervals of [16] are also reported.

4.4.5.1 Example 1

We consider the simulation settings with change points $a_1 = 400$ and $a_2 = 800$, and generate data from the model

$$\begin{aligned} y_t &= \cos(t\pi/61) + \sin(t\pi/61) + 0.1y_{t-1} \\ &+ (2\sin(t\pi/61) + 3\cos(t\pi/61) + 0.4y_{t-1})I_{(400,1200]}(t) \\ &+ (2\sin(t\pi/61) - 0.4y_{t-1})I_{(800,1200]}(t) + \varepsilon_t, \end{aligned} \tag{4.120}$$

where ε_t are generated from independent standard normal distributions. In this model, $\{y_t, t = 1, \ldots, 1200\}$ represents a periodic auto-correlated sequence with a period of 122 and an auto-correlation order of 1. According to [131], we set $c_n = 2\log(\log n)$ in (4.98).

We perform 1000 Monte Carlo simulations for multiple change points estimation and report the results in Table 4.4. The percentage of correct identification c_0 is used to record the proportion of obtaining the correct number of all change points among replications, while c_1 represents the frequency of obtaining the correct number of an individual change point. The means and standard errors of the estimated change points are computed across the replications. To examine the finite sample performance of the bootstrap method, Figure 4.6 displays the estimates of change points in model (4.120). We also report the bootstrap sample distributions in Figure 4.7. The corresponding bootstrap CIs and theoretical CIs with $B = 1000$ are given in Table 4.4. The $100(1-\alpha)\%$ theoretical CI is given by

$$[\hat{\zeta}_{j,k_1} - [c_{\alpha/2}/L_j] - 1, \hat{\zeta}_{j,k_1} + [c_{\alpha/2}/L_j] + 1],$$

where $c_{0.025} = 7$, $c_{0.05} = 11$ and $L_j = \boldsymbol{\delta}_j^\top \left(\sum_{i=n_j^{(l)}}^{n_j^{(r)}} \boldsymbol{x}_i \boldsymbol{x}_i^\top / N_j \right) \boldsymbol{\delta}_j / \sigma^2$.

Table 4.4: Methods for multiple change points detection

Method	c_0		a_1	a_2
$\text{TSP}_{\text{oga,wald}}$	0.974	c_1	1	0.999
		Mean	400	799
		SE	2.40	3.05
$\text{TSMCD}_{\text{lasso}}$	0.930	c_1	0.992	1
		Mean	399	798
		SE	6.04	3.84

From Table 4.4, we observe that $\text{TSP}_{\text{oga, wald}}$ outperforms $\text{TSMCD}_{\text{lasso}}$ in terms of the percentage of correct identification and estimation precision. In particular, $\text{TSMCD}_{\text{lasso}}$ sometimes misses the first change point a_1, while $\text{TSP}_{\text{oga, wald}}$ detects almost all change points successfully. It is clear from Figures 4.6–4.7 that the bootstrap approximation is clustered around the estimates 399 and 799, respectively. In the case of Example 1, it is preferred to choose a larger m, to obtain a larger

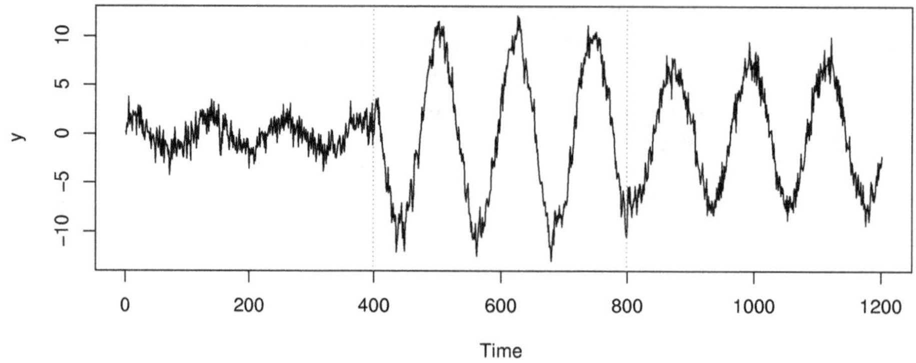

Figure 4.6: Change points estimates are displayed with dashed lines.

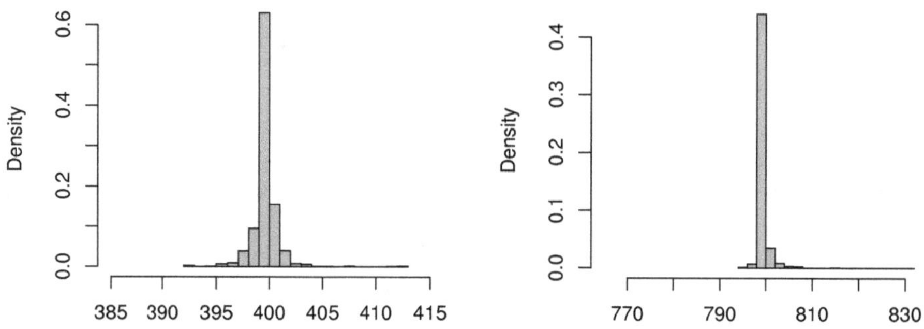

Figure 4.7: The bootstrap approximation of change points estimates

chunk interval in the two-stage procedures, leading to a more satisfactory estimation result. From Table 4.5, we note that bootstrap$_{\text{oga, wald}}$ and bootstrap$_{\text{wald}}$ have similar performance. Results in Table 4.5 demonstrate the validity of our bootstrap method since the bootstrap CIs are very close to the theoretical CIs.

Table 4.5: The bootstrap CIs for change points the estimates

$100(1-\alpha)\%$ CI	Method	$\hat{a}_1 = 399$	$\hat{a}_2 = 799$
95%	bootstrap$_{\text{oga,wald}}$	[397,401]	[798,803]
	bootstrap$_{\text{wald}}$	[398,402]	[799,803]
	Theoretical	[397,401]	[798,801]
90%	bootstrap$_{\text{oga,wald}}$	[397,401]	[799,801]
	bootstrap$_{\text{wald}}$	[398,402]	[799,801]
	Theoretical	[398,400]	[798,800]

4.4.5.2 Example 2

Consider the following linear model:

$$\begin{aligned} y_t &= \cos(t\pi/61) + \sin(t\pi/61) + 0.1y_{t-1} \\ &+ (\sin(t\pi/61) + 0.5\cos(t\pi/61) + 0.2y_{t-1})I_{(400,1200]}(t) \\ &+ (\sin(t\pi/61) + 0.2y_{t-1})I_{(800,1200]}(t) + \varepsilon_t. \end{aligned} \tag{4.121}$$

Table 4.6: Methods for multiple change points detection

Method	c_0		a_1	a_2
$\text{TSP}_{\text{oga,wald}}$	0.913	c_1	1	1
		Mean	402	797
		SE	10.74	7.23
$\text{TSMCD}_{\text{lasso}}$	0.839	c_1	0.958	1
		Mean	397	797
		SE	16.90	7.49

Table 4.7: The bootstrap CIs for change points the estimates

$100(1-\alpha)\%$ CI	Method	$\hat{a}_1 = 399$	$\hat{a}_2 = 801$
95%	$\text{bootstrap}_{\text{oga,wald}}$	[391,423]	[771,805]
	$\text{bootstrap}_{\text{wald}}$	[392,421]	[774,806]
	Theoretical	[390,408]	[795,807]
90%	$\text{bootstrap}_{\text{oga,wald}}$	[393,410]	[778,804]
	$\text{bootstrap}_{\text{wald}}$	[394,408]	[780,805]
	Theoretical	[393,405]	[797,805]

We basically adopt the same model setup as in the first study except for the parameter values. The results based on 1000 Monte Carlo simulations are summarized in Tables 4.6. Figures 4.6–4.7 show the change points estimates and the corresponding bootstrap approximations, respectively. Table 4.7 summarizes the performance measurements of the CIs.

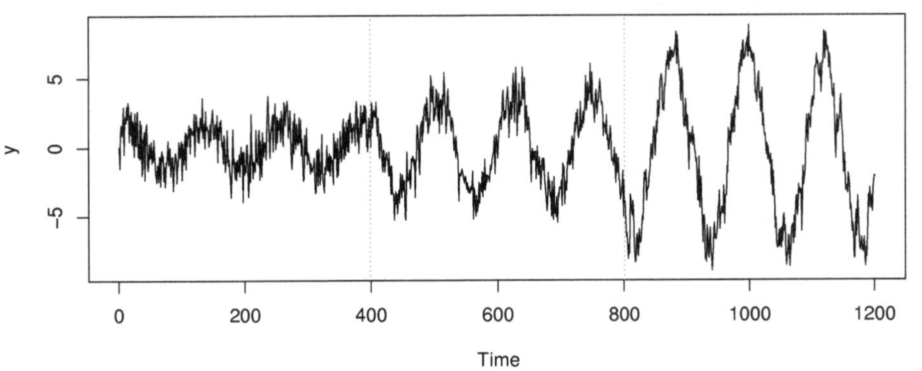

Figure 4.8: Change point estimates are displayed with dashed lines.

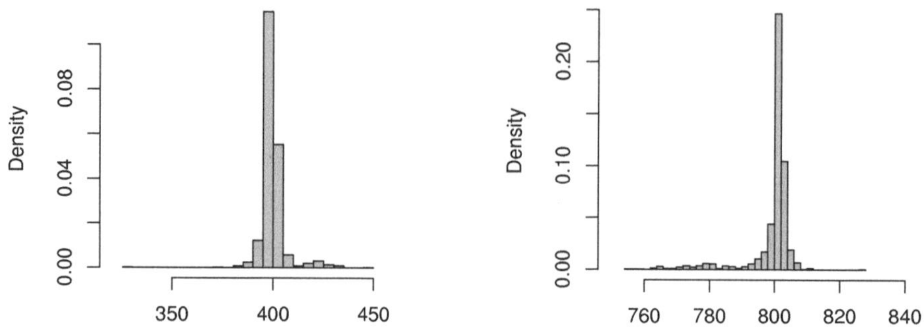

Figure 4.9: The bootstrap approximation of change points estimates

In view of Tables 4.4 and 4.6, the percentage of correct identification c_0 decreases and the SE of change point estimate increases with smaller changes. From Table 4.7, 90% bootstrap CIs are closer to the theoretical CIs than 95% bootstrap CI. Compared with Example 1, Example 2 prefers to choose a smaller one, so as to obtain finer interval chunks in the two-stage procedures, and obtain better estimation results. Compared with Table 4.5, it is obvious that smaller changes result in longer confidence intervals.

4.4.6 Earthquake data analysis

In this section, we illustrate the proposed method by applying it to the east-west component of seismograms recorded at the Iwanai station during the first foreshock of the 1982 Urakawa-Oki earthquake. This dataset was previously analyzed in [242].

Table 4.8: The bootstrap CIs for the estimated change points

Change point	95% bootstrap CIs	90% bootstrap CIs
3074	[3072, 3084]	[3073, 3080]
3914	[3877, 3952]	[3882, 3948]

The time series data is analyzed using an autoregressive (AR) model of order 5. The estimates and visualization of the 95% confidence level bootstrap CIs are presented in Table 4.8 and Figure 4.11, respectively.

It can be seen from Table 4.8 that two change points are detected at 3074 and 3914. In geography, these two change points represent the arrival times of P and S waves, which are two types of seismic waves. It is obvious from Figure 4.10 that the CI of the first change point is much narrower than that of the second change point.

This real example demonstrates the applicability and effectiveness of the proposed methodology in detecting change points in seismic data. By accurately identifying the locations of structural faults, we can gain valuable insights into the underlying geological processes and improve our understanding of seismic events.

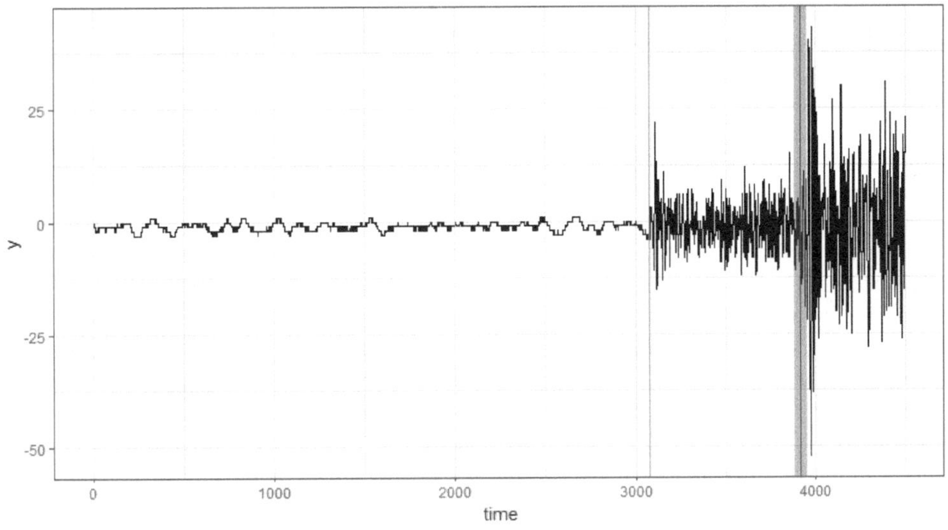

Figure 4.10: The change points estimated by our method (vertical lines), shaded areas representing the 95% confidence intervals around the change points

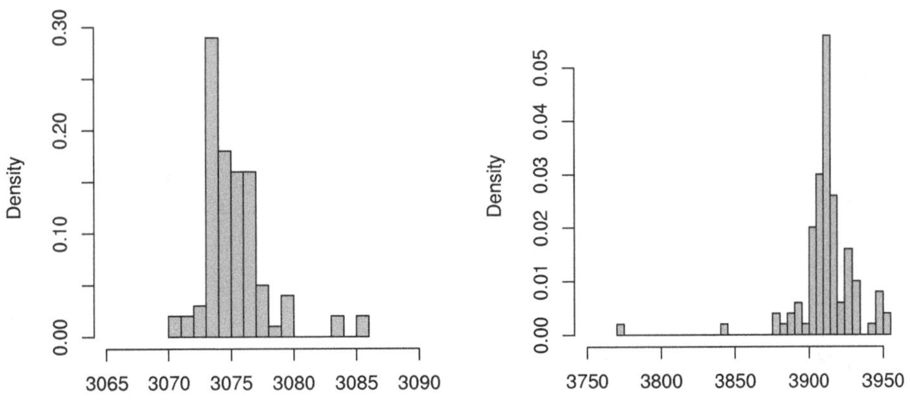

Figure 4.11: Bootstrap approximation for change points estimates

5

Regression models with change points

The application of change point detection methods in regression analysis is an important area of study. The class of threshold models is particularly valuable as it allows for the incorporation of multiple structural breaks in regression coefficients. Unlike constant coefficient models, threshold models offer greater flexibility and are thus applicable in a wide range of scenarios.

By identifying structural breaks in regression coefficients, threshold models can capture changes in relationships between variables over time or across different conditions. This flexibility makes them well-suited for modeling complex phenomena where relationships may evolve or shift abruptly.

Threshold models have been widely applied across various fields, including economics, finance, environmental science, and epidemiology. In economics and finance, they are used to analyze financial time series with regime shifts, capturing abrupt changes in market conditions or the impact of monetary policies. In environmental science, threshold models help investigate the effects of climate variables on environmental phenomena, while in epidemiology, they are employed to study how environmental or policy changes influence health outcomes. These models allow researchers to better understand complex systems where relationships between variables may experience a sudden change, usually triggered by external factors.

Permitting regression coefficients to vary has been explored extensively by researchers, leading to the development of various models. One prominent example is the varying coefficient model, in which regression coefficients change smoothly as a function of a continuous variable. In contrast, the threshold model maintains constant coefficients within specific regions but allows for abrupt non-smooth jumps when an underlying variable crosses a threshold. Due to this characteristic, the threshold model is also known as a piecewise linear model.

While both varying coefficient models and threshold models offer flexibility in capturing non-constant relationships between variables, they differ in their approach to modeling coefficient changes. Varying coefficient models assume that coefficients change continuously and smoothly, potentially reflecting gradual shifts in the relationship between variables. On the other hand, threshold models capture abrupt changes in coefficients at specific points, corresponding to distinct regimes or conditions. For example, in precision medicine, where the goal is to identify subgroups of patients that exhibit enhanced treatment effects in response to a drug, threshold models are often more appropriate. This is because they allow for the identification of sharp changes in treatment effects, which can pinpoint the thresholds at which certain subgroups benefit more significantly, providing clearer insights for personalized treatment strategies.

The distinction between these two classes of models is essential as it influences the interpretation of coefficient changes and the underlying dynamics of the relationships being modeled. Depending on the nature of the data and the research question at hand, researchers may choose between varying coefficient models and threshold models to effectively capture the complexities of the data-generating process.

Overall, the class of threshold models provides researchers with a powerful tool for capturing nonlinear and dynamic relationships in regression analysis, thereby enabling more accurate and insightful modeling of real-world phenomena.

DOI: 10.1201/9781032657073-5

5.1 Introduction

The threshold model, also referred to as segment regression or break-point regression, is widely utilized in fields such as economics and medicine. In many data analysis scenarios, the assumption of linearity in the regression effect may not be valid across the entire range of a covariate. To address this, a two-phase transformation model with unknown change points at of a covariate can significantly enhance the model fit, offering greater flexibility and accuracy in capturing the true relationship between the variables.

This approach enables the segmentation of data into distinct regions, each with its own unique regression relationship. By identifying the threshold point where the regression relationship shifts, researchers can effectively account for nonlinearity and heterogeneity within the data. This segmentation not only enhances model flexibility but also improves the accuracy of capturing complex patterns that may vary across different ranges of the covariate.

Applications of this approach are widespread. In economics, threshold models are used to analyze phenomena such as threshold effects in income levels on consumer behavior or investment decisions, where distinct economic regimes may emerge at certain thresholds. In medicine, these models are applied to study dose-response relationships, such as the effect of a treatment dosage on patient outcomes, or the impact of environmental factors on health, where the relationship may exhibit nonlinearity or abrupt changes at specific threshold values. This makes threshold models an invaluable tool for capturing complex dynamics that traditional linear models may fail to detect.

Overall, the use of such two-phase transformation models with threshold points offers a powerful tool for improving model fit and capturing complex relationships in data analysis across various fields. These models allow for the segmentation of data, effectively addressing nonlinearity and heterogeneity. Furthermore, theoretical developments, such as those ensuring convergence rates for split-based regression approaches, have been established to provide a solid foundation for these methods [19].

5.2 Threshold model

5.2.1 Multi-threshold model: two-stage estimation

Let T_i be the response variable and $\mathbf{X}_i$ be the p-dimensional regressor for the ith subject in the sample, $i = 1, \ldots, n$. We also observe a continuous thresholding variable Z_i. Assume $(T_i, Z_i, \mathbf{X}_i)$, $i = 1, \ldots, n$, satisfy the following regression model with s thresholds located at $a_1 < \ldots < a_s$:

$$
\begin{aligned}
T_i &= \sum_{j=1}^{s+1} \mathbf{X}_i^\top \beta_j^* 1_{\{a_{j-1} < Z_i \le a_j\}} + \varepsilon_i \\
&= \mathbf{X}_i^\top \left[\beta_1^* + \sum_{\ell=1}^{s} \mathbf{d}_\ell^* 1_{\{a_\ell < Z_i \leqslant a_{s+1}\}} \right] + \varepsilon_i, \\
&= \begin{cases} \mathbf{X}_i^\top \beta_1^* + \varepsilon_i & \text{if } a_0 < Z_i \le a_1, \\ \mathbf{X}_i^\top (\beta_1^* + \mathbf{d}_1^*) + \varepsilon_i & \text{if } a_1 < Z_i \le a_2, \\ \cdots & \cdots \\ \mathbf{X}_i^\top (\beta_1^* + \sum_{\ell=1}^{s} \mathbf{d}_\ell^*) + \varepsilon_i, & \text{if } a_s < Z_i \le a_{s+1}, \end{cases}
\end{aligned}
\tag{5.1}
$$

where $\beta_1^*, \ldots, \beta_{s+1}^*$ are unknown p-dimensional regression coefficients for $s+1$ subgroups, $\mathbf{d}_\ell^* = \beta_{\ell+1}^* - \beta_\ell^*, \ell = 1, \ldots, s$, are the increments of coefficients between two adjacent subgroups, $s \geq 0$ is the unknown number of thresholds, $a_1, \ldots, a_s$ are unknown threshold locations, $\varepsilon_1, \ldots, \varepsilon_n$ are independent random errors, and $\varepsilon_{a_{\ell-1}+1}, \ldots, \varepsilon_{a_\ell}$, the subgroup in the series $\{\varepsilon_i, i = 1, \ldots, n\}$ separated by the threshold locations $\{a_\ell, \ell = 1, \ldots, s\}$, are identically distributed with mean zero and variance $\sigma_\ell^2, \ell = 1, \ldots, s+1$.

Notice that if $s = 0$, the model (5.1) does not involve threshold. If $s \geq 1$, we denote $P(Z_i \leq a_j) = \tau_j, j = 1, \ldots, s$ where $0 < \tau_1 < \tau_2 < \ldots < \tau_s < 1$. If $Z_i = i$, we retain the usual change point model, and the condition $P(Z_i \leq a_j) = \tau_j$ will be replaced by $a_j/n \to \tau_j$ since change point locations actually depend on the sample size in this setting.

If s and all the change point locations are given, then model (5.1) is usually referred to as a piece-wise linear regression model. Fitting such a model can be painlessly carried out with a least squares approach. However, when s and all the change point locations are unknown, the model estimation is not straightforward. This problem is sometimes called the estimation of change-in-slope, as we exemplified in chapter 1.

This problem is particularly challenging for several reasons. First, a straightforward method for detecting changes in slope involves taking the first differences of the data, which transforms a change-in-slope into a change-in-mean. One could then apply standard methods for detecting changes in the mean. However, this approach loses a significant amount of information about the precise location of the changes and often performs poorly. For example, when comparing the raw data in the top-left plot of Figure 1, the approximate location of changes in slope is visually apparent, but detecting changes in mean from the differenced data becomes almost impossible. Running the PELT change-in-mean algorithm [140] on the first differenced data leads to poor estimates of the change locations.

Second, a common method for detecting multiple change points is binary segmentation [221] or its variants [102, 149], which repeatedly tests for a single change-in-slope. However, [20] shows that binary segmentation methods perform poorly for change-in-slope problems. Fitting a single change-in-slope model to data with multiple changes often results in detecting changes at locations that are near the middle of segments between actual changes.

Finally, dynamic programming algorithms that minimize an l_0-penalized cost, such as Optimal Partitioning [132] or PELT [140], cannot be applied to the change-in-slope problem. This is because of dependencies across change points due to the requirement for continuity of the mean at each change.

Despite these challenges, three classes of methods have been specifically developed for detecting changes in slope. The first is Trend Filtering [143, 246], which minimizes the residual sum of squares for the fit, combined with an l_1 penalty on changes in slope. The second method is NOT [20], which applies tests for single changes in slope on various subsets of the data and combines the results using the narrowest-over-threshold procedure. Lastly, CPOP [97] uses a novel variant of dynamic programming to minimize the residual sum of squares along with an l_0 penalty, which applies a constant penalty for adding each change.

The distinction between the l_1 penalty used in Trend Filtering and the l_0 penalty of CPOP is significant in practice. Trend Filtering's l_1 penalty allows the model to fit a single change in slope with multiple changes of the same sign, which can result in overfitting the number of changes or over-smoothing the mean function if the penalty is sufficiently large. This trade-off is illustrated in the bottom row of plots in Figure 1. CPOP, on the other hand, fits all changes simultaneously, providing a different approach from the localized nature of NOT. An empirical comparison of these methods can be found in [97].

The CPOP method is implemented efficiently in the `R` package `CPOP`. In addition to these aforementioned approaches, we focus on the two-stage estimation method introduced in [161], which we will discuss further in the following presentation.

When the response variable T is the logarithm of a survival time subject to right censoring, as in a time-to-event study, the model (5.1) is usually referred to as a multi-threshold accelerated failure time (AFT) model in the literature of survival analysis. In the following we will present the estimation of model (5.1) for this slightly more difficult situation. The estimation procedure for non-censored response can be easily modified. Let $C_i, i = 1, \dots, n$ be logarithm of the censoring time which are assumed to be independent and identically distributed (i.i.d.). In practice we only observe $(Y_i, \delta_i, \mathbf{X}_i), i = 1, \dots, n$, where $Y_i = \min(T_i, C_i)$ is the censored logarithm of the failure time and $\delta_i = 1_{\{T_i \le C_i\}}$ is the censoring indicator.

AFT model itself is a class of statistical models with wide applications in statistical research. Sometimes in order to achieve a sensible estimation researchers would assume a known parametric distribution for ε_i, such as the logistic distribution, Gumbel distribution or normal distribution. These distribution choices lend support to various parametric distributions for survival time. The estimation and inference can then follow from the maximum likelihood. However, such a parametric estimation approach may suffer mis-specification error when the true distribution differs from the assumed forms. As an alternative, researchers may opt for semiparametric estimation approaches to the traditional AFT model without thresholds. These approaches offer more flexibility and robustness against misspecification. Examples of semiparametric estimation methods include rank-based estimation, Buckley-James estimation, and Stute estimators[164]. The method introduced in this chapter is rooted in the classical Stute estimator [238].

Let $Y_{(1,\mathcal{I})} \le \dots \le Y_{(b,\mathcal{I})}$ be the ordered failure time responses in an index set $\mathcal{I}$ and $b = \sharp\mathcal{I}$. Let $\delta_{[\ell,\mathcal{I}]}$ be the concomitant of the ℓth-ordered $Y_{(\ell,\mathcal{I})}, \ell = 1, \dots, b$. Define the Kaplan-Meier weights as

$$w_{[1,\mathcal{I}]} = \frac{\delta_{[1,\mathcal{I}]}}{b}, w_{[\ell,\mathcal{I}]} = \frac{\delta_{[\ell,\mathcal{I}]}}{b-\ell+1} \prod_{k=1}^{\ell-1} (\frac{b-k}{b-k+1})^{\delta_{[k,\mathcal{I}]}}, \qquad \ell = 2, \dots, b.$$

Such weights are constructive elements of the Kaplan-Meier estimator for the survival function. Let $r_{Y_i,\mathcal{I}} \in [1, b]$ be the rank of Y_i among $\{Y_i : i \in \mathcal{I}\}$ i.e. $Y_i = Y_{(r_{Y_i,\mathcal{I}},\mathcal{I})}, i \in \mathcal{I}$. For simplicity, we denote $Y_{(r_{Y_i,\mathcal{I}})} = Y_{(r_{Y_i,\mathcal{I}},\mathcal{I})} = Y_i$ and $w_{[r_{Y_i,\mathcal{I}}]} = w_{[r_{Y_i,\mathcal{I}},\mathcal{I}]}$.

If $a_j, j = 1, \dots, s$ are known, define $\mathcal{I}_j^* = \{i : a_{j-1} < Z_i \le a_j\}$ and $b_j^* = \sharp\mathcal{I}_j^*$, we can use the Stute estimator [238] to fit the AFT model. The resulting estimator $\hat{\beta}_1^*, \hat{\mathbf{d}}_1^*, \dots, \hat{\mathbf{d}}_s^*$ may be represented by a weighted least squares estimator that minimizes

$$\sum_{j=1}^{s+1} \frac{b_j^*}{2n} \sum_{i \in \mathcal{I}_j^*} w_{[r_{Y_i,\mathcal{I}_j^*}]} \left(Y_i - \mathbf{X}_i^\top (\beta_1^* + \sum_{k=1}^{j-1} \mathbf{d}_k^*) \right)^2. \tag{5.2}$$

We note that the weights $\{w_{[r_{Y_i,\mathcal{I}_j^*}]}\}$ depend on the order of the random failure times and hence the related asymptotic argument is not as standard as the independent and identically distributed (i.i.d.) case in linear regression.

We intend to estimate s, a_1, ..., and a_s and then use the detected change point(s) to obtain the Stute estimator for regression coefficients. To this end, first we need to decide the total number of thresholds and the approximate distances between pairs of thresholds. After that we may proceed to determine the exact locations of thresholds. We thus adopt the two-stage procedure introduced in the following sections.

5.2.2 The splitting stage

First we split the data sequence into $q_n + 1$ segments where q_n tends to infinity as $n \to \infty$. Let $\{Z_i : \delta_i = 1, i = 1, \dots, n\} = \{\tilde{Z}_1, \dots, \tilde{Z}_{n^*}\}$ where $n^* = \sum_{i=1}^n \delta_i$ is the total number of events. In order to allow each segment to include enough failure event observations, the data sequence is split such that the first segment $\mathcal{I}_1 = \{i : Z_i \le \tilde{Z}_{(n^*-q_n m)}\}$ involves $n^* - q_n m$ events, and each of the other q_n segments $\mathcal{I}_j = \{i : \tilde{Z}_{(n^*-(q_n-j+2)m)} < Z_i \le \tilde{Z}_{(n^*-(q_n-j+1)m)}\}, j = 2, \ldots, q_n + 1$ involve m events where $m = \lceil n^*/q_n \rceil$ and $\tilde{Z}_{(1)} \le \tilde{Z}_{(2)} \cdots \le \tilde{Z}_{(n^*)}$ are the ordered threshold variables associated with subjects no failed. Define $b_j = \sharp\mathcal{I}_j, j = 1, \ldots, q_n + 1$.

To select these intervals covering thresholds $\{a_j\}$, we propose a concave 2-norm group selection method, such that an estimate $\hat{\theta} = ((\hat{\beta}_1)^\top, (\hat{\mathbf{d}}_1)^\top, \ldots, (\hat{\mathbf{d}}_{q_n})^\top)^\top$ is given by minimizing

$$\sum_{j=1}^{q_n+1} \frac{b_j}{2n} \sum_{i \in \mathcal{I}_j} w_{[r_{Y_i}, \mathcal{I}_j]} \left(Y_i - \mathbf{X}_i^\top (\beta_1 + \sum_{k=1}^{j-1} \mathbf{d}_k) \right)^2 + \sum_{k=1}^{q_n} p_{\lambda_n, \gamma_n}(\|\mathbf{d}_k\|), \tag{5.3}$$

where $\lambda_n > 0$, $\gamma_n > 1$ are tuning parameters, and the penalty function $p_{\lambda_n,\gamma_n}(|u|) > 0$ is concave in $|u|$. For the simplicity of presentation, we shall only consider two well-studied penalty functions in this book, namely the smoothly clipped absolute deviation (SCAD) penalty and the minimax concave penalty (MCP). SCAD was introduced in Fan and Li (2001) [91] and defined as $p_{\lambda,\gamma}(u) = \lambda u I_{[0,\, \lambda]}(u) + \frac{\gamma\lambda u - 0.5(u^2+\lambda^2)}{\gamma-1} I_{(\lambda,\, \gamma\lambda]}(u) + \frac{\lambda^2(\gamma^2-1)}{2(\gamma-1)} I_{(\gamma\lambda,\, \infty)}(u), \gamma > 1$. MCP was introduced in Zhang (2010) [281] and defined as $p_{\lambda,\gamma}(u) = (\lambda u - \frac{u^2}{2\gamma}) I_{[0,\, \gamma\lambda]}(u) + \frac{1}{2}\gamma\lambda^2 I_{(\gamma\lambda,\, \infty)}(u), \gamma > 2$.

Let $\mathbf{Y}_{(j)} = (Y_i, i \in \mathcal{I}_j)^\top$, $\mathbf{X}_{(j)} = (\mathbf{X}_i, i \in \mathcal{I}_j)^\top$, $\tilde{\mathbf{w}}_{(j)} = b_j(w_{[r_{Y_i}, \mathcal{I}_j]}, i \in \mathcal{I}_j)^\top$, $\tilde{\mathbf{y}}_{(j)} = diag(\tilde{\mathbf{w}}_{(j)})^{1/2}\mathbf{Y}_{(j)}$, $\tilde{\mathbf{X}}_{(j)} = diag(\tilde{\mathbf{w}}_{(j)})^{1/2}\mathbf{X}_{(j)}$, $j = 1, \ldots, q_n + 1$. Denote $\tilde{\mathbf{y}} = (\tilde{\mathbf{y}}_{(1)}^\top, \ldots, \tilde{\mathbf{y}}_{(q_n+1)}^\top)^\top$, $\tilde{\mathbf{X}} = (\tilde{\mathbf{X}}^{(1)}, \ldots, \tilde{\mathbf{X}}^{(q_n+1)})$ where $\tilde{\mathbf{X}}^{(1)} = (\tilde{\mathbf{X}}_{(1)}^\top, \ldots, \tilde{\mathbf{X}}_{(q_n+1)}^\top)^\top$ and $\tilde{\mathbf{X}}^{(j)} = (\mathbf{0}_{p \times n_{m,j-1}}, \tilde{\mathbf{X}}_{(j)}^\top, \ldots, \tilde{\mathbf{X}}_{(q_n+1)}^\top)^\top$, $j = 2. \ldots, q_n + 1$. The estimator $\hat{\theta}$ in (5.3) can be written as

$$\hat{\theta} = \arg\min_{\theta} \left\{ \frac{1}{2n} \|\tilde{\mathbf{y}} - \tilde{\mathbf{X}}\theta\|^2 + \sum_{j=1}^{q_n} p_{\lambda_n,\gamma_n}(\|\mathbf{d}_j\|) \right\}, \tag{5.4}$$

We apply the group coordinate descent (GCD) algorithm to estimate $\hat{\theta}_n$ from (5.4).

For simplicity, we write $\hat{\theta} = ((\hat{\theta}_1)^\top, \ldots, (\hat{\theta}_{q_n+1})^\top)^\top$ such that $\hat{\theta}_1 = \hat{\beta}_1$ and $\hat{\theta}_j = \hat{\mathbf{d}}_{j-1}$, $j = 2, \ldots, q_n + 1$. Let $\hat{\mathcal{A}} = \{j : \hat{\theta}_j \ne 0, j = 1, \ldots, q_n + 1\}$, and $\hat{\mathcal{A}}^*$ be a subset of $\hat{\mathcal{A}}$ such that $\hat{\mathcal{A}}^* = \{j : j \in \hat{\mathcal{A}}, j - 1 \notin \hat{\mathcal{A}}, j = 2, \ldots, q_n + 1\} = \{\hat{k}_1, \ldots, \hat{k}_{\hat{s}}\}$. If $\hat{s} = 0$, we declare there is no threshold. If $\hat{s} > 0$, by Theorem 5.2.2, the true threshold a_j is highly likely to be located in $(\tilde{Z}_{(n^*-(q_n-\hat{k}_j+3)m)}, \tilde{Z}_{(n^*-(q_n-\hat{k}_j+1)m)}]$, $j = 1, \ldots, \hat{s}$.

5.2.3 The refining stage

If $\hat{s} > 0$, by Theorem 5.2.2, we can estimate the threshold a_j in $(\tilde{Z}_{(n^*-(q_n-\hat{k}_j+3)m)}, \tilde{Z}_{(n^*-(q_n-\hat{k}_j+1)m)}]$. To obtain the estimates of the thresholds $\{a_j\}$, we denote $\hat{\mathcal{I}}_j = \{i : \tilde{Z}_{(n^*-(q_n-\hat{k}_j+3)m)} < Z_i \le \tilde{Z}_{(n^*-(q_n-\hat{k}_j+1)m)}\}$, $\hat{\mathcal{I}}_{j,\varsigma-} = \{i : \tilde{Z}_{(n^*-(q_n-\hat{k}_j+3)m)} < Z_i \le \varsigma\}$, $\hat{\mathcal{I}}_{j,\varsigma+} = \{i : \varsigma < Z_i \le \tilde{Z}_{(n^*-(q_n-\hat{k}_j+1)m)}\}$

$$Q_j(\varsigma^-, \beta) = \frac{\sharp\hat{\mathcal{I}}_{j,\varsigma-}}{\sharp\hat{\mathcal{I}}_j} \sum_{i \in \hat{\mathcal{I}}_{j,\varsigma-}} w_{[r_{Y_i}, \hat{\mathcal{I}}_{j,\varsigma-}]}(Y_i - \mathbf{X}_i^\top \beta)^2,$$

$$Q_j(\varsigma^+, \beta) = \frac{\sharp\hat{\mathcal{I}}_{j,\varsigma+}}{\sharp\hat{\mathcal{I}}_j} \sum_{i \in \hat{\mathcal{I}}_{j,\varsigma+}} w_{[r_{Y_i}, \hat{\mathcal{I}}_{j,\varsigma+}]}(Y_i - \mathbf{X}_i^\top \beta)^2,$$

and use the following method to estimate a_j:

$$\hat{a}_j = \text{argmin}_{\varsigma \in (\tilde{Z}_{(n^*-(q_n-\hat{k}_j+3)m)}, \tilde{Z}_{(n^*-(q_n-\hat{k}_j+1)m)}] \cap \{Z_{(1)}, \ldots, Z_{(n)}\}} \{Q_j(\varsigma)\}, \tag{5.5}$$

where $Q_j(\varsigma) = \min_\beta Q_j(\varsigma^-, \beta) + \min_\beta Q_j(\varsigma^+, \beta)$, $j = 1, \ldots, \hat{s}$, and $Z_{(1)} \le Z_{(2)} \cdots \le Z_{(n)}$ are the order of $Z_i, i = 1, \ldots, n$. The regions separated by the thresholds achieve the overall minimum least squares errors. The consistency of $\hat{a}_j$ will be provided in Theorem 6.3.3.

Remark 5.1. *After we obtain $\hat{a}_j, j = 1, \ldots, \hat{s}$ by (5.5), it is sensible to use the weighted least squares to obtain a final estimate of the coefficients in model (5.1). Let $\hat{\mathcal{I}}_j^* = \{i : \hat{a}_{j-1} < Z_i \le \hat{a}_j\}$ and $\hat{b}_j^* = \sharp\hat{\mathcal{I}}_j^*$, $j = 1, \ldots, \hat{s}+1$ where $\hat{a}_0 = -\infty$ and $\hat{a}_{\hat{s}+1} = +\infty$. Then the coefficient $\theta^* = ((\beta_1^*)^\top, (\mathbf{d}_1^*)^\top, \ldots, (\mathbf{d}_{\hat{s}}^*)^\top)^\top = (\theta_1^*, \ldots, \theta_{p(\hat{s}+1)}^*)^\top$ can be estimated by minimizing the following penalized least squares*

$$M(\theta) = \sum_{j=1}^{\hat{s}_n+1} \frac{\hat{b}_j^*}{2n} \sum_{i \in \hat{\mathcal{I}}_j^*} w_{[r_{Y_i}, \hat{\mathcal{I}}_j^*]} \left(Y_i - \mathbf{X}_i^\top (\beta_1 + \sum_{k=1}^{j-1} \mathbf{d}_k) \right)^2 + \sum_{i=1}^{p(\hat{s}_n+1)} p_{\lambda_n, \gamma_n}(|\theta_i|), \tag{5.6}$$

where the penalty function $p_{\lambda_n,\gamma_n}(|u|) > 0$ is the same as in (5.4) and $\theta = ((\beta_1)^\top, (\mathbf{d}_1)^\top, \ldots, (\mathbf{d}_{\hat{s}})^\top)^\top = (\theta_1, \ldots, \theta_{p(\hat{s}+1)})^\top$. The inclusion of the penalty functions may lead to a sparse solution since the number of parameters could be quite large for all the subgroups. The oracle property of $\hat{\theta}^* = \arg\min_\theta M(\theta)$ will be given below.

Similar to other related works [135], γ_n is usually set as 2.4 for SCAD and MCP penalties. The regularization parameter λ_n can be chosen by the Bayesian information criterion (BIC).

The performance of $\hat{\theta}^*$ is further dependent on the segment length m. The selection of an optimal m can be carried out as follows: Apply the splitting stage to the data sequence L times with the common segment length of events (excluding the first segment) m_ℓ, $\ell = 1, \ldots, L$. In order to meet the assumption, i.e., $m = \lfloor c\sqrt{n} \rfloor$ made in Theorem 5.2.2, we set $m_\ell = \lfloor \kappa_\ell \sqrt{n} \rfloor$, $\ell = 1, \ldots, L$, where κ_ℓ takes values from L grid-points in an interval, say [0.1, 2.0]. For each m_ℓ, applying the proposed two-stage procedure, we obtain the set of estimated thresholds $\hat{\mathcal{M}}_\ell = \{\hat{a}_{1,\ell}, \ldots, \hat{a}_{\hat{s}_\ell,\ell}\}$, $\ell = 1, \ldots, L$. We use the BIC to choose the best index

$$\hat{\ell} = \text{argmin}_{\ell=1,\ldots,L} \left\{ BIC_{\hat{\mathcal{M}}_\ell} \right\}, \tag{5.7}$$

where

$$BIC_{\hat{\mathcal{M}}_\ell} = n \log \left(\sum_{j=1}^{\hat{s}_\ell+1} \frac{\sharp\hat{\mathcal{I}}_{j,\ell}}{n} \left(\min_\beta \sum_{i \in \hat{\mathcal{I}}_{j,\ell}} w_{[r_{Y_i}, \hat{\mathcal{I}}_{j,\ell}]} (Y_i - \mathbf{X}_i^\top \beta)^2 \right) \right) + q(\hat{s}_\ell + 1) \log(n) \tag{5.8}$$

with $\hat{\mathcal{I}}_{j,\ell} = \{i : \hat{a}_{j-1,\ell} < Z_i \le \hat{a}_{j,\ell}\}$, $\hat{a}_{0,\ell} = -\infty$, $\hat{a}_{\hat{s}_\ell+1,\ell} = +\infty$. The optimal $m_{opt} = \lfloor \kappa_{\hat{\ell}} \sqrt{n} \rfloor$ and the optimal estimated thresholds $\hat{\mathcal{M}}_{opt} = \{\hat{a}_{1,\hat{\ell}}, \ldots, \hat{a}_{\hat{s}_{\hat{\ell}},\hat{\ell}}\}$.

We refer to the above two-stage procedure as *T*wo *S*tage *M*ultiple *C*hange-points *D*etection (TSMCD). Since two regularization methods, i.e. MCP, and SCAD, will be utilized in the splitting stage, we refer to the corresponding TSMCD as TSMCD(MCP) and TSMCD(SCAD), respectively. The detailed algorithm of TSMCD is described in the following.

5.2.4 Asymptotic theory

Under model (5.1), we further denote the distributions of $\{T_i, i \in \mathcal{I}_j^*\}$ and $\{Y_i, i \in \mathcal{I}_j^*\}$ to be F_j and H_j, $j = 1, \ldots, s+1$. Denote the distribution of log censoring times $\{c_1, \ldots, c_n\}$ by G. We define U_{F_j}, U_{H_j} and U_G to be the least upper bound for the support of F_j, H_j and G, respectively.

Algorithm 9 TSMCD

1: **for** $\ell = 1, 2, \ldots, 20$ **do**
    Step 1: Splitting stage
2:     Set $m = \lfloor 0.1\ell\sqrt{n^*} \rfloor$ and $q_n = \lfloor n^*/m \rfloor - 1$ where n^* is the number of events;
3:     Split the data sequence into $q_n + 1$ segments $\mathcal{I}_j, j = 1, \ldots, q_n + 1$ as Section 2.1;
4:     Estimate $\hat{\theta} = ((\hat{\theta}_1)^\top, \ldots, (\hat{\theta}_{q_n+1})^\top)^\top$ by minimizing (5.3) or (5.4);
5:     Compute the index sets $\hat{\mathcal{A}} = \{j : \hat{\theta}_j \neq 0, j = 1, \ldots, q_n + 1\}$ and $\hat{\mathcal{A}}^* = \{j : j \in \hat{\mathcal{A}}, j - 1 \notin \hat{\mathcal{A}}, j = 2, \ldots, q_n + 1\} \equiv \{\hat{k}_1, \ldots, \hat{k}_{\hat{s}}\}$ where $\hat{k}_1 < \hat{k}_2 < \ldots < \hat{k}_{\hat{s}}$ and $\hat{s} = \sharp\hat{\mathcal{A}}^*$;
    Step 2: Refining stage
6:     **if** $\hat{s} = 0$ **then** go to the step 9;
7:     **else if** $\hat{s} > 0$ **then** estimate the threshold a_j in $(\tilde{Z}_{(n^* - (q_n - \hat{k}_j + 3)m)}, \tilde{Z}_{(n^* - (q_n - \hat{k}_j + 1)m)}]$ by (5.5);
8:     **end if**
9:     Estimate the coefficient $\theta^* = ((\beta_1^*)^\top, (\mathbf{d}_1^*)^\top, \ldots, (\mathbf{d}_{\hat{s}}^*)^\top)^\top$ by minimizing (5.6);
10:     Return the set of estimated thresholds $\hat{\mathcal{M}}_\ell = \{\hat{a}_{1,\ell}, \ldots, \hat{a}_{\hat{s}_\ell,\ell}\}$ in step 7 and the estimator of the coefficient $\hat{\theta}_\ell^*$ in step 9, and compute $BIC_{\hat{\mathcal{M}}_\ell}$ by (5.8);
11: **end for**
12: Choose $\hat{\ell}$ that minimizes the $BIC_{\hat{\mathcal{M}}_\ell}$ and obtain the final estimators $\hat{\mathcal{M}}_{opt} = \hat{\mathcal{M}}_{\hat{\ell}}$ and $\hat{\theta}_{opt}^* = \hat{\theta}_{\hat{\ell}}^*$.

Throughout the section the following assumptions will be made:

(A1) $P(Z_i \leq a_j) = \tau_j$ where $0 < \tau_1 < \tau_2 < \ldots < \tau_s < 1$, $E(\varepsilon_i|\mathbf{X}_i) = 0$, $E(T_i^2) < \infty$, $E(\mathbf{X}_i\mathbf{X}_i^\top) = \Sigma_0$ is finite and nonsingular, and Z_i and ε_i are independent, $i = 1, \ldots, n$.

(A2) T_i and C_i are independent, H_j is continuous, and censoring mechanism is independent of covariates.

(A3) $U_{F_j} < U_G$ or $U_{F_j} = U_G = \infty$.

Assumption (A1) allows heteroscedastic error. For example, we may relax the error assumption to be $\varepsilon_i = \sigma(\mathbf{X}_i^\top \beta_j^*)\varepsilon_i^*$ where ε_i^* is independent with $\mathbf{X}_i$ and $\varepsilon_1^*, \ldots, \varepsilon_n^*$ are i.i.d. with mean zero and variance σ^2. (A2) assumes that δ_i is conditionally independent of $\mathbf{X}_i$ given T_i. (A3) implies that $U_{H_j} = U_{F_j}$.

We first present a result for the simple case where $s = 1$ is known. Tentatively we rewrite a_1 and $\mathbf{d}_1^*$ as a and $\mathbf{d}^*$, respectively, in model (5.1), and replace τ_1 with τ in assumption (A1) for simplicity.

Theorem 5.2.1. *Assume conditions (A1)–(A3) hold for model (5.1), $s = 1$ is known and $\|\mathbf{d}^*\| > 0$. Let $\hat{a} = argmin_{\varsigma \in \{Z_{(1)}, \ldots, Z_{(n)}\}}\{\min_\beta \frac{\sharp\hat{\mathcal{I}}_{\varsigma-}}{n} \sum_{i \in \hat{\mathcal{I}}_{\varsigma-}} w_{[r_{Y_i,\hat{\mathcal{I}}_{\varsigma-}}]}(Y_i - \mathbf{X}_i^\top\beta)^2 + \min_\beta \frac{\sharp\hat{\mathcal{I}}_{\varsigma+}}{n} \sum_{i \in \hat{\mathcal{I}}_{\varsigma+}} w_{[r_{Y_i,\hat{\mathcal{I}}_{\varsigma+}}]}(Y_i - \mathbf{X}_i^\top\beta)^2\}$ where $\hat{\mathcal{I}}_{\varsigma-} = \{i : Z_i \leq \varsigma\}$ and $\hat{\mathcal{I}}_{\varsigma+} = \{i : \varsigma < Z_i\}$, then we have $\frac{1}{n}\sum_{i=1}^n 1_{\{Z_i \leq \hat{a}\}} \rightarrow_{a.s.} \tau$ as $n \rightarrow \infty$. Furthermore, if Z_i is a continuous random variable, we have $\hat{a} \rightarrow_{a.s.} a$; if $Z_i = i$, we have $\hat{a}/n \rightarrow_{a.s.} \tau$.*

The proof of Theorem 5.2.1 is given in [161]. The result itself may be of interest since many econometric studies deal with a single change point.

If $s \geq 0$ is unknown, we use the TSMCD method to estimate s, and the change points $a_j, j = 1, \ldots, s$ when $s > 0$. The following technical conditions are also needed:

(A4) $m \rightarrow \infty$ and $m = O((n^*)^r)$, where $0 < r \leq 1/2$ is a constant.

(A5) The penalty function $p_{\lambda_n,\gamma_n}(|u|)$ of (5.4) satisfies $p_{\lambda_n,\gamma_n}(0) = 0,\ p^{\top}_{\lambda_n,\gamma_n}(u) = 0$ if $u > \gamma_n\lambda_n$, $p^{\top}_{\lambda_n,\gamma_n}(0) = \lambda_n$, $\lambda_n \to 0$ and $\lambda_n\sqrt{n}/\log n \to \infty$ as $n \to \infty$.

(A6) $\varepsilon_i = \sigma(\mathbf{X}_i^{\top}\beta_j^*)\varepsilon_i^*, i \in \mathcal{I}_j^*, j = 1, \ldots, s+1, \varepsilon_i^*$ is independent of $\mathbf{X}_i, \varepsilon_1^*, \ldots, \varepsilon_n^*$ are i.i.d. random errors with distribution F_0. Suppose $\tau_{F_0} < \tau_{F_j}$, and there exists some $U^*_{F_j} < U_{F_j} - U_{F_0}$, such that $\sigma(\mathbf{X}_i^{\top}\beta_j^*) = 0$ if $\mathbf{X}^{\top}\beta_j^* > U^*_{F_j}$, otherwise $\sigma(\mathbf{X}_i^{\top}\beta_j^*) = 1, j = 1, \ldots, s+1$.

By law of large numbers and Assumption (A1), $P(Z_i \le a_j) = \tau_j$ implies

$$\sum_{i=1}^{n} 1_{\{a_{j-1}<Z_i\le a_j\}}/n \to_{a.s.} \tau_j - \tau_{j-1} > 0.$$

Assumption (A4) implies with probability 1

$$\sum_{i=1}^{n} 1_{\{\tilde{Z}_{(n^*-(q_n-j+2)m)}<Z_i\le \tilde{Z}_{(n^*-(q_n-j+1)m)}\}}/n = m/n^* \to 0.$$

Thus Assumption (A1) and (A4) ensure that there is at most one threshold in each segment $\{Z_i : \tilde{Z}_{(n^*-(q_n-j+2)m)} < Z_i \le \tilde{Z}_{(n^*-(q_n-j+1)m)}\}$ for sufficiently large n where $\tilde{Z}_{(n^*-(q_n-j+2)m)}$ and $\tilde{Z}_{(n^*-(q_n-j+1)m)}$, $j = 1, \ldots, q_n+1$, are defined in Section 2.1. Both SCAD and MCD satisfy the penalty assumption (A5). (A6) implies $T_i = \mathbf{X}_i^{\top}\beta_j^*$ if $U^*_{F_j} + U_{F_0} < Y_i \le U_{F_j}$.

Define $\mathcal{A}^* = \{i_1, \ldots, i_a\}$ to be the group index set, $\tilde{\mathbf{X}}_{\mathcal{A}^*} = (\tilde{\mathbf{X}}^{(i_1)}, \ldots, \tilde{\mathbf{X}}^{(i_a)})$ and $\tilde{\mathbf{y}}_{\mathcal{A}^*} = (\tilde{\mathbf{y}}^{\top}_{(i_1)}, \ldots, \tilde{\mathbf{y}}^{\top}_{(i_a)})^{\top}$. Similarly we can define $\hat{\theta}_{\mathcal{A}^*}$ and $\hat{\theta}_{(\mathcal{A}^*)^c}$. Let $\mathcal{A} = \{1, k_{j,n}, k_{j,n}+1 : j = 1, \ldots, s\}$.

Theorem 5.2.2. *If Assumptions (A1)–(A6) hold,* $\min\{||\beta_1^*||, ||\mathbf{d}_1^*||, \ldots, ||\mathbf{d}_s^*||\} > 2\sqrt{p}\gamma\lambda_n$, *then with probability 1, (5.4) has a local minimizer* $\hat{\theta}$ *such that* $\hat{\theta}_{\mathcal{A}^*} = (\tilde{\mathbf{X}}^{\top}_{\mathcal{A}^*}\tilde{\mathbf{X}}_{\mathcal{A}^*})^{-1}\tilde{\mathbf{X}}_{\mathcal{A}^*}\tilde{\mathbf{y}}_{\mathcal{A}^*}$ *and* $\hat{\theta}_{(\mathcal{A}^*)^c} = \mathbf{0}$, *where*

$$\mathcal{A}^* \subseteq \mathcal{A} \text{ and } 1 \in \mathcal{A}^*, k_{\ell,n} \in \mathcal{A}^* \text{ or } k_{\ell,n}+1 \in \mathcal{A}^* \text{ for all } \ell = 1, \ldots, s. \tag{5.9}$$

The complete proof is provided in [161]. Theorem 5.2.2 guarantees the existence of the solution of (5.4). In this theorem, $\mathcal{A}^*$ is the group index set of no-zero group elements of $\hat{\theta}$. By (5.9), $\mathcal{A}^*$ is not unique and may take 2^s possible forms.

Let $\mathbf{a} = (a_1, \ldots, a_s)^{\top}$ be the vector of threshold locations. Similar to the definition of $\tilde{\mathbf{y}}_{\mathcal{A}^*}$ and $\tilde{\mathbf{X}}_{\mathcal{A}^*}$, we may define $\tilde{\mathbf{y}}_{(\mathbf{a})}$, $\tilde{\mathbf{X}}_{(\mathbf{a})}$. By Assumption (A1), we have $\frac{1}{n}\tilde{\mathbf{X}}^{\top}_{(\mathbf{a})}\tilde{\mathbf{X}}_{(\mathbf{a})} \to_{a.s} \Gamma > 0$.

To obtain the limiting property of $\hat{\theta}^*$ in (5.6), we will further need the following assumption:

(A7) $\max_{u\ge 0}\{p''_{\lambda_n,\gamma_n}(u)\} + \lambda_{(s+1)p}(\Gamma) > 0$ where $\lambda_{(s+1)p}(\Gamma)$ is the minimal eigenvalue of Γ.

For MCP, (A7) is equivalent to $\lambda_{(s+1)p}(\Gamma) > 1/\gamma$, and for SCAD, (A7) is equivalent to $\lambda_{(s+1)p}(\Gamma) > 1/(\gamma-1)$. Let $(\theta_1^*, \ldots, \theta^*_{(s+1)p})^{\top} = ((\beta_1^*)^{\top}, (\mathbf{d}_1^*)^{\top}, \ldots, (\mathbf{d}_s^*)^{\top})^{\top}$ and $\mathcal{S} = \{j : \theta_j^* \neq 0, j = 1, \ldots, (s+1)p\}$.

Let $\hat{\mathcal{A}} = \{j : \hat{\theta}_j \neq 0, j = 1, \ldots, q_n+1\}$, and $\hat{\mathcal{A}}^*$ be a subset of $\hat{\mathcal{A}}$ such that $\hat{\mathcal{A}}^* = \{j : j \in \hat{\mathcal{A}}, j-1 \notin \hat{\mathcal{A}}, j = 2, \ldots, q_n+1\} = \{\hat{k}_1, \ldots, \hat{k}_{\hat{s}}\}$ where $\hat{s}$ is the size of $\hat{\mathcal{A}}^*$.

Theorem 5.2.3. *Under the Assumptions (A1)–(A7),* $\min_{j\in\mathcal{S}}\{|\theta_j|\} > \gamma\lambda_n$, $\hat{\theta}$ *is the local minimizer in Theorem 5.2.2,* $\hat{\mathbf{a}} = (\hat{a}_1, \ldots, \hat{a}_{\hat{s}})^{\top}$ *is obtained by (5.5),* Z_i *is a continuous random variable and* $\hat{\theta}^*$ *is a minimizer of* $M(\theta)$ *given by (5.6), then we have, with probability 1, we have (1)* $\hat{s} = s$; *(2)* $\hat{\mathbf{a}} = \mathbf{a}$; *(3)* $\hat{\theta}^* = \hat{\theta}^o$, *where* $\hat{\theta}^o = \arg\min_{\theta}\{\|\tilde{\mathbf{y}}_{(\mathbf{a})} - \tilde{\mathbf{X}}_{(\mathbf{a})}\theta\|^2 : \theta = (\theta_1, \ldots, \theta_{(s+1)p})^{\top}, \theta_j = 0\,\forall\, j \notin \mathcal{S}\}$ *is the oracle estimator when thresholds and the set* $\mathcal{S}$ *are known.*

The proof of Theorem 5.2.3 is also given in [161]. We attain the consistency of the TSMCD estimators in this theorem. Our estimators work as well as the oracle estimators in large samples.

5.2.5 Simulation study

We generate random samples from model (5.1) with $s = 2$ thresholds in the following. Specifically, we generate the regressors $\mathbf{X}_i = (x_{1,i}, x_{2,i}, \ldots, x_{6,i})^\top$ with $x_{1,i} = 1, x_{j,i} \sim N(0,1), j = 2, \ldots, 6$, and $\varepsilon_i \sim N(0, 0.5)$. We specify the coefficients $(\theta_1^*, \ldots, \theta_{18}^*)^\top = ((\beta_1^*)^\top, (\mathbf{d}_1^*)^\top, (\mathbf{d}_2^*)^\top)^\top = (2,1,1,1,1,1,-1,0,0,-1,-1,-1,0,-1,1,0,0,0)^\top$, the threshold variable $Z_i = x_{2,i}$, and true thresholds $a_1 = -0.5244$, $a_2 = 0.2533$. The two thresholds are the 30% and 60% lower percentiles of the standard normal distribution. We first design the following four cases:

Example 1: $n = 150$, $x_{j,i}, j = 2, \ldots, 6$ are independent and censoring variable $C_i \sim N(2, 16)$;
Example 2: The same setting as Example 1 except $n = 300$;
Example 3: The same setting as Example 2 except $C_i \sim N(\sum_{j=2}^{6} x_{j,i}, 16)$;
Example 4: The same setting as Example 3 except covariance between $x_{j_1,i}$ and $x_{j_2,i}$ is $0.5^{|j_1 - j_2|}$.

The censoring rates are about 40% for all cases. Assumption (A2) stipulates that T_i and C_i must be independent. This assumption is satisfied in the first two examples. However, a more realistic assumption is that T_i and C_i are conditionally independent given $\mathbf{X}_i$. Example 3 is thus introduced to test the robustness of our method. Example 4 is the case where the regressors are moderately correlated.

In order to examine the performance of TSMCD when the model has no threshold. We design the following two cases:

Example 5: The same as Example 1 except $s = 0$ and $\beta_1^* = (1, 0, 2, 0, 0, 0)$;
Example 6: The same as Example 4 except $s = 0$ and $\beta_1^* = (1, 0, 2, 0, 0, 0)$;.

All numerical studies are performed on a computer(Intel(R) Core(TM) i7 930 2.80GHz 8 M Caches, 8GB Memory). The means and standard deviations of computing time by TSMCD are reported in Table 5.1 based on 100 simulations. The average computing time for MCP is slightly less than that for SCAD. With the increase of model complexity and sample size, the average computing time also tends to increase.

Table 5.1: The means and standard deviations of computation time (seconds).

		$s=2$				$s=0$	
	Example	1	2	3	4	5	6
TSMCD-	Mean	7.55	19.85	21.90	29.56	2.77	7.59
MCP	Stand deviation	1.97	3.15	3.56	5.14	1.59	3.34
TSMCD-	Mean	9.76	22.58	24.51	30.81	3.61	9.82
SCAD	Stand deviation	2.05	3.59	3.50	5.66	1.99	4.04

The estimation results for s are reported in Table 5.2 based on 1000 simulations. Our methods can correctly identify the number of thresholds with very high probability. While their performance is comparable, SCAD seems to identify slightly more unnecessary change points than MCP. Both methods improve as sample size increases.

Table 5.3 summarizes the estimation performance of the estimated thresholds for the cases with correct estimation of $\hat{s} = s$. Again the two penalty methods give quite similar results. The estimation bias and mean squared error for a_1 are relatively smaller than those for a_2. In fact, we note that the jumps at the two change points are $\|\mathbf{d}_1^*\|^2 = 4$ and $\|\mathbf{d}_2^*\|^2 = 2$, respectively, under our model. In general it is easier for our methods to detect a break point with greater jump.

It is noted in all simulations that results from Examples 3–4 and Example 6 are comparable to those from Example 2 and Example 5, respectively. Even though Examples 3–4 and Example 6

Table 5.2: Frequency of estimated $\hat{s}$ in 1000 simulations.

		TSMCD-MCP							TSMCD-SCAD					
	$\hat{s}$	0	1	2	3	4	5	6	0	1	2	3	4	5
	Example 1	2	150	778	65	4	1	0	0	153	773	70	4	0
$s=$	Example 2	0	13	919	64	3	0	1	0	12	869	116	3	0
2	Example 3	0	8	937	52	3	0	0	0	8	916	71	4	1
	Example 4	0	24	947	26	3	0	0	0	33	920	43	4	0
$s=$	Example 5	914	82	4	0	0	0	0	899	88	11	1	1	0
0	Example 6	945	49	6	0	0	0	0	936	58	6	0	0	0

Table 5.3: Estimation performance for the threshold estimation. Bias is the average of estimated parameter minus the true value. RMSE refers to the relative mean squared errors.

	TSMCD-MCP				TSMCD-SCAD			
	$\hat{a}_1$		$\hat{a}_2$		$\hat{a}_1$		$\hat{a}_2$	
	Bias	RMSE	Bias	RMSE	Bias	RMSE	Bias	RMSE
Example 1	−0.014	0.031	0.037	0.476	−0.013	0.037	0.039	0.428
Example 2	−0.007	0.007	0.010	0.171	−0.008	0.007	0.012	0.158
Example 3	−0.009	0.006	0.008	0.145	−0.007	0.006	0.010	0.159
Example 4	−0.006	0.005	0.013	0.166	−0.006	0.005	0.012	0.158

violate the independence censoring assumption, numerical results in this section suggest our methods are quite robust and may still work steadily under conditional independence.

5.2.6 Real data analysis

Follicular lymphoma is the second most common form of non-Hodgkins lymphoma, accounting for about 22% of all documented cases. Dave et al. (2004) [74] investigated whether the survival risks of patients with follicular lymphoma can be predicted by the gene-expression profiles of tumors and standard clinical risk factors at diagnosis. Fresh-frozen tumor-biopsy specimens and clinical data from 191 untreated patients who had received a diagnosis of follicular lymphoma between 1974 and 2001 were obtained. The median age at diagnosis was 51 years (range 23–81), and the median follow-up time was 6.6 years (range: less than 1.0–28.2). The median follow-up time among patients alive at last follow-up was 8.1 years.

A total of 156 subjects are included in analysis after excluding cases with missing information. Many authors analyzed this data in earlier works. We re-visit this data set in this section and consider an AFT model between the failure time and 5 most significant genetic markers selected in Yu et al. (2012) [276]. According to [276], genes 357, 2345, 6267, 6271, and 3653 in the original sample are the most important markers for the survival risk prediction when clinical information is adjusted. We thus set $x_{1,i} = 1$ as the intercept, chose the five gene expressions as the regressors $x_{2,i}, \ldots, x_{6,i}$. We pick gene 357 as the index variable in the following analysis and set $Z_i = x_{2,i}$ since it is recognized as the most predictive marker for the failure time. We have also attempted to detect change points for other covariates but our numerical program returned an estimate $\hat{s} = 0$ for them.

Applying TSMCD(MCP) and TSMCD(SCAD), we obtain the same estimation results, yielding $\hat{s} = 2$ and two thresholds $\hat{a}_1 = -0.483$ and $\hat{a}_2 = 0.907$. These two change points divide the sample into three groups with cumulative group sizes $\hat{\tau}_1 = \sum_{i=1}^{156} 1_{\{Z_i \le -0.483\}}/156 = 0.263$, and $\hat{\tau}_2 = \sum_{i=1}^{156} 1_{\{Z_i \le 0.907\}}/156 = 0.814$. Figure 5.1 displays the Kaplan-Meier survival curves for the three groups separated by genes 357. The survival curves for Group 3 drops rapidly since the baseline and represents a high-risk group in this sample. Such an observation is not available without considering

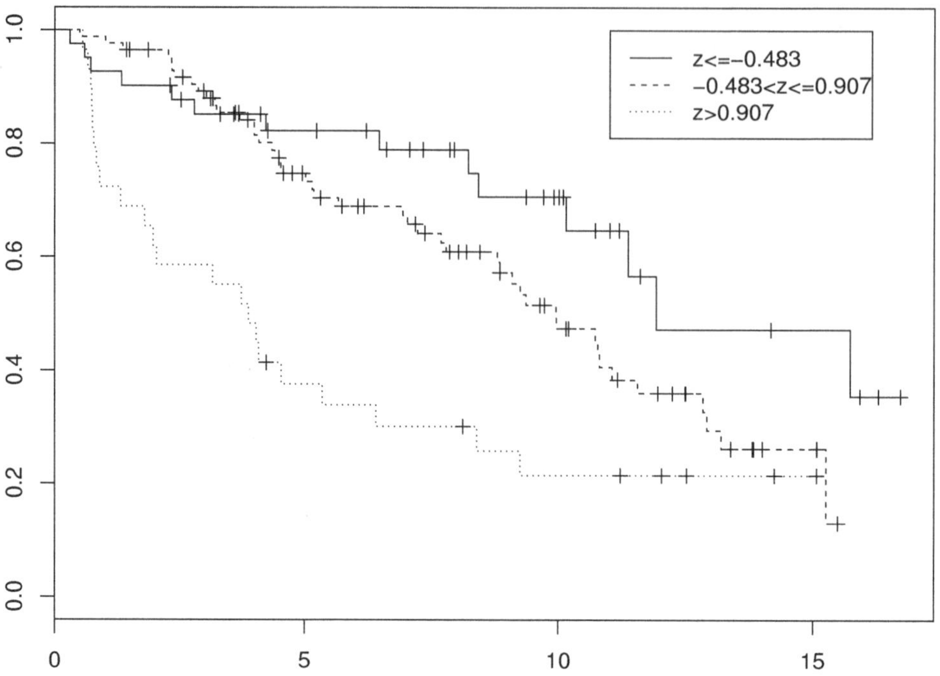

Figure 5.1: Survival curves for the three groups separated by the two thresholds of Gene 357 (Z). $+$ indicates a censored observation. Groups 1, 2 and 3 refer to $Z \leq -0.482$, $-0.482 < Z \leq 0.907$ and $Z > 0.907$, respectively.

the threshold AFT model in this book. Practitioners may adapt our model easily to discover more meaningful sub-populations with defining features.

Traditionally when the goal is to identify subgroups from the empirical sample one may resort to clustering analysis. Such unsupervised learning methods produce segmentation or grouping based on the variation of covariates and ignore the censored response variable. One drawback is that those learned groupings may not be relevant to the survival response. We have implemented a hierarchical clustering procedure (`R` function `hclust`) using only the covariate matrix and the identified clusters are quite different from our change point analysis. The comparison of cluster sizes between clustering analysis and our method is given in Table 5.4.

Table 5.4: Comparison of cluster sizes of real data analysis using two methods.

	Cluster analysis		
TSMCD	Group 1	Group 2	Group 3
Group 1	32	5	4
Group 2	20	46	20
Group 3	0	14	15

There is only slight agreement between the two methods with Cohen's kappa value equal to 0.36. Furthermore, it is hard to interpret the groups resulted from the hierarchical clustering algorithm. Specifically, it remains unclear whether the three identified groups reflect any practical differences in survival probability. Upon examining the Kaplan-Meier curves for these groups (which are not included in the book but are available from the authors), we observe a significant overlap in the survival functions of groups 2 and 3 over the observation period. Such overlap implies that the clustering results may have limited clinical relevance, as the differences between these groups do not appear to provide meaningful insights into differentiating survival experiences.

The estimated regression coefficients are $\hat{\theta}^* = ((\hat{\beta}_1^*)^\top, (\hat{\mathbf{d}}_1^*)^\top, (\hat{\mathbf{d}}_2^*)^\top)^\top$ where $\hat{\beta}_1^* = (0.791, -0.492, 0.291, 0.990, -0.589, 1.115)^\top$, $\hat{\mathbf{d}}_1^* = (1.166, 0, -0.451, -1.055, 0.540, -0.991)^\top$, $\hat{\mathbf{d}}_2^* = (0,\ -0.381,\ 0,\ 0,\ 0,\ 0.770)^\top$. Such results suggest at the first threshold $\hat{a}_1 = -0.483$, only the coefficient of $x_{2,i}$ does not change; and at the second threshold $\hat{a}_2 = 0.907$, only coefficients of $x_{2,i}$ and $x_{6,j}$ change and all other coefficients remain the same.

Using the identified break point structure, we refitted the AFT model and reported the estimation results for the three sub-groups in Table 5.5, along with 95% bootstrap confidence intervals. The effects of genes on the survival outcome are quite different for the three groups. For example, Gene 2345 is not significant for the first group but are negatively associated with the failure time for the other two groups. Using our empirical findings, investigators may evaluate the effects of various genetic biomarkers on mortality more specifically for different sub-populations.

Table 5.5: Estimates of regression coefficients for the three groups, along with bootstrap standard errors (S.E.) and P-values by Wald test.

	$Z \leq -0.482$			$-0.482 < Z \leq 0.907$			$Z > 0.907$		
Covariates	Coef.	S.E.	P-value	Coef.	S.E.	P-value	Coef.	S.E.	P-value
Intercept	0.791	0.507	0.056	1.957	0.083	0	1.957	0.083	0
Gene 357	−0.492	0.179	0.003	−0.492	0.179	0.003	−0.874	0.144	6e−10
Gene 2345	0.291	0.627	0.321	−0.159	0.084	0.029	−0.159	0.084	0.029
Gene 6267	0.990	0.725	0.086	−0.065	0.110	0.277	−0.065	0.110	0.277
Gene 6271	−0.589	0.568	0.150	−0.049	0.120	0.341	−0.049	0.120	0.341
Gene 3653	1.115	0.556	0.022	0.124	0.078	0.056	0.894	0.250	2e−4

5.2.7 Discussion

The Stute estimator for Accelerated Failure Time (AFT) models is well-suited for right-censored survival data. However, when dealing with interval-censored data, the estimation process becomes more challenging due to the nature of the censoring. In such cases, a likelihood-based objective function is typically constructed for regression estimation. Incorporating a change point structure into this framework allows for potential improvements in identifying shifts in the relationship between covariates and the outcome over time. However, the likelihood function for interval-censored data tends to be more complex compared to right-censored data, making numerical solutions non-trivial. While the two-step estimation approach, similar to what we propose, can be adopted, further theoretical and computational advancements are necessary to develop a robust and efficient solution. These include better handling of the non-standard functional forms that arise in the likelihood function under interval censoring and devising efficient algorithms to handle the added computational complexity.

In this chapter, we focus on independent survival times. In many practical applications, individuals may experience multiple events of the same type (recurrent events) or a number of events of different types (competing risks). Such correlated survival time data require more advanced modeling approaches and additional technical assumptions to achieve efficient results. Two widely used approaches for handling correlated survival times are the marginal model and the frailty model: Marginal models focus on estimating the average effect across all individuals, accounting for within-subject correlation, typically using generalized estimating equations (GEE) or other marginal methods; Frailty models introduce random effects (frailties) to capture unobserved heterogeneity between individuals, allowing for dependence between event times. In both approaches, one can still consider the application of the Stute estimator for the AFT model. Moreover, the two-stage multiple change-point detection (TSMCD) method we propose can be directly adapted to these models. However, deriving asymptotic results for such analyses is non-trivial. This complexity arises due to the need to properly account for the inter-dependence of repeated measures or event times across individuals, which can introduce additional variability and complicate the estimation and inference process. Hence, further investigation into the theoretical properties is required to extend the robustness and efficiency of these models to dependent survival data.

The TSMCD involves the penalization approaches and thus can be easily extended to incorporate high-dimensional data analysis. Not many authors examined the change point problem under such settings and therefore the results in this chapter could contribute significantly towards this goal. However, when the dimension of feature space is ultra high with an exponential order of the sample size, usually a screening step must be conducted before the application of penalized estimation. In particular, one may adopt TSMCD at the screening step and estimate the effects of individual markers using the change point structure. This differs from the existing literature where the marginal effects are usually modeled as a smooth function at the initial screening. The detailed methodology construction is beyond the scope of this book.

5.3 Multi-threshold structural equation model

5.3.1 Model and estimation

As a further application of the threshold model (5.1) introduced in the preceding section, we now extend the estimation framework to a multi-threshold structural equation model (SEM), incorporating endogeneity in the regressors. Addressing endogeneity is crucial in observational studies where causal effects are confounded by unmeasured variables or measurement errors. To tackle these issues, instrumental variable (IV) methods are frequently employed, allowing researchers to obtain unbiased causal estimates even in the presence of confounders. The theory and methods for estimating causal effects in the presence of unmeasured confounders have expanded significantly in recent years. Although IV methods originated in economics research, they have gained traction in epidemiological and statistical research over the past few decades, offering robust tools for causal inference in observational data.

The instrumental variable estimation is usually implemented within a two-stage least squares (2SLS) framework to deal with the endogeneity issue and this approach has been widely practiced in causal inference studies (cf. [261, 32, 159]). An alternative procedure is the classical Wald approach [253], also known as the ratio method, which agrees with the 2SLS under the simple linear regression setting.

The standard SEM is usually assumed as

$$Y = \beta_e X_e + \mathbf{X}_o^\top \boldsymbol{\beta}_o + U, \tag{5.10}$$

where Y is the response variable, X_e is the exposure whose regression coefficient β_e is of major interest, $\mathbf{X}_o$ include observed covariates with regression coefficients $\boldsymbol{\beta}_o$ and U is a disturbance involving unmeasured confounders correlated with X_e and thus invoking the endogeneity issue. For example in a medical study, X_e is usually a considered treatment for a disease outcome Y and β_e is the causal treatment effect. Without randomization, the treatment variable might be potentially correlated with many observed or unobserved factors. While observed factors can be easily included in $\mathbf{X}_o$ as adjustment, unobserved factors in U can never be adjusted directly. We thus have to seek the IV approach.

Usually at the same time, a reduced form equation for X_e is assumed as

$$X_e = \alpha_I X_I + \mathbf{X}_o^\top \boldsymbol{\alpha}_o + \varepsilon, \tag{5.11}$$

where X_I is the instrumental variable, and ε is an error. It is assumed that $E(U|X_I, \mathbf{X}_o) = 0$ and $E(\varepsilon|X_I, \mathbf{X}_o) = 0$. The usual 2SLS approach is to fit model (5.11) first using the least squares, use the fitted response $\hat{X}_e$ to replace X_e in model (5.10), and then carry out another least squares estimation for the modified model (5.10). The two least squares are both valid since the covariates are exogenous. The resulting estimate for β_e is proved to be consistent. We will leave this approach for now and specifically consider the ratio method in this section.

Substituting equation (5.11) into model (5.10), we arrive at a synthetic model form

$$Y = \eta_I X_I + \mathbf{X}_o^\top \boldsymbol{\eta}_o + E, \tag{5.12}$$

where $\eta_I = \beta_e \alpha_I$, $\boldsymbol{\eta}_o = \beta_e \boldsymbol{\alpha}_o + \boldsymbol{\beta}_o$, and $E = \beta_e \varepsilon + U$. Once we obtain the ordinary least squares (OLS) estimators $\hat{\alpha}_I$ and $\hat{\eta}_I$ by fitting models (5.11) and (5.12), we obtain the following ratio estimator of β_e:

$$\hat{\beta}_e = \hat{\eta}_I / \hat{\alpha}_I. \tag{5.13}$$

It has been well-known that the estimated coefficient $\hat{\beta}_e$ is consistent and asymptotically normally distributed under mild technical conditions [114, 32, 18]. We focus on this type of estimator in the following development.

Until recently researchers in many fields, including medicine and business, are more and more aware of the importance of treating subgroups of the population with special attention, especially when facing substantial heterogeneous effects. There are abundant discussions of recursive-partitioning approaches for subgroup analysis in the statistical literature, such as the classification and regression tree (CART) [38], the generalized unbiased interaction detection and estimation (GUIDE) [177], and the interaction tree [240], among many others. Change point detection (CPD) method is an alternative strategy for subgroup identification. Recently, [161] proposed a penalization method to identify multiple change points in the linear accelerated failure time model while [259] and [163] further considered the selection of threshold variables. We intend to apply the CPD technique ([243]) to study the estimation of heterogeneous causal effects in SEM.

For the ith subject in an empirical sample of size n, let Y_i be the response variable, X_{ei} be the endogenous variable, $\mathbf{X}_{oi}$ be a vector of observed exogenous variables, and U_i be the model error in the SEM. Assume that U_i are i.i.d. with mean zero and variance σ_u^2. Let T_i be the threshold variable that divides the sample into different subgroups. In this work, we assume that the threshold variable is given and available in the sample. We acknowledge that there is also a large literature concerning the latent grouping variable where the threshold variable is not directly observed and has to be constructed by some statistical methods. See for example, [31, 25, 169, 33, 259, 163]. In many applications, the threshold variable is a time variable or has a continuous distribution [122, 43, 118, 119]. We thus assume that T is continuous throughout the book.

We consider an SEM with s unknown thresholds $\boldsymbol{a} = (a_1, \cdots, a_s)^\top$ and assume that $a_0 \geq -\infty$ and $a_{s+1} \leq \infty$ are known interval bounds for the support of T_i. The number of threshold locations s is also unknown. While partitioning the sub-regions of the variable T_i, the s change points naturally divide the entire sample into $s+1$ subgroups. Specifically, let $\mathcal{I}_j^* = \{i : a_{j-1} < T_i \leq a_j\}$ be the subsets for the j-th subgroup, for $j = 1, \cdots, s+1$, and let $n_j^* > 0$ be the size of subset $\mathcal{I}_j^*$ such that $\sum_{j=1}^{s+1} n_j^* = n$.

We propose the following multi-threshold SEM model:

$$
\begin{aligned}
Y_i &= \sum_{j=1}^{s+1}(X_{ei}\beta_{ej} + \mathbf{X}_{oi}^\top \boldsymbol{\beta}_{oj})1_{\{a_{j-1}<T_i\le a_j\}} + U_i, \qquad (5.14)\\
&= \sum_{j=1}^{s+1}\mathbf{X}_i^\top \boldsymbol{\beta}_j 1_{\{a_{j-1}<T_i\le a_j\}} + U_i\\
&= \mathbf{X}_i^\top(\boldsymbol{\beta}_1 + \sum_{j=1}^{s}\boldsymbol{d}_j 1_{\{a_j<T_i\le a_{s+1}\}}) + U_i\\
&= \begin{cases}
\mathbf{X}_i^\top \boldsymbol{\beta}_1 + U_i, & \text{if } i\in \mathcal{I}_1^*,\\
\mathbf{X}_i^\top(\boldsymbol{\beta}_1 + \boldsymbol{d}_1) + U_i, & \text{if } i\in \mathcal{I}_2^*,\\
\cdots & \cdots\\
\mathbf{X}_i^\top(\boldsymbol{\beta}_1 + \sum_{j=1}^{s}\boldsymbol{d}_j) + U_i, & \text{if } i\in \mathcal{I}_{s+1}^*,
\end{cases}
\end{aligned}
$$

where $\mathbf{X}_i = (X_{ei}, \mathbf{X}_{oi}^\top)^\top$, β_{ej} is the coefficient for the endogenous variable and $\boldsymbol{\beta}_{oj}$ is the coefficient for the observed exogenous variables for the jth subgroup. When $s = 0$, model (5.14) reduces to the standard SEM (5.10). We write $\boldsymbol{\beta}_j = (\beta_{ej}, \boldsymbol{\beta}_{oj}^\top)^\top$, $j = 1,\cdots,s+1$, in which β_{ej} is of primary interest. Under this model, the causal effects for the exposure variable may thus be different for subgroups. Denote the coefficient difference for two adjacent regions to be $\boldsymbol{d}_j = \boldsymbol{\beta}_{j+1} - \boldsymbol{\beta}_j$, $j = 1,\cdots,s$, usually referred to as the threshold effect ([122]). To ensure the models for $s+1$ regimes are indeed different, we assume that $\|\boldsymbol{d}_j\| > 0$, $j = 1,\cdots,s$, where $\|\cdot\|$ is the L_2 norm.

Because U_i is unobserved and correlated with X_{ei}, simply fitting a model with covariates X_{ei} and $\mathbf{X}_{oi}$ may lead to inconsistent estimation of $\beta_{ej}, j = 1,\cdots,s+1$. Suppose we have also observed an instrumental variable $\mathbf{X}_{Ii}$ for every subject, satisfying $E(\mathbf{X}_{Ii}X_{ei}|\mathbf{X}_{oi}) \neq 0$ and $E(\mathbf{X}_{Ii}U_i|\mathbf{X}_{oi}) = 0$. Then we may impose the following segment reduced form model

$$
\begin{aligned}
X_{ei} &= \sum_{j=1}^{s+1}(X_{Ii}\alpha_{Ij} + \mathbf{X}_{oi}^\top \boldsymbol{\alpha}_{oj})1_{\{a_{j-1}<T_i\le a_j\}} + \varepsilon_i \qquad (5.15)\\
&= \sum_{j=1}^{s+1}\mathbf{Z}_i^\top \boldsymbol{\alpha}_j 1_{\{a_{j-1}<T_i\le a_j\}} + \varepsilon_i\\
&= \mathbf{Z}_i^\top(\boldsymbol{\alpha}_1 + \sum_{j=1}^{s}\boldsymbol{g}_j 1_{\{a_j<T_i\le a_{s+1}\}}) + \varepsilon_i\\
&= \begin{cases}
\mathbf{Z}_i^\top \boldsymbol{\alpha}_1 + \varepsilon_i, & \text{if } i\in \mathcal{I}_1^*,\\
\mathbf{Z}_i^\top(\boldsymbol{\alpha}_1 + \boldsymbol{g}_1) + \varepsilon_i, & \text{if } i\in \mathcal{I}_2^*,\\
\cdots & \cdots\\
\mathbf{Z}_i^\top(\boldsymbol{\alpha}_1 + \sum_{j=1}^{s}\boldsymbol{g}_j) + \varepsilon_i, & \text{if } i\in \mathcal{I}_{s+1}^*,
\end{cases}
\end{aligned}
$$

where $\mathbf{Z}_i = (X_{Ii}, \mathbf{X}_{oi}^\top)^\top$, $\alpha_{Ij}, \boldsymbol{\alpha}_{oj}, j = 1,\cdots,s+1$, are unknown coefficients for $s+1$ subgroups, $\boldsymbol{\alpha}_j = (\alpha_{Ij}, \boldsymbol{\alpha}_{oj}^\top)^\top$, $\boldsymbol{g}_j = \boldsymbol{\alpha}_{j+1} - \boldsymbol{\alpha}_j, j = 1,\cdots,s$, ε_i are i.i.d. random errors with mean zero and variance σ_ε^2. We assume ε_i is independent of X_{Ii} and $\mathbf{X}_{oi}$, but correlated with U_i with $cov(U_i, \varepsilon_i) = \sigma_{u\varepsilon}$. Also assume that the threshold effect $\|\boldsymbol{g}_j\| > 0$, for $j = 1,\cdots,s$. Denote $\boldsymbol{\zeta} = (\boldsymbol{\zeta}_1^\top,\cdots,\boldsymbol{\zeta}_{s+1}^\top)^\top$, $\boldsymbol{\zeta}_1 = \boldsymbol{\alpha}_1$, $\boldsymbol{\zeta}_j = \boldsymbol{g}_{j-1}$ for $j = 2,\cdots,s+1$.

We consider the classical Wald method to estimate β_{ej}. To this end, we substitute (5.15) into (5.14) and obtain the following synthetic segment regression model:

$$
\begin{aligned}
Y_i &= \sum_{j=1}^{s+1} X_{Ii}\alpha_{Ij}\beta_{ej}1_{\{a_{j-1}<T_i\le a_j\}} + \sum_{j=1}^{s+1} \mathbf{X}_{oi}^\top(\boldsymbol{\beta}_{oj} + \beta_{ej}\boldsymbol{\alpha}_{oj})1_{\{a_{j-1}<T_i\le a_j\}} + E_i \\
&= \sum_{j=1}^{s+1} X_{Ii}\eta_{Ij}1_{\{a_{j-1}<T_i\le a_j\}} + \sum_{j=1}^{s+1} \mathbf{X}_{oi}^\top\boldsymbol{\eta}_{oj}1_{\{a_{j-1}<T_i\le a_j\}} + E_i \\
&= \sum_{j=1}^{s+1} \mathbf{Z}_i^\top\boldsymbol{\eta}_j1_{\{a_{j-1}<T_i\le a_j\}} + E_i \\
&= \mathbf{Z}_i^\top(\boldsymbol{\eta}_1 + \sum_{j=1}^{s} \boldsymbol{h}_j1_{\{a_j<T_i\le a_{s+1}\}}) + E_i \\
&= \begin{cases} \mathbf{Z}_i^\top\boldsymbol{\eta}_1 + E_i, & \text{if } i \in \mathcal{I}_1^*, \\ \mathbf{Z}_i^\top(\boldsymbol{\eta}_1 + \boldsymbol{h}_1) + E_i, & \text{if } i \in \mathcal{I}_2^*, \\ \cdots & \cdots \\ \mathbf{Z}_i^\top(\boldsymbol{\eta}_1 + \sum_{j=1}^{s}\boldsymbol{h}_j) + E_i, & \text{if } i \in \mathcal{I}_{s+1}^*, \end{cases}
\end{aligned} \tag{5.16}
$$

where $\eta_{Ij} = \alpha_{Ij}\beta_{ej}$, $\boldsymbol{\eta}_{oj} = \boldsymbol{\beta}_{oj} + \beta_{ej}\boldsymbol{\alpha}_{oj}$, $E_i = U_i + \sum_{j=1}^{s+1}\beta_{ej}\varepsilon_i 1_{\{i\in\mathcal{I}_j^*\}}$, $\boldsymbol{\eta}_j = (\eta_{Ij}, \boldsymbol{\eta}_{oj}^\top)^\top$, for $j = 1, \cdots, s+1$, $\boldsymbol{h}_1 = \boldsymbol{\eta}_2 - \boldsymbol{\eta}_1, \cdots, \boldsymbol{h}_s = \boldsymbol{\eta}_{s+1} - \boldsymbol{\eta}_s$. Denote $\boldsymbol{\xi}_1 = \boldsymbol{\eta}_1$, $\boldsymbol{\xi}_2 = \boldsymbol{h}_1, \cdots, \boldsymbol{\xi}_{s+1} = \boldsymbol{h}_s$, $\boldsymbol{\xi} = (\boldsymbol{\xi}_1^\top, \cdots, \boldsymbol{\xi}_{s+1}^\top)^\top = (\xi_{11}, \xi_{21}\cdots, \xi_{\varrho,1}, \xi_{12}, \cdots, \xi_{\varrho,2}, \cdots, \xi_{\varrho(s+1)})^\top$, where ϱ is the length of $\mathbf{Z}_i$. For model (5.16), we may apply a multi-threshold CPD method TSMCD introduced in the preceding section to estimate the thresholds and fit the resulting piecewise linear model regressing $\mathbf{Y}$ on $\mathbf{X}_I$ and $\mathbf{X}_o$. See [258] for more details.

The estimated coefficients for model (5.16) are $\hat{\boldsymbol{\eta}}_j = \sum_{k=1}^{j}\hat{\boldsymbol{\xi}}_k^*, j = 1, \cdots, \hat{s}+1$, and

$$
\hat{\eta}_{Ij} = \{\text{the first element of } \hat{\boldsymbol{\eta}}_j\}, j = 1, \cdots, \hat{s}. \tag{5.17}
$$

Now we consider the estimation of model (5.15). Using the estimated threshold parameters $\hat{a}_1, \cdots, \hat{a}_{\hat{s}}$, we carry out a segment regression of X_{ei} on $\mathbf{Z}_i$. The coefficient estimator can be attained by:

$$
\hat{\zeta} = \text{argmin}_\zeta \sum_{j=1}^{\hat{s}+1} \frac{\hat{n}_j^*}{2n^2} \sum_{i\in\hat{\mathcal{I}}_j^*} (X_{ei} - \sum_{k=1}^{j} \mathbf{Z}_i^\top\zeta_k)^2. \tag{5.18}
$$

Then the estimated coefficients for the instrumental variable is given by

$$
\hat{\alpha}_{Ij} = \{\text{the first element of } \hat{\boldsymbol{\alpha}}_j\}, j = 1, \cdots, \hat{s}+1, \tag{5.19}
$$

where $\hat{\boldsymbol{\alpha}}_j = \sum_{k=1}^{j}\hat{\zeta}_k$. Finally, the estimated causal coefficients for the multi-threshold SEM are

$$
\hat{\beta}_{ej} = \hat{\eta}_{Ij}/\hat{\alpha}_{Ij}, \qquad j = 1, \cdots, \hat{s}+1. \tag{5.20}
$$

5.3.2 Theoretical results

We establish the asymptotic results for the proposed SEM estimators in this section. The following technical conditions are needed.

(C1): Assume that U_i are i.i.d. with mean zero and variance σ_u^2; ε_i are i.i.d. with mean zero and variance σ_ε^2; U_i and ε_i are uncorrelated with X_{Ii} and $\mathbf{X}_{oi}$.

(C2): $\alpha_{Ij} \neq 0$ for $j = 1, \cdots, s+1$.

(C3): $E(\mathbf{X}_{oi}\mathbf{X}_{oi}^\top 1_{\{a_{j-1}<T_i\le a_j\}})$ is finite and nonsingular.

(C4): $E(X_{Ii}^4)$ is nonzero and finite.

(C5): The covariance matrix of $\begin{pmatrix} U_i \\ \varepsilon_i \end{pmatrix}$ is $\mathbf{W} = \begin{pmatrix} \sigma_u^2 & \sigma_{u\varepsilon} \\ \sigma_{u\varepsilon} & \sigma_\varepsilon^2 \end{pmatrix}$ which is a finite positive definite matrix.

(C6): As $n \to \infty$, $n_j^*/n \to C_j^*$ and $0 < C_j^* \le 1$ for $j = 1, \cdots, s+1$.

(C7): $\mathbf{X}_I^{(j)\top}\mathbf{P}_j\mathbf{X}_I^{(j)}/n_j^* \xrightarrow{a.s.} Q_{Ij} \neq 0$ and $\mathbf{X}_I^{(j)\top}\mathbf{P}_j\mathbf{X}_e^{(j)}/n_j^* \xrightarrow{a.s.} Q_{Iej} \neq 0$ where $\mathbf{P}_j = \mathbf{I}_{n_j^*} - \mathbf{X}_o^{(j)}(\mathbf{X}_o^{(j)\top}\mathbf{X}_o^{(j)})^{-1}\mathbf{X}_o^{(j)\top}$, $\mathbf{X}_I^{(j)} = \{X_{Ii} : i \in \mathcal{I}_j^*\}$, $\mathbf{X}_e^{(j)} = \{X_{ei} : i \in \mathcal{I}_j^*\}$ and $\mathbf{X}_o^{(j)} = \{\mathbf{X}_{oi} : i \in \mathcal{I}_j^*\}$ for $j = 1, \cdots, s+1$.

In condition (C7), $\mathbf{P}_j$ is a projection matrix for the jth group such that $\mathbf{P}_j^\top = \mathbf{P}_j$, $\mathbf{P}_j^\top\mathbf{P}_j = \mathbf{P}_j$ and $\mathbf{P}_j\mathbf{X}_o^{(j)} = \mathbf{0}$. The projection matrix is used to derive an explicit form for the estimator of the causal coefficients. This condition also ensures the identification of our model, following the generic argument for nonlinear IV regression in [8].

We first consider an ideal case where the number of subgroups $s \ge 0$ and the thresholds positions $a_0, a_1, \cdots, a_{s+1}$ are known in advance. A complete knowledge of all the thresholds may not be empirically plausible for most applications. We provide such distribution results primarily to serve as a guidance in theoretical discussions and numerical references. Similar to the estimator (5.18), the coefficients can be simply estimated by the following least squares:

$$\tilde{\boldsymbol{\xi}} = \operatorname{argmin}_{\boldsymbol{\xi}} \sum_{j=1}^{s+1} \frac{n_j^*}{2n^2} \sum_{i\in\mathcal{I}_j^*} \Big(Y_i - \sum_{k=1}^{j} \mathbf{Z}_i^\top \boldsymbol{\xi}_k\Big)^2, \tag{5.21}$$

and

$$\tilde{\boldsymbol{\zeta}} = \operatorname{argmin}_{\boldsymbol{\zeta}} \sum_{j=1}^{s+1} \frac{n_j^*}{2n^2} \sum_{i\in\mathcal{I}_j^*} (X_{ei} - \sum_{k=1}^{j} \mathbf{Z}_i^\top \boldsymbol{\zeta}_k)^2. \tag{5.22}$$

Then for $j = 1, \cdots, s+1$, we can write the estimated $\tilde{\eta}_{Ij}$ and $\tilde{\alpha}_{Ij}$ as follows:

$$\begin{aligned} \tilde{\eta}_{Ij} &= (\mathbf{X}_I^{(j)\top}\mathbf{P}_j\mathbf{X}_I^{(j)})^{-1}\mathbf{X}_I^{(j)\top}\mathbf{P}_j\mathbf{X}_e^{(j)}, \\ \tilde{\alpha}_{Ij} &= (\mathbf{X}_I^{(j)\top}\mathbf{P}_j\mathbf{X}_I^{(j)})^{-1}\mathbf{X}_I^{(j)\top}\mathbf{P}_j\mathbf{Y}^{(j)}. \end{aligned}$$

And the estimator of the causal coefficients $\beta_{ej}, j = 1, \cdots, s+1$, can be expressed in the following closed-form

$$\tilde{\beta}_{ej} = \tilde{\eta}_{Ij}/\tilde{\alpha}_{Ij} = (\mathbf{X}_I^{(j)\top}\mathbf{P}_j\mathbf{X}_e^{(j)})^{-1}\mathbf{X}_I^{(j)\top}\mathbf{P}_j\mathbf{Y}^{(j)}. \tag{5.23}$$

Theorem 5.3.1. *Assume s and $a_1, \cdots, a_s$ in models (5.14) and (5.15) are known. Under conditions (C1)–(C7), as $n \to \infty$, we have*

$$\tilde{\beta}_{ej} \xrightarrow{a.s.} \beta_{ej},$$

for $j = 1, \cdots, s+1$. And $\tilde{\boldsymbol{\beta}}_e = (\tilde{\beta}_{e1}, \cdots, \tilde{\beta}_{e,s+1})^\top$ is asymptotically normally distributed

$$\sqrt{n}(\tilde{\boldsymbol{\beta}}_e - \boldsymbol{\beta}_e) \xrightarrow{d} N(\mathbf{0}, \boldsymbol{\Sigma}),$$

where $\boldsymbol{\Sigma} = diag(\Sigma_1 C_1^{*-1}, \cdots, \Sigma_{s+1}C_{s+1}^{*-1})$, $\Sigma_j = \sigma_u^2 Q_{Ij} Q_{Iej}^{-2}$, for $j = 1, \cdots, s+1$.

The proof of the theorem is given in [258]. We acknowledge that the conclusion of this theorem agrees with Theorem 5 of [275] when $s = 1$. It is easy to see that the following marginal distribution result holds.

Corollary 5.1. *Under the same conditions of theorem 5.3.1, for $j = 1, \cdots, s+1$, we have*

$$\sqrt{n_j^*}(\tilde{\beta}_{ej} - \beta_{ej}) \xrightarrow{d} N(0, \Sigma_j).$$

When the number of subgroups and thresholds are unknown, we have to use the penalization methods to detect multiple thresholds. The following conditions are further required to consistently identify the unknown number and location of thresholds.

(C8): The penalty function $\Gamma_{\lambda_n,\gamma_n}$ satisfies: $\Gamma_{\lambda_n,\gamma_n}(0) = 0$, the first order derivative $\Gamma'_{\lambda_n,\gamma_n}(\theta) = 0$ when $\theta > \lambda_n\gamma_n$, and $\Gamma'_{\lambda_n,\gamma_n}(0) = \lambda_n$.

(C9): $r \to \infty$ and $r = O(n^l)$, where $0 < l \le 1/2$.

Condition (C8) is often assumed for the non-concave penalty function. Both SCAD and MCP can satisfy this condition. The following theorem justifies the strong consistency of the estimated threshold parameters.

Theorem 5.3.2. *Suppose that conditions (C1) to (C9) hold. Assume that* $\min\{|\alpha_{Ij}\beta_{ej}|, j = 1, \cdots, s+1\} > \gamma_n\lambda_n$. $\lambda_n \to 0$ *and* $\lambda_n\sqrt{n}/\log n \to \infty$ *when* $n \to \infty$. *We have*

$$\hat{s} \xrightarrow{a.s.} s$$

and

$$\hat{a} \xrightarrow{a.s.} \boldsymbol{a}.$$

Furthermore, $\hat{\beta}_e$ *is asymptotically identically distributed as* $\tilde{\beta}_e$.

This theorem further lends support to the inference for the estimated causal parameters. One may thus perform hypothesis tests or construct confidence intervals using the distribution results.

If there is no threshold, that is, $s = 0$, by theorem 5.3.2, we have the following corollary:

Corollary 5.2. *Under conditions (C1), (C3)–(C5), (C7)–(C9), if there is no break point, we have* $\hat{s} \xrightarrow{a.s.} 0$ *and* $\hat{\beta}_e = (\mathbf{X}_I^\top \mathbf{X}_e)^{-1}\mathbf{X}_I^\top \mathbf{Y} \xrightarrow{a.s.} \beta_e$.

5.3.3 Households data analysis

To illustrate the practical application of the change point detection (CPD) method, we re-analyze the Philippine farm households data from surveys conducted in Bukidnon province [34, 35, 232]. This data set includes information on rural households and has been used in previous studies to explore the causal relationship between income changes and food expenditure. Initially, 406 households were sampled through a stratified random sampling method, but after removing a few outliers, we carried out the analysis on the remaining 403 households. Similar to [232], we take the household's food expenditure as response Y, the worth of assets as instrumental variable X_I, and the household's log income as the endogenous variable X_e. Other observed confounders ($\mathbf{X}_o$) include mother's education, father's education, mother's age, father's age, mother's nutritional knowledge, price of corn, and price of rice. For more details of the data, we refer to [34] and [35]. The justification of the instrument X_I may follow [34] since the land availability is assumed to be a constraint in the short run, and therefore exogenous to the household decision making process. However, arguably, the confounding variables involved in $\mathbf{X}_o$ may not always be exogenous, even though they were used this way in the earlier publications. To explore more possibilities, we consider the following two types of analyses: (1) use mother's nutritional knowledge as the threshold variable T; (2) use the household's log income as the T. We can also consider some other analyses, for example, first use the mother's nutritional knowledge as the threshold variable to obtain subgroups and then reapply the proposed method with log income as the threshold variable to obtain more subgroups; and take the mother's nutritional knowledge as the endogenous variable X_e and threshold variable T at the same time. Those additional analyses can be found in [258]. We use the SCAD penalty when applying our CPD method with $\gamma_n = 2.4$ and λ_n chosen by BIC.

Analysis 1: When mother's nutritional knowledge is the threshold variable and log income is the endogenous, applying our proposed CPD approach, we end up with two subgroups

Table 5.6: Analysis results for Philippine households study with mother's nutritional knowledge as the threshold variable (Analysis 1). "CPD": the proposed method; "Wald": the Wald (ratio) method; "Est": estimated coefficients; "SE": standard error for estimated coefficients; "P": p value for the estimated coefficients. The tuning parameter chosen by BIC for SCAD is $\lambda_n = 0.0431$.

		Subgroup 1			Subgroup 2		
	Method	Est	SE	P	Est	SE	P
log income	Wald	0.992	0.016	0.000	–	–	–
	CPD	0.754	0.101	0.000	1.494	0.481	0.000
mother's education	Wald	0.078	0.073	0.284	–	–	–
	CPD	0.079	0.073	0.282	−0.027	0.183	0.882
father's education	Wald	−0.103	0.071	0.147	–	–	–
	CPD	−0.201	0.071	0.005	0.081	0.173	0.642
mather's age	Wald	0.021	0.095	0.026	–	–	–
	CPD	−0.249	0.089	0.005	−0.034	0.286	0.906
father's age	Wald	0.014	0.095	0.879	–	–	–
	CPD	0.092	0.091	0.309	−0.451	0.268	0.092
mother's nutritional knowledge	Wald	−0.005	0.064	0.934	–	–	–
	CPD	0.036	0.098	0.705	−0.280	0.328	0.393
price of corn	Wald	0.072	0.060	0.232	–	–	–
	CPD	0.104	0.056	0.063	−0.104	0.188	0.581
price of rice	Wald	0.020	0.061	0.737	–	–	–
	CPD	0.047	0.059	0.421	0.096	0.158	0.544

($\hat{s}+1 = 2$). The first group (group size $n_1^* = 292$) involves households where mother's nutritional knowledge is less than or equal to 9, while the rest are in another group with size $n_2^* = 111$. CH method also leads to the same two subgroups. After implementing the HHB method introduced in [118], we fail to identify any subgroup and thus obtain the same results as the Wald method. We present the estimated coefficients, their standard errors and P-values based on the classical Wald method, and our proposed CPD method in Table 5.6. The estimated coefficient for X_e, representing the household's log income, is 0.754 for subgroup 1, indicating a moderate positive effect. When the mother's nutritional knowledge exceeds 9, the estimated coefficient for the endogenous variable increases to 1.494, nearly doubling the effect observed in subgroup 1. Both of these effects are highly significant for this dataset. Generally, household income is positively associated with food expenditure. However, when mothers possess more nutritional knowledge, they may prioritize their families' diets, which strengthens the causal relationship between income and food expenditure. In contrast, the naive Wald method estimates the household's log income coefficient to be 0.992 for the entire sample, effectively averaging the effects from both subgroups. Coefficients for all other confounders (treated as exogenous variables) are also estimated and reported in Table 5.6. These should be interpreted with caution, as some confounders may themselves be endogenous, requiring further adjustment.

Analysis 2: Using the household's log income as the threshold variable T and the endogenous variable X_e at the same time, we present the estimation results in Table 5.7. In this case we specify the coefficients in the reduced form equation to be constant without threshold, similar to that in [118]. Both the CH method and the proposed CPD method can split the data into two groups while the HHB method in [118] fails to detect any subgroup. For the proposed method, the first group is of size 96 (log income $\leq$ 3.093) while the second group is of size 307. For CH method, the first group is of size 358 (log

Table 5.7: Analysis results for Philippine households study with the household's log income as the threshold variable (Analysis 2). "CPD": the proposed method; "CH": the method proposed by [43]; "Est": estimated coefficients; "SE": standard error for estimated coefficients; "P": p value for the estimated coefficients. The tuning parameter chosen by BIC for SCAD is $\lambda_n = 0.0316$.

		Subgroup 1			Subgroup 2		
	Method	Est	SE	P	Est	SE	P
log income	CPD	0.136	0.432	0.000	1.153	0.178	0.000
	CH	1.829	0.34	0.000	1.595	0.611	0.013
mother's education	CPD	0.079	0.060	0.187	0.014	0.083	0.864
	CH	0.149	0.052	0.005	−0.110	0.215	0.611
father's education	CPD	−0.086	0.195	0.100	−0.041	0.079	0.600
	CH	−0.179	0.071	0.012	0.085	0.196	0.671
mother's age	CPD	−0.349	0.087	0.000	−0.155	0.107	0.146
	CH	−0.315	0.067	0.000	0.558	0.309	0.079
father's age	CPD	0.262	0.099	0.008	−0.076	0.104	0.464
	CH	0.088	0.067	0.185	−0.981	0.285	0.001
mother's nutritional knowledge	CPD	0.019	0.062	0.763	0.060	0.071	0.397
	CH	−0.087	0.064	0.173	0.243	0.219	0.274
price of corn	CPD	0.013	0.044	0.770	0.068	0.072	0.348
	CH	0.124	0.042	0.003	0.071	0.243	0.771
price of rice	CPD	−0.016	0.062	0.791	0.029	0.066	0.662
	CH	0.068	0.048	0.165	−0.071	0.165	0.672

income $\leq$ 4.827) while the second group is of size 45. According to the CPD method, the estimated coefficient for log income is 0.136 for group 1 and 1.153 for group 2, indicating a significant difference between the two groups. Specifically, low-income families exhibit a much weaker effect compared to high-income families. On the other hand, the CH method estimates the coefficient for log income to be 1.829 for the first group and 1.595 for the second group. These two groups, identified by the CH method, appear less distinct compared to the two groups detected by our method, where the disparity in the effects of income on the two subgroups is more pronounced.

5.4 Tree methods and recursive partitioning

Data partitioning refers to divide the domain space of covariates into subsets where a local regression function can be specified. This idea is very similar to the threshold model we discuss in this chapter.

The initial partition can be coarse, often just two subsets. However, it becomes more refined with each iteration as the sets within the partition are split—typically into two subsets (binary segmentation), though occasionally into more. In many cases, the local regression within a partition element is simply a constant, representing the mean response from the covariates corresponding to a set from the partition.

Trees, as a modeling and prediction technique, differ significantly from linear models in several key aspects:

- Region-Dependent Interactions: Trees encapsulate region-dependent interactions by employing decision thresholds. This approach contrasts with multiple regression, where interactions are typically represented as product terms that are not localized. Decision thresholds allow trees to capture complex interactions within specific regions of the feature space.

- Discontinuous Regression Functions: Trees fit regression functions that are discontinuous at the boundaries of regions. While this may be considered a drawback if the response is believed to be smooth, from a predictive standpoint, it often is not a serious concern. Trees excel at capturing non-linear relationships and can effectively model discontinuities in the data.
- Difficulty in Approximating Some Functions: Like any modeling method, trees may struggle to approximate certain functions. It is important to carefully consider when to use tree models, opting for situations where the relationship between predictors and response can be easily described by regions.
- Adaptability to High Dimensions: Recursive partitioning methods, such as tree-based models, are well-suited to high-dimensional data. The sparsity of data in high dimensions naturally leads to rougher models, and recursive partitioning methods can effectively capture these patterns. However, in low dimensions, tree-based methods may not be as competitive compared to additive smooth models.

In summary, tree-based models offer a unique approach to modeling complex interactions and non-linear relationships in data. While they may not be suitable for every scenario, they excel in high-dimensional data settings and are particularly effective when relationships can be easily described by distinct regions in the feature space.

5.4.1 Decision tree

We sketch a simple roadmap for growing a decision tree in this section. There are many variants in the literature. Suppose we have $p-$dimensional predictor (independent variable) $X = (\boldsymbol{x}_1, ..., \boldsymbol{x}_p)$ and response (dependent variable) Y. Given $X = x$, we aim at predicting

$$m(x) = E(T|X = x).$$

The main objective of the tree method is to model the multi-dimensional regression surface

$$m(x_1, ..., x_p) = E(Y|\boldsymbol{x}_1 = x_1, ...\boldsymbol{x}_p = x_p).$$

Instead of using a simple linear function, the surface is assumed to involve piece-wise constants. Mathematically, we assumes that the sample space can be partitioned into H regions $\{N_1, ..., N_H\}$ and we approximate the conditional mean function by

$$m(x_1, ..., x_p) \approx \sum_{h=1}^{H} c_h I(x \in N_h)$$

where $I(.)$ is the indicator function.

In practice, if we have samples $S = \{(\boldsymbol{x}_{i,1}, ..., \boldsymbol{x}_{i,p}, Y_i) : i = 1, ..., n\}$, then c_h is estimated as

$$\hat{c}_h = \frac{\sum_{i=1}^{n} Y_i I(X_i \in N_h)}{\sum_{i=1}^{n} I(X_i \in N_h)}.$$

The main computational challenge in tree-based models is partitioning the feature space effectively. A regression tree achieves this partitioning through a recursive binary splitting method. The key task involves deciding both the splitting variables (the features used for partitioning) and the split points (also called nodes or thresholds). The following is a short summary of the procedure.

- First, calculate the cross-validation (CV) squared error for the simplest model (no splitting)

$$Y = c + \xi, \tag{5.24}$$

where ξ is a generic notation for error in this section. After fitting the model we denote

$$CV^{(0)} = n^{-1}\sum_{i=1}^{n}(Y_i - \bar{Y}_i)^2$$

where $\bar{Y}_i = (Y_1 + ... + Y_{i-1} + Y_{i+1} + ... + Y_n)/(n-1)$.

- For each variable $\boldsymbol{x}_k$ and a node x'_k, we consider a model

$$Y = \begin{cases} \stackrel{\boldsymbol{x}_k \leq x'_k}{=} c_1 + \xi \\ \stackrel{\boldsymbol{x}_k > x'_k}{=} c_2 + \xi \end{cases} \tag{5.25}$$

where $c_1 \neq c_2$. Clearly, we can choose x_k to minimize

$$Q_k = \min_{c_1, c_2, x_k} \sum_{\boldsymbol{x}_{i,k} \leq x_k} (Y_i - c_1)^2 + \sum_{\boldsymbol{x}_{i,k} > x_k} (Y_i - c_2)^2.$$

It is easy to notice that for a fixed x_k, the estimated values of c_1 and c_2 are simply the group means of Y_i in the two divided regions $\{\boldsymbol{x}_{i,k} \leq x_k\}$ and $\{\boldsymbol{x}_{i,k} > x_k\}$, respectively. We further denote the optimal value of x_k from the above minimization by x'_k.

- For each partition (i.e. each chosen $\boldsymbol{x}_k$ and x'_k), calculate its leave-one-out CV value. For each observation j, find $x'_k(j), c_1(j), c_2(j)$ to minimize

$$\sum_{\boldsymbol{x}_{i,k} \leq x_k, i \neq j} (Y_i - c_1)^2 + \sum_{\boldsymbol{x}_{i,k} > x_k, i \neq j} (Y_i - c_2)^2$$

and predict Y_j by $\hat{Y}_j = I(\boldsymbol{x}_{j,k} \leq x'_k(j))c_1(j) + I(\boldsymbol{x}_{j,k} > x'_k(j))c_2(j)$, and calculate the squared prediction error

$$cv_k(j) = (Y_j - \hat{Y}_j)^2 = I(\boldsymbol{x}_{j,k} \leq x'_k(j))(Y_j - c_1(j))^2 + I(\boldsymbol{x}_{j,k} > x'_k(j))(Y_j - c_2(j))^2$$

and the CV value

$$CV_k = n^{-1}\sum_{j=1}^{n} cv_k(j)$$

Compare all the $CV_k, k = 1, ..., p$. The partition with the smallest CV is identified as the first splitting variable and the associated node. Suppose the selection is $\boldsymbol{x}_1$ with x'_1. Denote the corresponding CV values by $CV^{(1)}$.

- If $CV^{(1)} > CV^{(0)}$, we stop the procedure and the final model is (5.24). Otherwise if $CV^{(1)} < CV^{(0)}$ we have to continue the binary splitting. Consider the sets $S_1 = \{(\boldsymbol{x}_{i,1}, ..., \boldsymbol{x}_{i,p}, y_i) : \boldsymbol{x}_{i,1} > x'_1\}$ and $S_2 = \{(\boldsymbol{x}_{i,1}, ..., \boldsymbol{x}_{i,p}, y_i) : \boldsymbol{x}_{i,1} \leq x'_1\}$. Apply the same computational procedure (as to the full sample S) to S_1 and S_2. Repeat the computation procedure until no more splitting is needed (i.e. CV does not descend any more).

The outcome variable can also be categorical, in which case the resulting method is known as a classification tree. Together with regression trees, this forms the well-known classification and regression tree (CART) method, or simply a decision tree. Although various CART procedures may employ different loss functions, they all share the common goal of identifying the optimal splitting variable and splitting points to partition the data and improve predictive accuracy.

There are quite a few famous packages in `R` to implement the tree method, including `tree`, `rpart`, among others. The following is the syntax for the `tree` function documented in `R`.

R code: tree(...), plot(...), text(...), predict(...)

- object = tree(formula, subset =??, ...)
  - formula: y∼ x1+x2+..., or y∼ X (where X is a matrix of n×p)
  - To show the result (nodes, and values in each nodes): object
- plot(object): to plot the regression tree
- text(object): to put text in the plot
- predict(object,newdata, type="class"/"vector"): if it is for classification, "class" return the classified classes, "vector" probabilities of classification
- factor(y): if y is categorical but labeled by numbers, using factor() is essential

Section 3.2 of Fryzlewicz (2007) [101] explores the approximation of functions belonging to different smoothness classes using the Unbalanced Haar transform as basis vectors. These basis functions are piecewise constant by construction, making them particularly effective for approximating underlying functions that exhibit similar structural characteristics. Indeed, according to the result of Theorem 3.1 in that paper, when f is in the class of functions of bounded variation the expected integrated mean squared error (IMSE) of the Haar transform based estimate attains a rate of $n^{-2/3}$ up to a logarithmic factor that involves the sample size as well as certain features of the approximation basis. On the other hand, when f is in $S[0,1]$, the space of piecewise constant functions with finitely many jumps, the rate improves to n^{-1}, again up to logarithmic terms and the number of jumps of the function.

In addition to CART, another alternative and closely related procedure is GUIDE [176] which employs the local linear regression instead of computing the local mean in the above calculation.

5.4.2 Ensemble learning

When used for classification, tree methods usually do not perform very well and have relatively low prediction accuracy. In machine learning literature, weak classifiers such as the decision tree are also referred to as weak learners. In order to improve the performance, we may resort to an ensemble learning approach to evolve the weak learner to a stronger version.

Ensemble learning is a powerful technique in machine learning that improves predictive performance by combining multiple models, allowing for greater accuracy and stability compared to relying on a single model. The core idea is to train a set of classifiers, often referred to as "experts", and combine their predictions, typically through a voting mechanism. Two widely used ensemble learning techniques are Bagging and Boosting. Both approaches focus on reducing the variance of the model by combining outputs from multiple models, but they operate in different ways:

- Bagging (Bootstrap Aggregating): Bagging builds multiple independent models by training each model on different random subsets of the data. The models are trained in parallel and the predictions are combined (e.g., via averaging or majority voting). The goal is mainly to reduce variance and improve the overall accuracy by aggregating the results of weak learners. A typical example of bagging is the Random Forest.

- Boosting: Boosting works sequentially, where each model is trained to correct the mistakes of its predecessor. Each learner is weighted according to its performance, and subsequent learners focus on the instances that were misclassified by previous ones. Boosting gradually improves prediction accuracy by building a series of weak learners that adapt and improve based on previous models' mistakes. Common examples include AdaBoost, Gradient Boosting. A key feature for boosting is that it enhances model accuracy by focusing on the hard-to-predict cases and iteratively refining the model.

5.5 Change plane model

5.5.1 Why do we need this model?

In this section, we explore a more recent extension of the threshold model, known as the change plane model. While the traditional threshold model (5.1) uses a single variable as the thresholding variable, the change plane model allows multiple variables to serve as thresholding variables, leading to a more general and flexible modeling framework. This extension is particularly motivated by recent advances in personalized medicine, where identifying subgroups with enhanced treatment effects is a central focus. Defining such subgroups often necessitates a change plane structure, which accommodates more complex relationships than the simple change point model.

Nonetheless, the change plane model introduces additional computational challenges, making it more difficult to fit compared to the threshold model. To address these challenges, we employ the induced smoothing approach, where the indicator function is replaced with a cumulative distribution function (CDF). This smoothing technique facilitates the computation by providing a differentiable approximation of the thresholding mechanism, enabling more efficient estimation of the model parameters.

Seo and Linton [222] studied a linear regression model with regime switching, where the form of the linear regression changes based on whether a specific subset of covariates lies above or below a hyperplane, the parameters of which are also unknown. This problem is essentially a change-plane problem, extending the traditional change-point model by introducing multidimensional thresholding variables. To avoid the complexities introduced by non-standard distributions that typically arise under a regular least squares approach, the authors employ a smooth integrated kernel function (akin to a distribution function) to replace the indicator function in the least squares criterion. This approach follows the spirit of the smoothed maximum score estimator proposed by Horowitz [125], facilitating a more tractable estimation process while maintaining the flexibility of regime switching models.

Individualized learning and modeling have gained significant attention in statistics and computer science, particularly in addressing personalized medical treatment problems. The traditional "one size fits all" approach often fails to capture critical patterns in sub-populations, making it inadequate for personalized predictions tailored to individual patients. For instance, in the fight against cancer and other diseases, finding a universal treatment that works for all patients is challenging. This has spurred the growth of precision medicine and the analysis of electronic health record (EHR) data, encouraging researchers to identify meaningful subgroups. By doing so, they can model the relationships between response variables and predictors in a way that accounts for the unique characteristics of each subgroup, ultimately improving treatment recommendations and outcomes for specific individuals.

Earlier advancements in personalized medicine primarily focused on determining dynamic treatment regimes at multiple stages. Some of the most popular model-based approaches for estimating the optimal individualized treatment regimes include Q-learning [205, 109] and A-learning

[191, 214, 219]. A-learning models interactions between treatments and covariates and tends to be more robust to model misspecification than Q-learning. In addition, [287] introduced the framework of outcome weighted learning (O-learning), which directly identifies the optimal binary treatment rule from a classification perspective. Other notable works in this area include [279], [280], and [288]. More recently, [252] developed a forest-based method for treatment effect estimation, providing a tractable asymptotic theory for its results.

In addition to these optimization-based learning strategy, another burgeoning research direction in personalized medicine is categorizing patients into subgroups using appropriate algorithms and then consider the treatment effects for those subgroups. Many data-driven approaches for subgroup identification have been developed in the literature. One commonly used approach is the tree-based method. Early works include Automatic Interaction Detection (AID) [188] and theta automatic interaction detection (THAID) [186]. [172] developed the generalized unbiased interaction detection and estimation (GUIDE) to identify subgroups of subjects for whom the treatment has an enhanced effect. [98] proposed a virtual twins (VT) to obtain the subgroups with an enhanced treatment effect. [41] and [286] used a parametric scoring system to estimate subject-specific treatment differences and then identify a promising population who benefit more from the new treatment. [226] adapted a finite logistic-normal mixture model to subgroup analysis by a likelihood-based test. [52] propose a general framework for subgroup identification by weighting and A-learning approaches. All the aforementioned works used similar techniques to those in change point analysis [16] and can be justified rigorously using the traditional theory. Recently, [89] considered a change plane method to test the existence of subgroup using a doubly robust score statistic. The advantage of change plane over change point is that we may allow the underlying grouping variable to be a linear combination of covariates instead of a single covariate. However, the approach in [89] only allows a single threshold (and thus only two subgroups) and searching the supremum of squared score test statistics over a unit ball could be quite challenging, especially when extending to multiple groups. In addition, as more information per individual is being collected in clinical studies and not all of the information is relevant for subgroup detection.

To formally address the issue, Li et al. (2021) [163] considered a change plane model with unknown number of multiple thresholds which extends the familiar change point model. In fact, the change point model has wide applications in economics [247, 158, 148] where the underlying grouping variable is usually the time point or a chosen regressor and subgroups are identified as the grouping variable moves across thresholds. For the single threshold change point model, [122] developed the asymptotic results for the threshold parameter estimate based on the diminishing effect assumption. [222] proposed a smoothed least squares estimator and established the consistency and asymptotic normality following the well-known smoothed maximum score estimator ([125]). Detecting multiple thresholds is a much more challenging problem since we need to first estimate the number of thresholds and then determine their exact locations. [161] proposed a penalization-based framework for the accelerated failure time regression model. They formulated the threshold problem as a group model selection problem and applied the fast computing tool in [135].

The model to be introduced in this section allows multiple change planes which automatically generates subgroups with different covariate effects. The technical difficulty mainly lie in the following three aspects. First, instead of using only a pre-assigned index variable in change point model [161] introduced in preceding sections, the notion of change plane allows a linear combination of the covariates and may lead to more meaningful definition of subgroups. This framework may offer a more flexible tool for precision medicine than other similar proposals. The inference for plane-related parameters is not standard as the objective function is not smooth. Second, the change plane model may include multiple unknown structural changes. This is a non-trivial improvement from single threshold models because of the difficulty in determining the number of break points. A fast splitting strategy needs to be developed to convert the threshold identification problem into a model selection problem. Third, we need to investigate moderate- to high-dimensional covariates for all the subgroups in order to obtain sparse solutions[178, 270, 234]. This goal can be achieved through a penalized induced smoothing estimation approach. We will provide the consistency of subgroup detection and asymptotic theory for the penalized estimates.

The rest of this section is arranged as follows. In Section 5.5.2, a penalized induced smoothing estimation is proposed for the single threshold change plane model. In Section 5.5.3, the multi-threshold change plane regression for subgroup detection is formulated. We adopt an iterative two-stage procedure to detect the change planes and estimate model parameters. The theoretical properties of the estimation procedure are established rigorously under technical conditions in the following sections. Two empirical applications are presented in Section 5.5.6. A discussion concludes Section 5.5.7.

Throughout the section, $\mathbf{1}_q = (1, \ldots, 1)^{\mathrm{T}}$ is the q-dimensional vector of ones, $\mathbf{1}(\cdot)$ is an indicator function, I_q is the $q \times q$ identity matrix, and $\boldsymbol{X}^{\mathrm{T}}$ is the transpose of a matrix $\boldsymbol{X}$. For a vector $\boldsymbol{a}$, $\boldsymbol{a}^{\mathrm{T}}$ is its transpose, a_j is its jth component, and $|\boldsymbol{a}|$, $\|\boldsymbol{a}\|$ and $\|\boldsymbol{a}\|_\infty$ are respectively its L_1-norm, L_2-norm (Euclidean norm) and L_∞ norm. For any matrix $\mathbf{M}$, $\|\mathbf{M}\|_{\max} = \max\{|M_{ij}|\}$ denotes the matrix max norm. If $\mathcal{A}$ is a set, its complement and its size are respectively denoted by $\mathcal{A}^c$ and $|\mathcal{A}|$. In addition, "$\rightarrow_{a.s}$" denotes convergence with probability 1 and "$\xrightarrow{D}$" denotes convergence in distribution.

5.5.2 Single-threshold change plane (SCPL)

Let us denote $Y_i \in \mathbb{R}$ to be the response variable of interest for the ith subject in a sample of size n. We first consider the following single threshold change plane model:

$$Y_i = \boldsymbol{X}_i^{\mathrm{T}}\boldsymbol{\beta} + \boldsymbol{X}_i^{\mathrm{T}}\boldsymbol{\delta}\mathbf{1}(\boldsymbol{Z}_i^{\mathrm{T}}\boldsymbol{\theta} \geq 0) + \epsilon_i, \quad i = 1, \ldots, n, \tag{5.26}$$

where $\boldsymbol{X}_i = (X_{i1}, \ldots, X_{ip})^{\mathrm{T}}$ is a p-dimensional vector, regression coefficients $\boldsymbol{\beta} = (\beta_1, \ldots, \beta_p)^{\mathrm{T}}$ and $\boldsymbol{\delta} = (\delta_1, \ldots, \delta_p)^{\mathrm{T}}$ are the covariate effects for the baseline group and the effect differences between the two groups. We also observe the grouping variables $\boldsymbol{Z}_i \in \mathbb{R}^{d+1}$ where the first element of $\boldsymbol{Z}_i$ is assumed to be constant one. The corresponding coefficient $\boldsymbol{\theta} = (\theta_0, \theta_1, \ldots, \theta_d)^{\mathrm{T}}$ is a $(d+1)$-dimensional vector. For the sake of identifiability, we assume that $\|\boldsymbol{\theta}\| = 1$. We assume $E(\epsilon_i|\boldsymbol{X}_i) = 0$ and does not impose additional distribution assumption on the error terms. When $d = 0$, the threshold variable is univariate and thus model (5.26) reduces to the change point model (5.1).

Similar models have been considered in [222] and [89] where only two comparison groups are considered. This kind of model itself may be of interest in many clinical applications and therefore we provide a new yet relatively simple solution first. In the next section we will consider more general multi-threshold model which requires more sophisticated procedures.

Denote $\boldsymbol{\gamma} = (\boldsymbol{\beta}^{\mathrm{T}}, \boldsymbol{\delta}^{\mathrm{T}})^{\mathrm{T}}$. The unknown parameters $\boldsymbol{\gamma}$ and $\boldsymbol{\theta}$ in model (5.26) can be estimated by minimizing the following objective function

$$\frac{1}{n}\sum_{i=1}^{n}[Y_i - \boldsymbol{X}_i^{\mathrm{T}}\boldsymbol{\beta} - \boldsymbol{X}_i^{\mathrm{T}}\boldsymbol{\delta}\mathbf{1}(\boldsymbol{Z}_i^{\mathrm{T}}\boldsymbol{\theta} \geq 0)]^2 \tag{5.27}$$

with constraint $\|\boldsymbol{\theta}\| = 1$. In addition, when p is large we usually assume a sparse structure for $\boldsymbol{\gamma}$. Motivated by the idea of penalization such as [91], we obtain $(\tilde{\boldsymbol{\gamma}}^{\mathrm{T}}, \tilde{\boldsymbol{\theta}}^{\mathrm{T}})^{\mathrm{T}} = \arg\min_{\|\boldsymbol{\theta}\|=1} Q_n(\boldsymbol{\gamma}, \boldsymbol{\theta})$, where

$$Q_n(\boldsymbol{\gamma}, \boldsymbol{\theta}) = \frac{1}{n}\sum_{i=1}^{n}[Y_i - \boldsymbol{X}_i^{\mathrm{T}}\boldsymbol{\beta} - \boldsymbol{X}_i^{\mathrm{T}}\boldsymbol{\delta}\mathbf{1}(\boldsymbol{Z}_i^{\mathrm{T}}\boldsymbol{\theta} \geq 0)]^2 + p_{\lambda_n}(|\boldsymbol{\gamma}|), \tag{5.28}$$

and $p_{\lambda_n}(\cdot)$ is a penalty function with a regularization parameter λ_n. For the simplicity of presentation, we only consider two well-studied penalty functions in this book, namely the smoothly clipped absolute deviation (SCAD, [91]) penalty and the minimax concave-plus penalty (MCP, [281]).

We can consider an iterative estimation procedure. For a given $\boldsymbol{\theta}$, model (5.26) can be treated as a piecewise linear model, then the baseline coefficients $\boldsymbol{\beta}$ and enhanced effects $\boldsymbol{\delta}$ can be estimated by the penalized least squares. On the other hand, given γ, the objective function (5.27) or (5.28) is not continuous and finding its minimizer is difficult. One way to overcome this difficulty is to approximate the discontinuous objective function with a smooth function ([137, 222]). We can show that the estimated results of the smoothed objective function have very similar asymptotic properties as those from the original non-smooth version.

Denote $\Phi(\cdot)$ to be the distribution function of the standard normal variable. We use $\Phi(\cdot/h)$ as a smooth approximation to the indicator function, where the bandwidth h is chosen to converge to zero as the sample size n increases. Note that if $\boldsymbol{Z}_i^{\mathrm{T}}\boldsymbol{\theta} > 0$, $\Phi((\boldsymbol{Z}_i^{\mathrm{T}}\boldsymbol{\theta})/h) \to 1$ as $h \to 0$. Thus we may consider the following approximate penalized objective function for the estimation

$$Q_n^*(\gamma, \boldsymbol{\theta}) = \frac{1}{n}\sum_{i=1}^{n}[Y_i - \boldsymbol{X}_i^{\mathrm{T}}\boldsymbol{\beta} - \boldsymbol{X}_i^{\mathrm{T}}\boldsymbol{\delta}\Phi((\boldsymbol{Z}_i^{\mathrm{T}}\boldsymbol{\theta})/h)]^2 + p_{\lambda_n}(|\gamma|). \tag{5.29}$$

This becomes a relatively standard nonlinear least squares problem ([110]). Denote $(\tilde{\gamma}^{*\mathrm{T}}, \tilde{\boldsymbol{\theta}}^{*\mathrm{T}})^{\mathrm{T}} = \arg\min_{\|\boldsymbol{\theta}\|=1} Q_n^*(\gamma, \boldsymbol{\theta})$. The smoothed objective function, $Q^*(\gamma, \boldsymbol{\theta})$, is now continuously differentiable and standard numerical methods can be used to efficiently compute $\tilde{\boldsymbol{\theta}}^*$. Our estimation procedure is described in details as follows.

- *Step 0*: Given an initial estimate of $\boldsymbol{\theta}$, say $\tilde{\boldsymbol{\theta}}^*_{int}$, and set $\tilde{\boldsymbol{\theta}}^{*[0]} = \tilde{\boldsymbol{\theta}}^*_{int}/\|\tilde{\boldsymbol{\theta}}^*_{int}\|$. Then, obtain $\tilde{\gamma}^{*[0]}$by ordinary least squares.

- *Step 1*: Given $\tilde{\gamma}^{*[k]} = (\tilde{\boldsymbol{\beta}}^{*[k]\mathrm{T}}, \tilde{\boldsymbol{\delta}}^{*[k]\mathrm{T}})^{\mathrm{T}}$, estimate $\boldsymbol{\theta}$ by solving

$$\tilde{\boldsymbol{\theta}}^{*[k+1]} = \arg\min_{\|\boldsymbol{\theta}\|=1}\left\{\frac{1}{n}\sum_{i=1}^{n}[Y_i - \boldsymbol{X}_i^{\mathrm{T}}\tilde{\boldsymbol{\beta}}^{*[k]} - \boldsymbol{X}_i^{\mathrm{T}}\tilde{\boldsymbol{\delta}}^{*[k]}\Phi((\boldsymbol{Z}_i^{\mathrm{T}}\boldsymbol{\theta})/h)]^2)\right\}.$$

- *Step 2*: Given $\tilde{\boldsymbol{\theta}}^{*[k+1]}$, write $\mathbb{X}_i^{[k+1]} = (\boldsymbol{X}_i^{\mathrm{T}}, \boldsymbol{X}_i^{\mathrm{T}}\Phi(\boldsymbol{Z}_i^{\mathrm{T}}\tilde{\boldsymbol{\theta}}^{*[k+1]}/h))^{\mathrm{T}}$, estimate γ by minimizing the following regularized least squares with a SCAD or MC+ penalty

$$\tilde{\gamma}^{*[k+1]} = \arg\min\left\{\frac{1}{n}\sum_{i=1}^{n}[Y_i - \mathbb{X}_i^{[k+1]}\gamma]^2 + p_{\lambda_n}(|\gamma|)\right\}.$$

- *Step 3*: Repeat Step 1 and Step 2 until convergence.

Remark 5.2. *In Step* 1, *a modified Newton-Raphson algorithm can be used to estimate $\tilde{\boldsymbol{\theta}}$ by normalizing $\boldsymbol{\theta}$ in every iteration. In practice, one can adopt function* `BBoptim` *in R package* BB *to optimizing a high-dimensional nonlinear objective function. More detailed descriptions of separable nonlinear least squares problems and the convergence properties of related algorithms can be found in [110] and references therein. In Step 2, $\tilde{\gamma}^{*[k+1]}$ can be obtained by the efficient coordinate descent algorithms ([37]). Moreover, other penalty methods can also be applied, such as the weighted lasso ([156]). The tuning parameters λ_n can be chosen by the Bayesian information criterion (BIC) or generalized cross validation method (GCV).*

5.5.3 Multi-threshold change planes (MCPL)

With a slight abuse of notation, we still use $\boldsymbol{Z}_i$ in the following presentation to denote the $d-$vector of grouping variables without the intercept one. We now consider change plane model with multiple thresholds and assume $\{(Y_i, \boldsymbol{X}_i^{\mathrm{T}}, \boldsymbol{Z}_i^{\mathrm{T}})^{\mathrm{T}}, i = 1, \ldots, n\}$ follows the change-plane model with s thresholds located at $a_1 < a_2 < \cdots < a_s$:

$$Y_i = \boldsymbol{X}_i^{\mathrm{T}} \left[\boldsymbol{\beta} + \sum_{j=1}^{s} \boldsymbol{\delta}_j \mathbf{1}(a_j < \boldsymbol{Z}_i^{\mathrm{T}} \boldsymbol{\theta} \leq a_{s+1}) \right] + \epsilon_i, \quad i = 1, \ldots, n, \tag{5.30}$$

where $\boldsymbol{\theta}$ is the change-plane parameter, $\boldsymbol{\beta}$ is the vector of coefficients for the baseline group and $\boldsymbol{\delta}_j$ is the vector of enhanced effects for the jth subgroup relative to the baseline group. In this case $s \geq 0$ is also unknown and needs to be estimated and $a_1, \ldots, a_s$ are the locations. We set $a_0 = -\infty$, and $a_{s+1} = \infty$. ϵ_i's are independent random errors with mean zero and variance σ^2. To identify the model, we need to assume $\boldsymbol{\theta} \in \boldsymbol{\Theta} = \{\boldsymbol{\theta} \in \mathbb{R}^d : \|\boldsymbol{\theta}\| = 1, \theta_r > 0, 1 \leq r \leq d\}$ with the r-th element being positive. When $d = 1$, the threshold variable is univariate and thus model (5.30) reduces to the change point model (5.1).

Denote $\boldsymbol{\eta} = (\boldsymbol{\beta}^{\mathrm{T}}, \boldsymbol{\delta}^{\mathrm{T}}, \boldsymbol{a}^{\mathrm{T}}, \boldsymbol{\theta}^{\mathrm{T}})^{\mathrm{T}}$. If $a_j, j = 1, \ldots, s$ are known, then the unknown parameters $\boldsymbol{\beta}, \boldsymbol{\delta}$, and $\boldsymbol{\theta}$ can be estimated by minimizing the following objective function with constraint $\|\boldsymbol{\theta}\| \in \boldsymbol{\Theta}$,

$$L_n(\boldsymbol{\eta}) = \frac{1}{n} \sum_{i=1}^{n} \left[Y_i - \boldsymbol{X}_i^{\mathrm{T}} \boldsymbol{\beta} - \boldsymbol{X}_i^{\mathrm{T}} \sum_{j=1}^{s} \boldsymbol{\delta}_j \mathbf{1}(a_j < \boldsymbol{Z}_i^{\mathrm{T}} \boldsymbol{\theta} \leq a_{s+1}) \right]^2. \tag{5.31}$$

In general, however, the number of change-planes s and the locations are all unknown. Estimation and establishing the relevant limiting distribution for $(\boldsymbol{a}, \boldsymbol{\theta})$ may be non-trivial. Moreover, locating the global minimum of the least squares criterion requires a multi-parameter grid search over all possible values of the s threshold parameters, which is typically computational infeasible. In fact, when s is unknown, [111] suggested a sequential estimation procedure for choosing s, under the homoscedasticity assumption and without the change plane parameter $\boldsymbol{\theta}$. We are not aware of any results for more general models.

Li et al. (2021) [163] proposed an iterative two-stage procedure for multi-threshold change plane estimation. Observe that once $(\boldsymbol{\theta}, \boldsymbol{a})$ is fixed, $(\boldsymbol{\beta}, \boldsymbol{\delta})$ can be estimated by a least square estimation. Therefore, given a consistent estimation $\hat{\boldsymbol{\theta}}$ in the first stage we can obtain a consistent estimation of s using a change point detection algorithm. After we obtain $\hat{s}$, we can use the induced smoothing approach introduced in section 5.5.2 to estimate $(\boldsymbol{\beta}, \boldsymbol{\delta}, \boldsymbol{\theta}, \boldsymbol{a})$ in the second stage. The details are as follows.

The Splitting Stage. For a given estimator $\hat{\boldsymbol{\theta}}$, we denote $\hat{W}_i = \boldsymbol{Z}_i^{\mathrm{T}} \hat{\boldsymbol{\theta}}, i = 1, \ldots, n$. We then generate the rank mapping $\{\iota_{(i)} : 1 \leq i \leq n\}$ such that $\hat{W}_{\iota_{(i)}}$ is the i-th smallest value in $\{\hat{W}_i : 1 \leq i \leq n\}$, and can be arranged in the ascending order, that is, $\hat{W}_{\iota_{(1)}} \leq \hat{W}_{\iota_{(2)}} \leq \cdots \leq \hat{W}_{\iota_{(n)}}$. First we split the data sequence into $q_n + 1$ segments based on $\hat{W}_{\iota_{(i)}}$ where q_n tends to infinity as $n \to \infty$. The data sequence is split such that the first segment $\mathcal{I}_1 = \{i : \hat{W}_i \leq \hat{W}_{\iota_{(n - q_n m)}}\}$ involves $n - q_n m$ observations, and each of the other q_n segments $\mathcal{I}_j = \{i : \hat{W}_{\iota_{(n-(q_n - j+2)m)}} < \hat{W}_i \leq \hat{W}_{\iota_{(n-(q_n-j+1)m)}}\}$, $j = 2, ..., q_n + 1$ involves m observations where $m = \lceil n/q_n \rceil$.

Let $\boldsymbol{Y}_{(j)} = (Y_i, i \in \mathcal{I}_j)^{\mathrm{T}}$, $\boldsymbol{X}_{(j)} = (\boldsymbol{X}_i, i \in \mathcal{I}_j)^{\mathrm{T}}$. Denote $\tilde{\boldsymbol{Y}} = (\boldsymbol{Y}_{(1)}^{\mathrm{T}}, \ldots, \boldsymbol{Y}_{(q_n+1)}^{\mathrm{T}})^{\mathrm{T}}$, $\tilde{\boldsymbol{X}} = (\boldsymbol{X}^{(1)}, \ldots, \boldsymbol{X}^{(q_n+1)})$ where $\boldsymbol{X}^{(1)} = (\boldsymbol{X}_{(1)}^{\mathrm{T}}, \ldots, \boldsymbol{X}_{(q_n+1)}^{\mathrm{T}})^{\mathrm{T}}$ and $\boldsymbol{X}^{(j)} = (\mathbf{0}_{p \times \sum_{i=1}^{j-1} b_i}, \boldsymbol{X}_{(j)}^{\mathrm{T}}, \ldots, \boldsymbol{X}_{(q_n+1)}^{\mathrm{T}})^{\mathrm{T}}, j = 2, \ldots, q_n + 1$. The estimator $\tilde{\gamma}^* = (\tilde{\boldsymbol{\beta}}_1^{\mathrm{T}}, \tilde{\boldsymbol{\delta}}_1^{\mathrm{T}}, \ldots, \tilde{\boldsymbol{\delta}}_{q_n}^{\mathrm{T}})^{\mathrm{T}}$ can be written as

$$\tilde{\gamma}^* = \arg\min_{\gamma} \left\{ \frac{1}{n} \|\tilde{\boldsymbol{Y}} - \tilde{\boldsymbol{X}} \gamma^* \|^2 + \sum_{j=1}^{q_n} p_{\lambda_n}(\|\boldsymbol{\delta}_j\|) \right\}, \tag{5.32}$$

We apply the group coordinate descent (GCD) algorithm to estimate $\tilde{\gamma}^*$ from (5.32). For simplicity, we write the estimator $\tilde{\gamma}^* = (\tilde{\gamma}_1^{*\mathrm{T}}, \ldots, \tilde{\gamma}_{q_n+1}^{*\mathrm{T}})^{\mathrm{T}}$. Denote $\hat{\mathcal{A}} = \{j : \tilde{\gamma}_j^* \neq 0, j = 1, \ldots, q_n + 1\}$,

and

$$\hat{\mathcal{A}}^* = \{j : j \in \hat{\mathcal{A}}, j-1 \notin \hat{\mathcal{A}}, j = 2, \ldots, q_n+1\} = \{\hat{k}_1, \ldots, \hat{k}_{\hat{s}}\}, \quad \hat{k}_1 < \ldots < \hat{k}_{\hat{s}}, \tag{5.33}$$

which is a subset of $\hat{\mathcal{A}}$. It is obvious that if $j-1 \notin \hat{\mathcal{A}}$, $j \in \hat{\mathcal{A}}$ and $j+1 \in \hat{\mathcal{A}}$, then $j \in \hat{\mathcal{A}}^*$ and $j+1 \notin \hat{\mathcal{A}}^*$. Therefore, with each estimator $\hat{\boldsymbol{\theta}}$, we obtain the estimated number of change planes $\hat{s} = |\hat{\mathcal{A}}^*|$. If the given estimator $\hat{\boldsymbol{\theta}}$ is consistent, then the estimated $\hat{s}$ in the splitting stage will also convergence with high probability. The consistency of $\hat{s}$ can be guaranteed by Theorem 5.5.1 in the next section. If $\hat{s} = 0$, we declare there is no subgroup. If $\hat{s} > 0$, according to the proof of Theorem 5.5.1, the true threshold a_j is highly likely to be located in $(\hat{W}_{\iota_{(n-(q_n-\hat{k}_j+3)m)}}, \hat{W}_{\iota_{(n-(q_n-\hat{k}_j+1)m)}}]$, $j = 1, \ldots, \hat{s}$. In the following step, we can refine the estimated thresholds $\hat{\boldsymbol{a}}$ and obtain all the regression coefficient estimates by an induced smoothing method.

The Smoothing Stage. Given an estimated number of change planes $\hat{s}$, we can estimate the parameters $\boldsymbol{\theta}$, $\boldsymbol{\gamma} = (\boldsymbol{\beta}_1^{\mathrm{T}}, \boldsymbol{\delta}_1^{\mathrm{T}}, \ldots, \boldsymbol{\delta}_{\hat{s}}^{\mathrm{T}})^{\mathrm{T}}$ and $\boldsymbol{a}$ in the model by minimizing the following smoothed objective function

$$L_n^*(\boldsymbol{\eta}) = \frac{1}{n}\sum_{i=1}^{n}\left[Y_i - \boldsymbol{X}_i^{\mathrm{T}}\boldsymbol{\beta} - \boldsymbol{X}_i^{\mathrm{T}}\sum_{k=1}^{\hat{s}}\boldsymbol{\delta}_k\Phi\left(\frac{\boldsymbol{Z}_i^{\mathrm{T}}\boldsymbol{\theta} - a_k}{h}\right)\right]^2. \tag{5.34}$$

Write $\tilde{\boldsymbol{\eta}}^* = \arg\min_{\boldsymbol{\theta}\in\boldsymbol{\Theta}}\{L_n^*(\boldsymbol{\eta})\}$. For a non-sparse problem, to minimize (5.34) we can use the familiar Newton-type algorithm. For a spares problem, similar to the single change plane model, a penalty function can be added in (5.34) to deal with the sparse parameterization among the coefficients $\boldsymbol{\gamma}$ such as Lasso, SCAD and MC+. Then $\boldsymbol{\eta}$ can be estimated by minimizing the following penalized objective function

$$Q_n^*(\boldsymbol{\eta}) = \frac{1}{n}\sum_{i=1}^{n}\left[Y_i - \boldsymbol{X}_i^{\mathrm{T}}\boldsymbol{\beta} - \boldsymbol{X}_i^{\mathrm{T}}\sum_{k=1}^{\hat{s}}\boldsymbol{\delta}_k\Phi\left(\frac{\boldsymbol{Z}_i^{\mathrm{T}}\boldsymbol{\theta} - a_k}{h}\right)\right]^2 + p_{\lambda_n}(|\boldsymbol{\gamma}|). \tag{5.35}$$

where $p_{\lambda_n}(\cdot)$ is the penalty function. Denote $\hat{\boldsymbol{\eta}}^* = \arg\min_{\boldsymbol{\theta}\in\boldsymbol{\Theta}}\{Q_n^*(\boldsymbol{\eta})\}$, which can be obtained by the iterative penalized induced smoothing procedure similar to that used in section 2.

We may iterate the splitting and smoothing stages many rounds until some convergence criterion is met. In particular, the iteration stops when the estimated number of change planes remains unchanged. The detailed algorithm is described in the following:

- *Step 0*: Given an initial estimate of $\boldsymbol{\theta}$, say $\hat{\boldsymbol{\theta}}_{int}^*$, and set $\hat{\boldsymbol{\theta}}^* = \hat{\boldsymbol{\theta}}_{int}^*/\|\hat{\boldsymbol{\theta}}_{int}^*\|$.
- *Step 1*: Implement the Splitting stage. Minimizing (5.32) and compute the index sets $\hat{\mathcal{A}}^*$ defined in (5.33), and obtain the number of thresholds by $\hat{s} = |\hat{\mathcal{A}}^*|$.
- *Step 2*: Given $\hat{s}$, update $(\hat{\boldsymbol{\theta}}^*, \hat{\boldsymbol{a}}^*, \hat{\boldsymbol{\gamma}}^*)$ by minimizing objective function (5.34) or (5.35) in the smoothing stage.
- *Step 3*: Continue Steps 1 and 2 until convergence.

Remark 5.3. *In the smoothing stage, the algorithm proposed in section 2 can be similarly adopted. The inclusion of the penalty functions may lead to a sparse solution since the number of parameters could be quite large with large number of subgroups. The oracle property of $\hat{\gamma}^*$ will be given in Theorem 6.3.3. We note that the objective function derived from the nonlinear model often has multiple local minima. Moreover, the computational efficiency might depend on the dimension of parameter space. The tuning parameter λ_n can be chosen by the BIC criterion under moderate- or high-dimensional situations ([94]). Finally, the performance of splitting stage depends on the segment length m, and the selection of an optimal m may follow the BIC method in [161].*

5.5.4 Asymptotic Properties

In this subsection, we study the theoretical properties of the estimation methods given in the preceding subsections. To establish the asymptotic theory, we impose the following necessary conditions to facilitate the technical proofs.

Condition 1. (a) $E(\boldsymbol{X}_i\boldsymbol{X}_i^{\mathrm{T}}) = \Sigma_0$ is finite and positive definite. $E(\boldsymbol{Z}_i\boldsymbol{Z}_i^{\mathrm{T}})$ is positive definite. $\boldsymbol{Z}_i$ and ϵ_i are independent, $i = 1, \ldots, n$. $E(\epsilon_i|\boldsymbol{X}_i) = 0$ almost surely. (b) Let $\boldsymbol{V}_i = (\boldsymbol{X}_i^{\mathrm{T}}, \boldsymbol{Z}_i^{\mathrm{T}})^{\mathrm{T}}$ be the vector of all the regressors, and for some $\xi > 1$, $0 < E\|\boldsymbol{V}_i\boldsymbol{V}_i^{\mathrm{T}}\|^{\xi} < \infty$, and $E\|\boldsymbol{V}_i\epsilon_i\|^{\xi} < \infty$. Furthermore, $E(\boldsymbol{X}_i\boldsymbol{X}_i^{\mathrm{T}}|\boldsymbol{Z}_i) > 0$ almost surely.

Condition 2. The parameter space of true parameter $\boldsymbol{\eta}$ are compact with $\min_{1\leq l<k\leq s}\{|a_l - a_k|\}$ and $\min\{\|\boldsymbol{\beta}/\sqrt{p}\|, \|\boldsymbol{\delta}_1/\sqrt{p}\|, \ldots, \|\boldsymbol{\delta}_s/\sqrt{p}\|\}$ bounded away from zero.

Let $\rho(t) = \lambda_n^{-1}p_{\lambda_n}(t)$ and $\bar{\rho}(t) = \rho'(|t|)\,\mathrm{sgn}(t)$. We assume that the penalty function $p_{\lambda_n}(\cdot)$ satisfies the following condition:

Condition 3. $p_{\lambda_n}(\cdot)$ is a symmetric function and it is nondecreasing and concave on $[0, \infty)$. There exists a constant $\nu > 0$ such that $\rho(t)$ is a constant for all $|t| \geq \nu\lambda_n$, and $\rho(0) = 0$. $\rho'(t)$ exists and is continuous except for a finite number of t, and $\rho'(0+) = 1$.

Denote $\mathcal{I}_j^0 = \{i : a_{j-1} < \boldsymbol{Z}_i^{\mathrm{T}}\boldsymbol{\theta} \leq a_j\}$ for $j = 1, \ldots, s+1$, with true vector of threshold locations $\boldsymbol{a}$ and change-plane $\boldsymbol{\theta}$. Similar to the definition of $\tilde{\boldsymbol{X}}$ and $\tilde{\boldsymbol{Y}}$, replacing $\mathcal{I}_j$ with the true split $\mathcal{I}_j^0$ and we can define $\tilde{\boldsymbol{X}}_{\boldsymbol{a}}$ and $\tilde{\boldsymbol{Y}}_{\boldsymbol{a}}$. By Condition 1, we have $\tilde{\boldsymbol{X}}_{\boldsymbol{a}}^{\mathrm{T}}\tilde{\boldsymbol{X}}_{\boldsymbol{a}}/n \rightarrow_{a.s} \Upsilon$, where Υ is a positive definite matrix. To obtain the asymptotic property of $\hat{\gamma}^*$ in (5.35), we assume the following condition:

Condition 4. $\max_{u\geq 0}\{p''_{\lambda_n}(u)\} + \Lambda_{(s+1)p}(\Upsilon) > 0$ where $\Lambda_{(s+1)p}(\Upsilon)$ is the minimal eigenvalue of Υ.

Condition 5. Let $W_i = \boldsymbol{Z}_i^{\mathrm{T}}\boldsymbol{\theta}$ be the true threshold, and $f_{w|V}(\cdot|V)$ denote the conditional density of W_i given $\boldsymbol{V}_i = V$ and $f(\cdot)$ the density of W_i, where given V, $f(\cdot|V)$ is of compact support and has a bounded second derivative. $P(W_i \leq a_j) = \tau_j$ where $0 < \tau_1 < \ldots < \tau_s < 1$. Furthermore, $E(\epsilon_i^4|\boldsymbol{V}_i) < M$ for some $M < \infty$.

Condition 6. $h^2\log(n) \rightarrow 0$ and $nh^2 \rightarrow \infty$.

Remark 5.4. *Condition 1 for the design matrix reduces to the Assumption* 1 *of [222] allowing for a regime specific heteroscedasticity. The error assumption can be relaxed to $\epsilon_i = \sigma(\boldsymbol{X}_i^{\mathrm{T}}\boldsymbol{\beta}_j)e_i$ where e_i is independent with $\boldsymbol{X}_i$ and $e_1, \ldots, e_n$ are i.i.d. with mean zero and variance σ^2. Condition 2 is about the parameter space, it excludes the possibility of a reduced model with less than $s+1$ subgroups by requiring $a_{j-1} < a_j$, $j = 1, \ldots, s$. Conditions* 3 *and Condition* 4 *are common assumptions in penalized regression in high-dimensional settings. The concave penalties such as MC+ and SCAD satisfy Condition 3. For the MC+ penalty, Condition 4 is equivalent to $\Lambda_{(s+1)p}(\Upsilon) > 1/\nu$, and for the SCAD penalty, Condition 4 is equivalent to $\Lambda_{(s+1)p}(\Upsilon) > 1/(\nu-1)$, which ensures the objective function (5.35) is globally convex. Condition 5 and 6 are common in estimation of threshold model for the consistency of smoothed LSE of a two-regime threshold regression model, see [126] and [222].*

When $\boldsymbol{\theta}$ is either known or estimated to be consistent, we have $\hat{W}_i = W_i + o_p(1)$ where $W_i = \boldsymbol{Z}_i^{\mathrm{T}}\boldsymbol{\theta}$. By law of large numbers and Condition 5, $P(\hat{W}_i \leq a_j) = P(W_i \leq a_j) = \tau_j$ implies $\sum_{i=1}^{n} 1(a_{j-1} < \hat{W}_i \leq a_j)/n \rightarrow_{a.s} \tau_j - \tau_{j-1} > 0$. Suppose that $m \rightarrow \infty$, $m/n \rightarrow 0$. By the Condition 5, it follows that with probability 1, $\sum_{i=1}^{n} 1(\hat{W}_{\iota(n-(q_n-j+2)m)} < \hat{W}_i \leq \hat{W}_{\iota(n-(q_n-j+1)m)})/n = m/n \rightarrow 0$. Thus there is at most one threshold locate in each segment $\{\hat{W}_{\iota(n-(q_n-j+2)m)} < \hat{W}_i \leq$

$\hat{W}_{\iota(n-(q_n-j+1)m)}\}$ for large n where $\hat{W}_{\iota(n-(q_n-j+2)m)}$ and $\hat{W}_{\iota(n-(q_n-j+1)m)}$, $j = 1, \ldots, q_n + 1$ are defined in Section 5.5.3. Then a consistent selection estimate in the splitting stage can be guaranteed by the following theorem.

Theorem 5.5.1. *Give a consistent $\hat{\boldsymbol{\theta}}^*$, suppose $m \to \infty$ and $m = O(n^r)$, where $0 < r \leq 1/2$ is a constant, $\lambda_n \to 0$ and $\lambda_n\sqrt{n}/\log n \to \infty$ as $n \to \infty$. If Conditions 1–5 hold, then we have $\lim_{n\to\infty} P(\hat{s} = s) = 1$.*

Let $\boldsymbol{\gamma} = (\gamma_1, \ldots, \gamma_{(s+1)p})^{\mathrm{T}} = ((\boldsymbol{\beta})^{\mathrm{T}}, (\boldsymbol{\delta}_1)^{\mathrm{T}}, \ldots, (\boldsymbol{\delta}_s)^{\mathrm{T}})^{\mathrm{T}}$ and $(\boldsymbol{a}, \boldsymbol{\theta})$ denote the true parameters in (5.30) and $\mathcal{G} = \{j : \gamma_j \neq 0, j = 1, \ldots, (s+1)p\}$ be the true set of important variables for the subgroup detection. For a given consistent estimate $\hat{s}$, the consistency of smoothed least square estimator $\tilde{\boldsymbol{\eta}}^*$ which minimizing the unregularized objective function (5.34) can be obtained directly from Theorem 1 in [222]. Here, we consider the estimator $\hat{\boldsymbol{\eta}}^*$ which minimizing the penalized smooth objective function (5.35). The following theorem guarantees the consistency of the penalized smoothed LSE.

Theorem 5.5.2. *Given a consistent $\hat{s}$. Under Conditions 1–6, and $\lambda_n \to 0$ as $n \to \infty$, then there is a local minimizer $\hat{\boldsymbol{\eta}}^*$ of $L_n^*(\boldsymbol{\eta})$ such that $\|\hat{\boldsymbol{\gamma}}^* - \boldsymbol{\gamma}\| = O_p(\sqrt{1/n})$, $\|\hat{\boldsymbol{a}}^* - \boldsymbol{a}\| = O_p(\sqrt{h/n})$ and $\|\hat{\boldsymbol{\theta}}^* - \boldsymbol{\theta}\| = O_p(\sqrt{h/n})$, where $\|\hat{\boldsymbol{\theta}}^*\| = \|\boldsymbol{\theta}\| = 1$.*

Without loss of generality, rewrite $\mathcal{G} = \{g_1, \ldots, g_{s+1}\}$, where $g_{j+1} = \{j_1, \ldots, j_{p_j}\}$ be the p_j nonzero covariates set in each subgroup with $j = 0, 1, \ldots, s$. We can also rewrite $\boldsymbol{\gamma} = (\boldsymbol{\gamma}_{(1)}^{\mathrm{T}}, \boldsymbol{\gamma}_{(2)}^{\mathrm{T}})^{\mathrm{T}}$ where $\boldsymbol{\gamma}_{(1)} = (\boldsymbol{\gamma}_{g_1}^{\mathrm{T}}, \ldots, \boldsymbol{\gamma}_{g_{s+1}}^{\mathrm{T}})^{\mathrm{T}}$ with $\boldsymbol{\gamma}_{g_{j+1}} = (\gamma_{j_1}, \ldots, \gamma_{j_{p_j}})^{\mathrm{T}}$ and $\boldsymbol{\gamma}_{(2)} = \mathbf{0}$. For $j = 0, 1, \ldots, s$, denote $\boldsymbol{X}_{i,g_{j+1}} = (X_{i,j_1}, \ldots, X_{i,j_{p_j}})^{\mathrm{T}}$. Denote $\Sigma_1 = (\sigma_{1,jk})_{0\leq j,k\leq s}$ be the $\sum_{j=0}^{s} p_j \times \sum_{j=0}^{s} p_j$ block matrix, where the block $\sigma_{1,jk} = 4\sigma^2 E\boldsymbol{X}_{i,g_{j+1}}\boldsymbol{X}_{i,g_{k+1}}^{\mathrm{T}}\mathbf{1}(\boldsymbol{Z}_i^{\mathrm{T}}\boldsymbol{\theta} > a_j)\mathbf{1}(\boldsymbol{Z}_i^{\mathrm{T}}\boldsymbol{\theta} > a_k)$, $\Sigma_2 = (\sigma_{2,kk})_{k=1,\ldots,s}$ be the $s \times s$ diagonal matrix, where $\sigma_{2,kk} = \frac{2\sigma^2}{\sqrt{\pi}}A_k$, and $A_k = E\{\boldsymbol{\delta}_{g_{k+1}}^{\mathrm{T}}\boldsymbol{X}_{i,g_{k+1}}\boldsymbol{X}_{i,g_{k+1}}^{\mathrm{T}}\boldsymbol{\delta}_{g_{k+1}}|W_i = a_k\}f(a_k)$, and $\Sigma_3 = (\sigma_{3,uv})_{0\leq u,v\leq d}$ be the $d \times d$ matrix with $\sigma_{3,uv} = \frac{2\sigma^2}{\sqrt{\pi}}\sum_{j=1}^{s}(G_{juv} + \sum_{k\neq j} e^{-\frac{(a_k-a_j)^2}{4h^2}} G_{kjsv})$, where $G_{jsv} = E\{Z_{is}Z_{iv}\boldsymbol{\delta}_{g_{j+1}}^{\mathrm{T}}\boldsymbol{X}_{i,g_{j+1}}\boldsymbol{X}_{i,g_{j+1}}^{\mathrm{T}}\boldsymbol{\delta}_{g_{j+1}}|W_i = a_j\}f(a_j)$ and $G_{kjsv} = E\{Z_{is}Z_{iv}\boldsymbol{\delta}_{g_{k+1}}^{\mathrm{T}}\boldsymbol{X}_{i,g_{k+1}}\boldsymbol{X}_{i,g_{j+1}}^{\mathrm{T}}\boldsymbol{\delta}_{g_{j+1}}|W_i = \bar{a}_{kj}\}f(\bar{a}_{kj})$. Let $V_{11} = (v_{k,l})_{0\leq k,l\leq s}$, where $v_{k,l} = 2E\boldsymbol{X}_{i,g_{k+1}}\boldsymbol{X}_{i,g_{l+1}}^{\mathrm{T}}\mathbf{1}(\boldsymbol{Z}_i^{\mathrm{T}}\boldsymbol{\theta} > a_k)\mathbf{1}(\boldsymbol{Z}_i^{\mathrm{T}}\boldsymbol{\theta} > a_l)$, $V_{12} = (v_{rj})_{0\leq r\leq s,1\leq j\leq s}$ be a block matrix, where $v_{rj} = 2E\boldsymbol{X}_{i,g_{r+1}}\boldsymbol{X}_{i,g_{j+1}}^{\mathrm{T}}\gamma_{g_{j+1}}\mathbf{1}(\boldsymbol{Z}_i^{\mathrm{T}}\boldsymbol{\theta} > a_r)f_{w|V}(a_j|V)$, $V_{13} = (v_{rj})_{0\leq r\leq s,1\leq j\leq d}$ be a block matrix, where $v_{rj} = 2EZ_{ij}\boldsymbol{X}_{i,g_{r+1}}\sum_{j=1}^{s}\boldsymbol{X}_{i,g_{j+1}}^{\mathrm{T}}\gamma_{g_{j+1}}\mathbf{1}(\boldsymbol{Z}_i^{\mathrm{T}}\boldsymbol{\theta} > a_r)f_{w|V}(a_j|V)$, $V_{33} = \frac{1}{\sqrt{\pi}}\sum_{j=1}^{s}(\boldsymbol{G}_j + \sum_{k\neq j} e^{-\frac{(a_k-a_j)^2}{4h^2}}\boldsymbol{G}_{kj})$, where $\boldsymbol{G}_j = (G_{j,uv})_{1\leq u,v\leq d}$ and $\boldsymbol{G}_{kj} = (G_{kj,uv})_{1\leq u,v\leq d}$ be the $d \times d$ matrix; $V_{22} = \mathrm{diag}(\frac{A_k}{\sqrt{\pi}}, k = 1, \ldots, s)$, $V_{23} = (v_{vk})_{1\leq v\leq d,1\leq k\leq s}$ where $v_{vk} = -\frac{1}{\sqrt{\pi}}E[Z_{iv}\boldsymbol{\delta}_{g_{k+1}}^{\mathrm{T}}\boldsymbol{X}_{i,g_{k+1}}f_{w|V}(a_k|V) + \sum_{j\neq k} Z_{iv}\boldsymbol{\delta}_{g_{j+1}}^{\mathrm{T}}\boldsymbol{X}_{i,g_{j+1}}\boldsymbol{X}_{i,g_{k+1}}^{\mathrm{T}}\boldsymbol{\delta}_{g_{k+1}}f_{w|V}(\bar{a}_{jk}|V)]$.
Denote

$$\begin{aligned}
\Gamma_{\lambda_n} =& \mathrm{diag}\{p_{\lambda_n}''(|\gamma_{0_1}|), \ldots, p_{\lambda_n}''(|\gamma_{0_{p_0}}|), \ldots, p_{\lambda_n}''(|\gamma_{s_1}|), \ldots, p_{\lambda_n}''(|\gamma_{s_{p_s}}|)\};\\
\boldsymbol{b}_n =& \{p_{\lambda_n}'(|\gamma_{0_1}|)\,\mathrm{sgn}(\gamma_{0_1}), \ldots, p_{\lambda_n}'(|\gamma_{0_{p_0}}|)\,\mathrm{sgn}(\gamma_{0_{p_0}}), \ldots,\\
& p_{\lambda_n}'(|\gamma_{s_1}|)\,\mathrm{sgn}(\gamma_{s_1}), \ldots, p_{\lambda_n}'(|\gamma_{s_{p_s}}|)\,\mathrm{sgn}(\gamma_{s_{p_s}})\}^{\mathrm{T}}.
\end{aligned}$$

With the smoothing techniques, the limiting distribution of the estimators are developed based on the standard Taylor approximation argument.

Theorem 5.5.3. *Given a consistent $\hat{s}$, under Conditions 1–6, $\lambda_n \to 0$ and $\lambda_n\sqrt{n}/\log n \to \infty$ as $n \to \infty$, with probability tending to 1, the penalized smooth estimator $\hat{\boldsymbol{\eta}}^* = (\hat{\boldsymbol{\gamma}}_{(1)}^*, \hat{\boldsymbol{\gamma}}_{(2)}^*, \hat{\boldsymbol{a}}^*, \hat{\boldsymbol{\theta}}^*)$ in Theorem 5.5.2 satisfy*

(a) Sparsity: $\hat{\gamma}^*_{(2)} = \mathbf{0}$.

(b) Asymptotic normality:

$$\sqrt{n}(\hat{\gamma}^*_{(1)} - \gamma_{(1)} + (V_{11} + \Gamma_{\lambda_n})^{-1}\boldsymbol{b}_n) \xrightarrow{D} N(0, (V_{11} + \Gamma_{\lambda_n})^{-1}\Sigma_1(V_{11} + \Gamma_{\lambda_n})^{-1}),$$
$$\sqrt{n/h}\tilde{\boldsymbol{V}}\begin{pmatrix} \hat{\boldsymbol{a}}^* - \boldsymbol{a} \\ \hat{\boldsymbol{\theta}}^* - \boldsymbol{\theta} \end{pmatrix} \xrightarrow{D} N(0, \boldsymbol{\Omega}).$$

where $\tilde{\boldsymbol{V}} = \begin{pmatrix} V_{22} & V_{23} \\ \boldsymbol{P_\theta} V_{23}^{\mathrm{T}} & \boldsymbol{P_\theta} V_{33} \end{pmatrix}$, $\boldsymbol{\Omega} = \mathrm{diag}(\Sigma_2, \boldsymbol{P_\theta}\Sigma_3\boldsymbol{P_\theta})$, *and* $\boldsymbol{P_\theta} = I - \boldsymbol{\theta}\boldsymbol{\theta}^{\mathrm{T}}$. *Furthermore,* $\sqrt{n/h}\tilde{\boldsymbol{V}}\begin{pmatrix} \hat{\boldsymbol{a}}^* - \boldsymbol{a} \\ \hat{\boldsymbol{\theta}}^* - \boldsymbol{\theta} \end{pmatrix}$ *and* $\sqrt{n}(\hat{\gamma}^*_{(1)} - \gamma_{(1)})$ *are asymptotically independent.*

For the penalty functions satisfy Conditions 3 and 4, if $\lambda_n \to 0$ and $\lambda_n\sqrt{n}/\log n \to \infty$ as $n \to \infty$, $\boldsymbol{b}_n = \mathbf{0}$, by Theorem 5.5.3, the corresponding penalized smoothed estimators possess the oracle property for estimating $\hat{\gamma}^*_{(1)}, \hat{\boldsymbol{a}}^*, \hat{\boldsymbol{\theta}}^*$ knowing $\hat{\gamma}^*_{(2)} = \mathbf{0}$. Hence, our proposed MCPL estimation can be used to estimate parameters and select variables simultaneously without losing any efficiency compared to the smoothed LSE when the true sub-model is known.

Considering the sparse single change-plane model (5.26), with the asymptotic normality given in Theorems 5.5.2 and 5.5.3, we can directly obtain the asymptotic distribution of $(\hat{\gamma}^*, \hat{\boldsymbol{\theta}}^*)$ which minimizing the objective function (5.29). Results are presented in the following corollary.

Corollary 5.3. *Let* $(\hat{\gamma}^*, \hat{\boldsymbol{\theta}}^*)$ *be the single change-plane estimator. Let Conditions* 1–6 *hold, then* $(\hat{\gamma}^*, \hat{\boldsymbol{\theta}}^*)$ *are consistent and*

$$\sqrt{n}(\hat{\gamma}^*_{(1)} - \gamma_{(1)} - (V_{11} - \Gamma_{\lambda_n})^{-1}\boldsymbol{b}_n) \xrightarrow{D} N(0, (V_{11} - \Gamma_{\lambda_n})^{-1}\Sigma_1(V_{11} - \Gamma_{\lambda_n})^{-1}),$$
$$\sqrt{n/h}(\hat{\boldsymbol{\theta}}^* - \boldsymbol{\theta}) \xrightarrow{D} N(0, V_{33}^{-1}\Sigma_3 V_{33}^{-1}),$$

and they are asymptotically independent.

Considering the threshold variables, we treat the one-dimensional case ($d = 1$), for which $\boldsymbol{\theta} = 1$ and $\boldsymbol{Z}_i^{\mathrm{T}}\boldsymbol{\theta} = Z_i$, separately. Then we can estimate $\hat{\boldsymbol{\eta}}^*$ by a one-step estimation, which has the advantage of being non-iterative in the smoothing stage, we also provide its distribution theory in the following corollary.

Corollary 5.4. *Let* $\hat{s}$ *and* $(\hat{\boldsymbol{a}}^*, \hat{\gamma}^*)$ *be the one-step estimate with* $\boldsymbol{\theta} = 1$. *Let Conditions* 1–6 *hold, we have* $\lim_{n\to\infty} P(\hat{s} = s) = 1$ *and furthermore* $\sqrt{n}(\hat{\gamma}^* - \gamma)$ *and* $\sqrt{n/h}(\hat{\boldsymbol{a}}^* - \boldsymbol{a})$ *are asymptotically independent, and*

$$\sqrt{n}(\hat{\gamma}^*_{(1)} - \gamma_{(1)} - (V_{11} - \Gamma_{\lambda_n})^{-1}\boldsymbol{b}_n) \xrightarrow{D} N(0, (V_{11} - \Gamma_{\lambda_n})^{-1}\Sigma_1(V_{11} - \Gamma_{\lambda_n})^{-1}),$$
$$\sqrt{n/h}(\hat{\boldsymbol{a}}^* - \boldsymbol{a}) \xrightarrow{D} N(0, V_{22}^{-1}\Sigma_2 V_{22}^{-1}).$$

Remark 5.5. *The proof of Theorems 5.5.1–5.5.3 are given in [163], and both Corollary 5.3 and 5.4 are direct consequence of the theorems. Moreover, Theorem 5.5.3 provides support for using the asymptotic normality in conducting statistical inference, however, it is only proved for the fixed* λ_n *values satisfying the conditions in Theorem 5.5.3 and requires the consistent model selection for the sparse regression coefficients in each subgroup.*

5.5.5 Asymptotic Variance Estimation

The construction of the asymptotic confidence set is straightforward by inverting the t or $Wald$ statistic given the asymptotic normality. Ways to estimate the asymptotic variances are described below.

Given the asymptotic normality of the penalized smoothed LSE, we have to estimate the variances of the limiting distributions for inference. Since $\hat{\gamma}^*$ is asymptotically independent of $(\hat{\boldsymbol{a}}^*, \hat{\boldsymbol{\theta}}^*)$, we can estimate its asymptotic variance first, which involves the expectation of $\mathbf{1}(a_j < \boldsymbol{Z}_i^{\mathrm{T}}\boldsymbol{\theta}), j = 0, 1, ..., s$. Since $\hat{\boldsymbol{a}}^*$ and $\hat{\boldsymbol{\theta}}^*$ are consistent, we can input them into the indicator function as if they were the true values. Then $\hat{\gamma}^*$ is the penalized LSE and estimation of its asymptotic variance is standard [91].

For the variance estimate of $\hat{\boldsymbol{a}}^*$ and $\hat{\boldsymbol{\theta}}^*$ in Theorem 5.5.3, we restricted $\|\boldsymbol{\theta}\| = 1$ as only the direction of $\boldsymbol{\theta}$ is identifiable. To avoid theoretical difficulty, we now suppose the first element of $\boldsymbol{\theta}$, i.e. θ_1, to be 1 for identification. We can divide $\hat{\boldsymbol{\theta}}^*$ by $\hat{\theta}_1^*$ to obtain the new estimator $\widehat{\tilde{\boldsymbol{\theta}}}^*$. Denote $\tilde{\boldsymbol{Z}}_i = (Z_{i2}, \dots, Z_{id})^{\mathrm{T}}$ and $\tilde{\boldsymbol{\theta}} = (\theta_2, \dots, \theta_d)^{\mathrm{T}}$, by the proof of Theorem 5.5.3, given the consistent $\hat{s}$, we have

$$\sqrt{n/h}\begin{pmatrix} \hat{\boldsymbol{a}}^* - \boldsymbol{a} \\ \widehat{\tilde{\boldsymbol{\theta}}}^* - \boldsymbol{\theta} \end{pmatrix} \xrightarrow{D} N(0, \boldsymbol{V}^{-1}\boldsymbol{\Omega}\boldsymbol{V}^{-1}).$$

where $\boldsymbol{V} = \begin{pmatrix} V_{22} & V_{23} \\ V_{23} & V_{33} \end{pmatrix}$ and $\boldsymbol{\Omega} = \mathrm{diag}(\Sigma_2, \Sigma_3)$. Let

$$\varphi_i^{a_j}(\boldsymbol{\eta}) = 2e_i \frac{\boldsymbol{\delta}_j^{\mathrm{T}}\boldsymbol{X}_i}{h}\phi\left(\frac{\tilde{\boldsymbol{Z}}_i^{\mathrm{T}}\tilde{\boldsymbol{\theta}} - a_j}{h}\right), \; j = 1, \dots, \hat{s}$$

$$\varphi_i^{\tilde{\boldsymbol{\theta}}}(\boldsymbol{\eta}) = 2e_i\tilde{\boldsymbol{Z}}_i \sum_{j=1}^{\hat{s}} \frac{\boldsymbol{\delta}_j^{\mathrm{T}}\boldsymbol{X}_i}{h}\phi\left(\frac{\tilde{\boldsymbol{Z}}_i^{\mathrm{T}}\tilde{\boldsymbol{\theta}} - a_j}{h}\right)$$

where $e_i = Y_i - \boldsymbol{X}_i^{\mathrm{T}}\boldsymbol{\beta} - \boldsymbol{X}_i^{\mathrm{T}}\sum_{k=1}^{\hat{s}}\delta_k\Phi\left(\frac{\tilde{\boldsymbol{Z}}_i^{\mathrm{T}}\tilde{\boldsymbol{\theta}} - a_k}{h}\right)$. Then $\boldsymbol{V}$ and $\boldsymbol{\Omega}$ can be estimated by $\hat{\boldsymbol{V}}^{-1}\hat{\boldsymbol{\Omega}}\hat{\boldsymbol{V}}^{-1}$, where $\widehat{\boldsymbol{\eta}}^* = (\hat{\gamma}^*, \hat{\boldsymbol{a}}^*, \widehat{\tilde{\boldsymbol{\theta}}}^*)$ and

$$\hat{\Sigma}_{2,jk} = \frac{1}{n}\sum_{i=1}^{n}\varphi_i^{a_j}(\widehat{\boldsymbol{\eta}}^*)\varphi_i^{a_k}(\widehat{\boldsymbol{\eta}}^*), \; j, k = 1, \dots, \hat{s}; \qquad \hat{\Sigma}_3 = \frac{1}{n}\sum_{i=1}^{n}\varphi_i^{\tilde{\boldsymbol{\theta}}}(\widehat{\boldsymbol{\eta}}^*)\varphi_i^{\tilde{\boldsymbol{\theta}}}(\widehat{\boldsymbol{\eta}}^*)$$

and for $j, k = 1, \dots, \hat{s}$ and $v = 2, \dots, d$,

$$\hat{V}_{22,jk} = \frac{1}{n}\sum_{i=1}^{n}\frac{\hat{\boldsymbol{\delta}}_j^{*\mathrm{T}}\boldsymbol{X}_i\boldsymbol{X}_i^{\mathrm{T}}\hat{\boldsymbol{\delta}}_k^*}{h^2}\phi\left(\frac{\tilde{\boldsymbol{Z}}_i^{\mathrm{T}}\widehat{\tilde{\boldsymbol{\theta}}}^* - \hat{a}_j^*}{h}\right)\phi\left(\frac{\tilde{\boldsymbol{Z}}_i^{\mathrm{T}}\widehat{\tilde{\boldsymbol{\theta}}}^* - \hat{a}_k^*}{h}\right),$$

$$\begin{aligned}\hat{V}_{23,kv} = -\frac{2}{nh^2}\sum_{i=1}^{n}\Bigg\{ & Z_{iv}\hat{\boldsymbol{\delta}}_j^{*\mathrm{T}}\boldsymbol{X}_i\phi\left(\frac{\tilde{\boldsymbol{Z}}_i^{\mathrm{T}}\widehat{\tilde{\boldsymbol{\theta}}}^* - \hat{a}_k^*}{h}\right)\sum_{j=1}^{\hat{s}}\boldsymbol{X}_i^{\mathrm{T}}\hat{\boldsymbol{\delta}}_j^*\phi\left(\frac{\tilde{\boldsymbol{Z}}_i^{\mathrm{T}}\widehat{\tilde{\boldsymbol{\theta}}}^* - \hat{a}_j^*}{h}\right) \\ & -\hat{e}_i^*\tilde{Z}_{iv}\hat{\boldsymbol{\delta}}_j^{*\mathrm{T}}\boldsymbol{X}_i\phi'\left(\frac{\tilde{\boldsymbol{Z}}_i^{\mathrm{T}}\widehat{\tilde{\boldsymbol{\theta}}}^* - \hat{a}_k^*}{h}\right)\Bigg\},\end{aligned}$$

$$\begin{aligned}\hat{V}_{33} = \frac{2}{nh^2}\sum_{i=1}^{n}\Bigg\{ & \tilde{\boldsymbol{Z}}_i\tilde{\boldsymbol{Z}}_i^{\mathrm{T}}\left(\sum_{j=1}^{\hat{s}}\boldsymbol{X}_i^{\mathrm{T}}\hat{\delta}_j^*\phi\left(\frac{\tilde{\boldsymbol{Z}}_i^{\mathrm{T}}\widehat{\tilde{\boldsymbol{\theta}}}^* - \hat{a}_j^*}{h}\right)\right)^2 \\ & -\hat{e}_i^*\tilde{\boldsymbol{Z}}_i\tilde{\boldsymbol{Z}}_i^{\mathrm{T}}\sum_{j=1}^{\hat{s}}\hat{\boldsymbol{\delta}}_j^{*\mathrm{T}}\boldsymbol{X}_i\phi'\left(\frac{\tilde{\boldsymbol{Z}}_i^{\mathrm{T}}\widehat{\tilde{\boldsymbol{\theta}}}^* - \hat{a}_j^*}{h}\right)\Bigg\}.\end{aligned}$$

Similarly, the variance estimate in Corollary 5.3 can be obtain in the same way as [222]. For the one-step estimate in Corollary 5.4, the variances V_{22} and Σ_2 that appear in limiting distribution of $\hat{a}^*$ are also limits of certain elements of the score and Hessian matrix of $L_n^*(\boldsymbol{\eta})$. An alternative approach to variance estimate is based on the bootstrap for threshold model, see [126] and [222], where they established the asymptotically correctness of bootstrap and the details are omit here.

5.5.6 Real data applications

5.5.6.1 Bovine collagen clinical trial (BCCT)

We illustrate model (5.30) using clinical data from a 3-year NIH-sponsored randomized Bovine Collagen Trial for Scleroderma patients conducted at 12 centers in the USA which contains 297 samples. This data set was analyzed by many authors using parametric and semiparametric models ([198, 165]). Patients were randomized to receive oral native collagen at a dose of 500 μg/day or a similar appearing placebo. Patients were evaluated clinically by the use of the Modified Rodnan Skin Score (MRSS) (the primary outcome variable), disability index of the Health Assessment Questionnaire (HAQ), patient's global assessment, patients pain assessment and physicians global assessment. To implement the proposed method, we consider 11 predictor variables: X_1 = haq (health assessment questionnaire); X_2 = pga (patient self-assessment of disease progression); X_3 = dlcop (lung performance measurement 3); X_4 = fvcp (lung performance measurement 1); X_5 = over (disease progression); X_6 = pain (index of pain); X_7 = fev1p (lung performance measurement 2); X_8 = durdis (duration of disease); X_9 = age (in years) X_{10} = ethnic (0 = hispanic, 1 = non-hispanic); X_{11} = sex (0 = female, 1 = male); which aim at predicting MRSS. Variables are standardized with mean zero and unit variance.

We first fit a linear regression model with $\boldsymbol{X}_i = (1, X_{i1}, \ldots, X_{i,11})^{\mathrm{T}}$ without considering subgroups, and denote $\hat{\boldsymbol{\beta}}^{\mathrm{ols}}$ the OLS estimation. Then, for subgroup identification, we choose $\boldsymbol{Z}_i = (X_{i1}, X_{i2}, X_{i3})^{\mathrm{T}}$ as the threshold variables and fit the multiple change-plane model for this dataset. The tuning parameters in (5.35) were chosen via generalized cross-validation (GCV). We detect one change by MCPL method with the estimated threshold $\hat{a}^* = -0.125$ and the change-plane parameter $\hat{\boldsymbol{\theta}}^* = (0.801, -0.206, 0.562)^{\mathrm{T}}$. The two subgroup sizes are 139 and 155 respectively and we report the estimated coefficients $\boldsymbol{\beta}$ and $\boldsymbol{\delta}$ in Table 5.8 with their standard errors (S.E.), and the p-values for testing the significance of the coefficients using our proposed method. We also report the OLS estimation by assuming a common intercept. We compared our method with the single change-plane (SCPL) model (5.26), the two stage multiple change-points (MCPT) detection proposed in [161] with single threshold covariate with X_1, X_2, X_3 respectively and the multiple

Table 5.8: Estimated results for Bovine Collagen Clinical Trial (BCCT), along with standard errors (S.E.) and P-values by t test. X_1 = haq; X_2 = pga; X_3 = dlcop; X_4 = fvcp; X_5 = over; X_6 = pain; X_7 = fev1p; X_8 = durdis; X_9 = age X_{10} = ethnic (non-hispanic); X_{11} = sex (male).

	$\boldsymbol{\beta}$			$\boldsymbol{\delta}$			$\boldsymbol{\beta}^{\mathrm{ols}}$		
Covariates	Coef.	S.E.	P-value	Coef.	S.E.	P-value	Coef.	S.E.	P-value
Intercept	0.022	0.104	0.830	−0.525	0.238	0.028	−0.383	0.188	0.042
X_1	0	–	–	0.443	0.125	< 0.001	0.220	0.065	< 0.001
X_2	0.416	0.057	< 0.001	0	–	–	0.369	0.059	< 0.001
X_3	0.332	0.102	0.001	−0.261	0.131	0.048	0.113	0.058	0.051
X_4	0.124	0.081	0.127	0	–	–	0.118	0.086	0.170
X_5	−0.296	0.128	0.022	0.547	0.173	0.002	0.065	0.092	0.480
X_6	0.245	0.122	0.045	−0.602	0.160	< 0.001	−0.095	0.087	0.276
X_7	−0.322	0.099	0.001	0.307	0.111	0.006	−0.121	0.084	0.153
X_8	−0.187	0.068	0.006	0.313	0.105	0.003	−0.063	0.055	0.251
X_9	−0.104	0.050	0.037	0	–	–	−0.115	0.053	0.031
X_{10}	0	–	–	0.367	0.209	0.080	0.386	0.195	0.049
X_{11}	−0.376	0.172	0.029	0.771	0.247	0.002	0.138	0.130	0.287

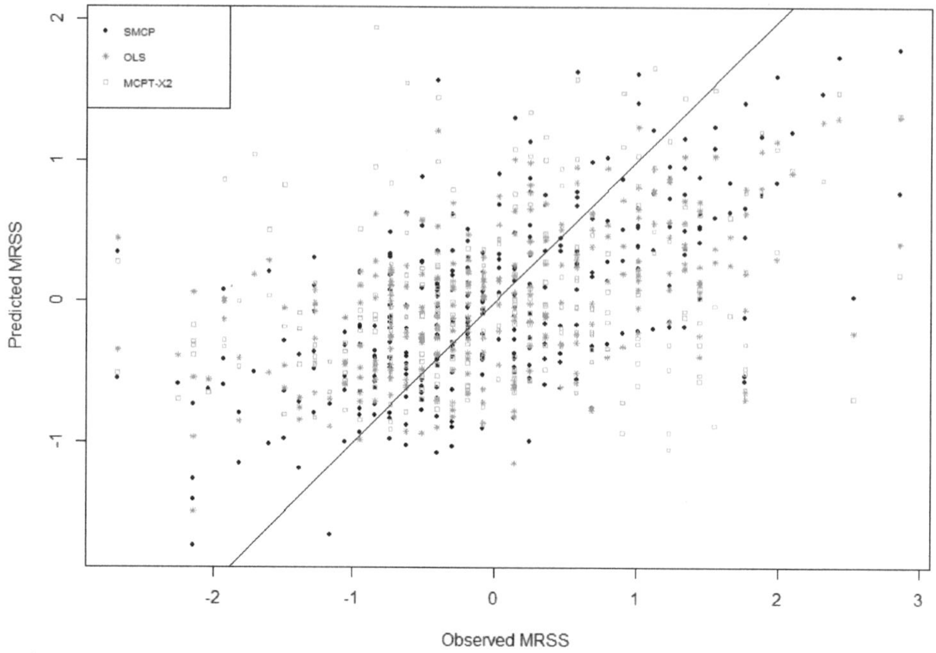

Figure 5.2: The scatter plot of the fitted Y by each method for BCCT data. MCPL stands for multiple change-plane, MCPT-X_2, stands for the MCPT method with threshold X_2 and OLS stands for the ordinary least square estimate.

change-plane variables $\boldsymbol{Z}_i$ with equal weight (E-MCPL). From Table 5.9, the first three models can split the dataset into two groups, the MCPT with X_2 as the threshold variable and E-MCPL methods both can detect 3 groups but all with larger MSE. Moreover, we present the bar plot of all the covariates for the subgroup detected by the methods mentioned before in Figure 5.3. Eyeballing the plots we can see that the mean difference of all the covariates between the two subgroups detected by MCPL are quite distinct compared to the other methods. We also plot the kernel density estimation of the thresholding variables for all methods in Figure 5.4.

5.5.6.2 AIDS clinical trials

We next apply model (5.30) to the AIDS Clinical Trials Group Study 175 (ACTG175), which contains 2139 HIV-infected subjects. ACTG175 was a randomized clinical trial to compare zidovudine (ZDV) monotherapy (treatment 0) with other three therapies including ZDV and didanosine (ddI) (treatment 1), ZDV and zalcitabine (zal) (treatment 2), and ddI monotherapy (treatment 3) in adults infected with the human immunodeficiency virus type I ([248, 178]). Our interest is to conduct subgroup analysis for the fitted value of CD4 counts (cells/mm^3) at 20 ± 5 weeks as the continuous response Y after adjusting for the effects of the covariates: $X_1 =$ hemophilia (0 =no, 1 =yes); $X_2 =$ gender (0 =female, 1 =male); $X_3 =$ CD4 counts at baseline; $X_4 =$ antiretroviral history (0 =naive, 1 =experienced); $X_5 =$ age (years); $X_6 =$ weight (kg); $X_7 =$ Karnofsky score; $X_8 =$ CD8 counts at baseline; $X_9 =$ homosexual activity (0 =no, 1 =yes); $X_{10} =$ history of intravenous drug use (0 =no, 1 =yes); $X_{11} =$ race (0 =white, 1 =white); $X_{12} =$ symptomatic status (0 =asymptomatic, 1 =symp-

Table 5.9: Estimated Comparison for Bovine Collagen Clinical Trial (BCCT). MCPL stands for multiple change-plane, SCPL stands for single change-plane, MCPT-X_1, MCPT-X_2, MCPT-X_3 stands for the MCPT method with threshold X_1, X_2 and X_3 respectively, E-MCPL stands for multiple change plane with equal weight and OLS stands for the ordinary least square estimate.

Model	MSE	$\hat{s}$	threshold $\hat{a}$	Group Sizes
MCPL	0.613	1	-0.125	$138 : 156$
SCPL	0.669	1	0	$155 : 139$
MCPT-X_1	0.695	1	-0.695	$72 : 222$
MCPT-X_2	0.672	2	$(-0.079, 0.916)$	$150 : 81 : 63$
MCPT-X_3	0.786	1	-0.996	$42 : 252$
E-MCPL	0.707	2	$(-0.259, 0.029)$	$103 : 51 : 140$
OLS	0.739	0	-	-

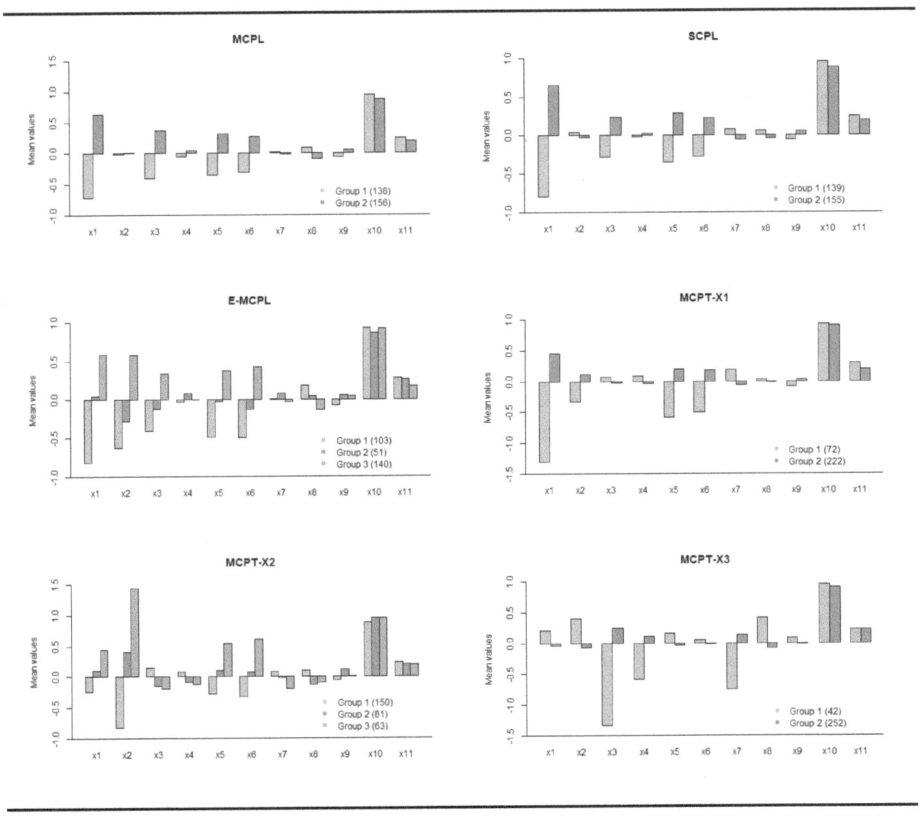

Figure 5.3: The mean value of predictors for different detected subgroups for BCCT data. MCPL stands for multiple change-plane model, SCPL stands for single change-plane model, E-MCPL stands for multiple change plane with equal weight, MCPT-X_1, MCPT-X_2, MCPT-X_3 stands for the MCPT method with threshold X_1, X_2 and X_3 respectively. X_1 = haq; X_2 = pga; X_3 = dlcop; X_4 = fvcp; X_5 = over; X_6 = pain; X_7 = fev1p; X_8 = durdis; X_9 = age X_{10} = ethnic (non-hispanic); X_{11} = sex (male). Group sizes are given in parentheses.

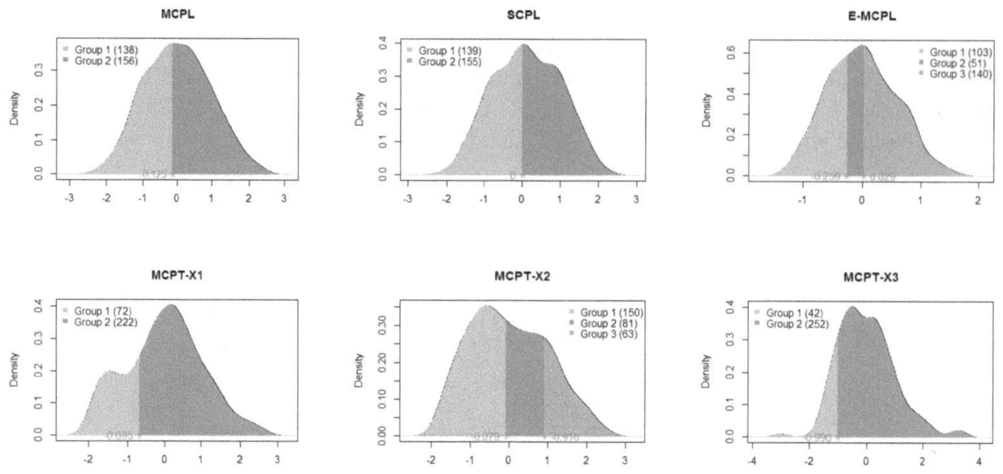

Figure 5.4: The density plots of thresholding variables estimated by each method for BCCT data, the red mark points in x-axis display the cut-off points. MCPL stands for multiple change-plane, SCPL stands for single change-plane, E-MCPL stands for multiple change plane with equal weight, MCPT-X_1, MCPT-X_2, MCPT-X_3 stands for the MCPT method with threshold X_1, X_2 and X_3 respectively. Group sizes are given in parentheses.

tomatic) and X_{13} = treatment arm (0=zidovudine, 1=zidovudine and didanosine, 2=zidovudine and zalcitabine, 3=didanosine).

We consider all 2139 patients to analyze the treatment ZDV vs. the other three treatments. We first fit a linear regression model with $\boldsymbol{X}_i = (1, X_{i1}, \ldots, X_{i,14})^{\mathrm{T}}$ without assuming subgroups, and denote $\hat{\boldsymbol{\beta}}^{\mathrm{ols}}$ the OLS estimation. Considering subgroups among all the patients, we fit the change-plane model (5.30) with the linear baseline mean function, and choose $\boldsymbol{Z}_i = (X_{i1}, X_{i2}, X_{i3}, X_{i4}, X_{i5})^{\mathrm{T}}$ as the threshold variables. Similarly, the tuning parameters in (5.35) were chosen via generalized cross-validation (GCV). The estimated change-plane parameter $\hat{\boldsymbol{\theta}}^* = (-0.268, -0.199, 0.876, -0.268, 0.223)^{\mathrm{T}}$. We detect two change by our method, the estimated threshold $\hat{\boldsymbol{a}}^* = (-0.309, 0.201)^{\mathrm{T}}$, and the three group size are 1162, 394, and 583 respectively. Table 5.10 reports the estimated coefficients $\boldsymbol{\beta}$ and $\boldsymbol{\delta}$, their standard errors (S.E.), and the p-values for testing the significance of the coefficients using our proposed method, comparing with OLS estimate by assuming a common intercept. We also compared our method with the single change-plane (SCPL) model (5.26), the two stage multiple change-points (MCPT) detection method proposed in [161] with single threshold covariate X_1, X_2, X_3, X_4 and X_5 respectively and the multiple change-plane variables $\boldsymbol{Z}_i$ with equal weight (E-MCPL), where only the first three models can detect the subgroups. Both the MCPL and the MCPT with threshold variable X_2 can detect 2 subgroups while the later has a larger MSE. The analysis results are summarized in table 5.11. Similarly, we present the bar plot of all the covariates for the subgroup detected by the methods mentioned before in Figure 5.6 from which we can see that the mean difference of all the covariates between the two subgroups detected by MCPL are quite distinct compared to the other methods. We also plot the kernel density estimation of the thresholding variables for all methods in Figure 5.7.

Table 5.10: Estimated results for AIDS Clinical Trials Group Study 175 (ZDV vs. the other three treatments), along with standard errors (S.E.) and P-values by t test. X_1 = hemophilia (yes); X_2 = gender (male); X_3 = CD4 counts at baseline; X_4 = antiretroviral history (experienced); X_5 = age; X_6 = weight; X_7 = Karnofsky score; X_8 = CD8 counts at baseline; X_9 = homosexual activity (yes); X_{10} = history of intravenous drug use (yes); X_{11} = race (white); X_{12} = symptomatic status (symptomatic), X_{13} = treatment arm[1] (zidovudine and didanosine), X_{14} = treatment arm[2] (zidovudine and zalcitabine) and X_{15} = treatment arm[3] (didanosine).

	β			δ_1			δ_2			β^{ols}		
Covariates	Coef.	S.E.	P-value	Coef.	S.E.	P-value	Coef.	S.E.	P-value	Coef.	S.E.	P-value
Intercept	−0.087	0.059	0.143	0.171	0.078	0.029	0.286	0.092	0.002	−0.001	0.061	0.981
X_1	−0.169	0.062	0.007	0	–	–	0	–	–	−0.184	0.080	0.021
X_2	0	–	–	0	–	–	0	–	–	−0.032	0.066	0.632
X_3	0.568	0.041	< 0.001	0	–	–	−0.269	0.058	< 0.001	0.571	0.017	< 0.001
X_4	−0.247	0.035	< 0.001	0	–	–	0	–	–	−0.281	0.035	< 0.001
X_5	0	–	–	−0.117	0.025	< 0.001	0	–	–	0.021	0.017	0.234
X_6	0	–	–	0	–	–	0	–	–	−0.003	0.018	0.846
X_7	0.085	0.021	< 0.001	−0.233	0.047	< 0.001	0.187	0.055	0.001	0.040	0.017	0.019
X_8	−0.109	0.024	< 0.001	0.085	0.034	0.012	0	–	–	−0.066	0.017	< 0.001
X_9	0	–	–	−0.064	0.057	0.257	0	–	–	0.003	0.059	0.957
X_{10}	0.092	0.058	0.111	0	–	–	−0.167	0.109	0.128	0.052	0.052	0.319
X_{11}	−0.101	0.043	0.020	0	–	–	−0.128	0.083	0.123	−0.128	0.040	0.002
X_{12}	−0.135	0.055	0.015	0.103	0.120	0.390	−0.110	0.145	0.450	−0.130	0.045	0.004
X_{13}	0.452	0.061	< 0.001	0.164	0.107	0.125	−0.124	0.121	0.309	0.488	0.048	< 0.001
X_{14}	0.251	0.062	< 0.001	0.008	0.086	0.927	0	–	–	0.252	0.048	< 0.001
X_{15}	0.228	0.052	< 0.001	0	–	–	0.261	0.096	0.007	0.294	0.047	< 0.001

5.5.7 Discussion

In model (5.30), the change plane variables $\boldsymbol{Z}_i$ are pre-specified, which introduces a degree of subjectivity. To enhance subgroup detection, a forward selection method could be employed to systematically identify the most relevant threshold variables, which warrants further investigation. Additionally, the minimum in the smoothing stage is influenced by both stochastic and numerical errors when optimizing the nonlinear parameters, as the objective function derived from the nonlinear model often exhibits multiple local minima. Therefore, in practical applications, more efficient optimization techniques could be beneficial in improving the overall performance of this method.

In the theoretical results, we allow the coefficients of the covariates to be sparse, but require their dimension to be much smaller than n. Addressing high or ultra-high dimensional settings can be further explored, as discussed in [229]. Additionally, the proposed estimation method can be extended to other frameworks, including generalized linear models and hazard regression models. While these extensions may seem conceptually straightforward, developing efficient computational algorithms and establishing the corresponding theoretical properties for such more complex models remains a challenging task.

Table 5.11: Estimated Comparison for AIDS Clinical Trials Group Study 175 data (ZDV vs. the other three treatments). MCPL stands for multiple change-plane, SCPL stands for single change-plane, MCPT-X_3, MCPT-X_5stands for the MCPT method with threshold X_3 and X_5 respectively, E-MCPL stands for multiple change plane with equal weight and OLS stands for the ordinary least square estimate.

Method	MSE	$\hat{s}$	Threshold $\hat{a}$	Group Sizes
MCPL	0.567	2	$(-0.309, 0.201)$	1162 : 394 : 583
SCPL	0.578	1	0	489 : 1650
MCPT-X_3	0.575	2	$(-1.758, 1.210)$	21 : 1889 : 229
MCPT-X_5	0.595	0	-	-
E-MCPL	0.595	0	-	-
OLS	0.595	0	-	-

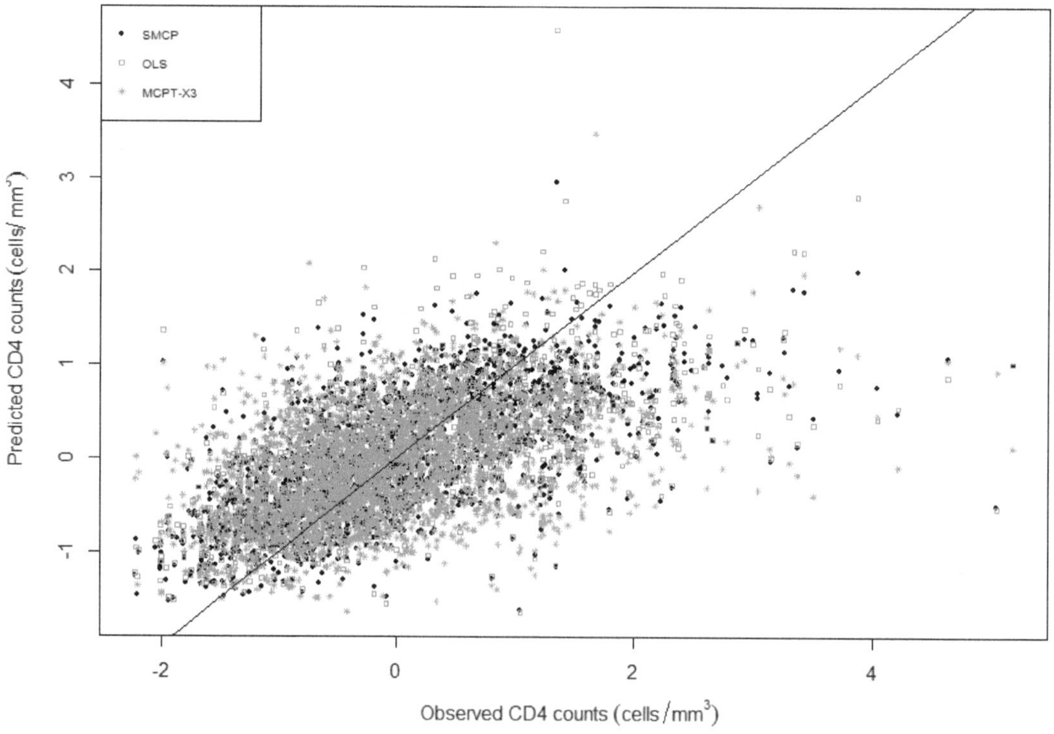

Figure 5.5: The scatter plot of the fitted Y by each method for ACTG 175 data. MCPL stands for multiple change-plane, MCPT-X_3, stands for the MCPT method with threshold X_3 and OLS stands for the ordinary least square estimate.

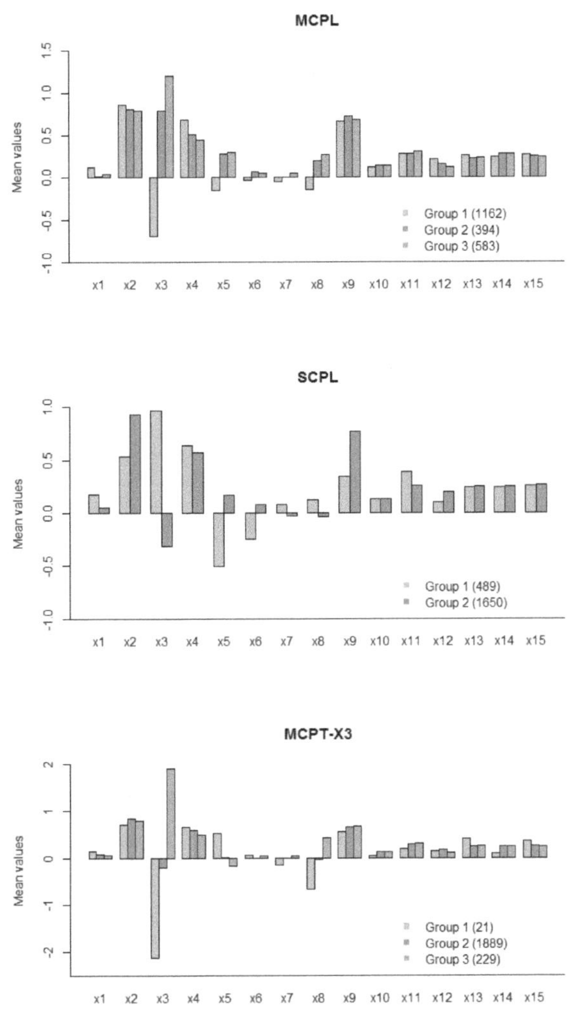

Figure 5.6: The mean value of predictors for different detected subgroups for ACTG 175 data. MCPL stands for multiple change-plane, SCPL stands for single change-plane, MCPT-X_3 stands for the MCPT method with threshold X_3. X_1 = hemophilia (yes); X_2 = gender (male); X_3 = CD4 counts at baseline; X_4 = antiretroviral history (experienced); X_5 = age; X_6 = weight; X_7 = Karnofsky score; X_8 = CD8 counts at baseline; X_9 = homosexual activity (yes); X_{10} = history of intravenous drug use (yes); X_{11} = race (white); X_{12} = symptomatic status (symptomatic), X_{13} = treatment arm^1 (zidovudine and didanosine), X_{14} = treatment arm^2 (zidovudine and zalcitabine) and X_{15} = treatment arm^3 (didanosine). Group sizes are given in parentheses.

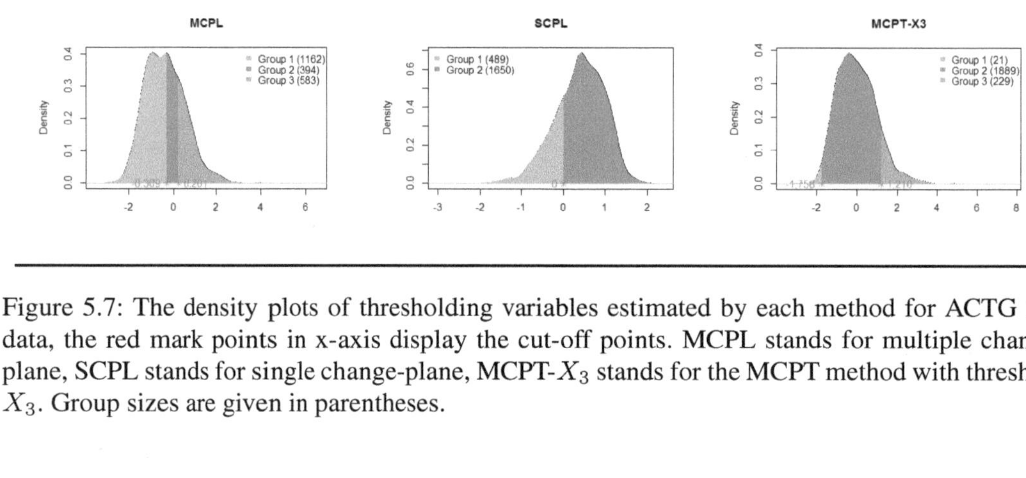

Figure 5.7: The density plots of thresholding variables estimated by each method for ACTG 175 data, the red mark points in x-axis display the cut-off points. MCPL stands for multiple change-plane, SCPL stands for single change-plane, MCPT-X_3 stands for the MCPT method with threshold X_3. Group sizes are given in parentheses.

6

Further Applications

Detecting patterns and similarities between observations is a key topic in modern statistics, especially in large datasets from which we can extract information. In the medical field, change point models have long been applied, particularly in constructing patient subgroups using decision tree methods. These methods yield piecewise constant or piecewise linear models, fundamentally characterized by change points of certain important features. In economics, structural break models have been extensively used to detect change points in time series data. Econometricians often conduct tests to identify these breaks, allowing for the analysis of different homogeneous periods with greater precision.

We point out a few meaningful applications of change point detection in this last chapter. Each section in the following is a field that is rapidly developing and growing in these days. We offer some efforts especially relevant to the theme of this book. Some practical examples are included to illustrate these methods.

6.1 Functional data analysis with discontinuity

6.1.1 Functional data

As mentioned in chapter 5, threshold detection and subgroup identification play an important role in personalized medicine. Such segmentation is needed in other fields as well. In this section we consider segmentation of functional data.

Functional data are of the form $\{X_i(t), t \in [a,b], i = 1, \cdots, n\}$, where $[a,b]$ is an interval on the line. Each observation is thus a curve from a functional space. These data can be viewed as an extension of the time series data from two perspectives. In a typical financial study, for each fixed i, the records $X_i(t)$ may be an observed time series for the stock price of a particular company over the time period $t \in [a,b]$. In fact, when the number of repeated observations is small and the data are collected at a few discrete time points, this type of data is usually called longitudinal data or repeated measures. Financial data can be collected on a frequent resolution, daily, hourly or event every minute. Such dense observations lead to the belief that we actually have an observation of a continuous function and we thus call them panel data or functional data. On the other hand, when n is small, say a couple of gigantic technology companies including Apple, Google and Amazon, such data may be analyzed as a multivariate time series data. Many univariate time series models can be generalized into vector version such as the vector autoregressive model. When n is very large, say, the complete set of all stocks listed on the NASDAQ, such data cannot be treated as a vector any more and must be best regarded as functions. The research questions usually concern the population characteristics of the random functional data, such as their distribution means and covariances.

Nonparametric techniques for functional data analysis (FDA) are well established and widely applied in practice. See [209], [127], and [128] for an overview. To deal with the non-standard function estimation, usually basis approximation or local polynomial techniques are employed. Furthermore, various complicated models were proposed to cope with the functional feature of the data,

DOI: 10.1201/9781032657073-6

such as the single index models, the varying coefficient models, the additive models, among many other works [160, 45]. For identification of meaningful subgroups and distinct patterns in functional data, earlier works include [57], [58] and [56]. In this section we attempt to identify different subregions of the functional process delineated by discontinuities of the mean function. This focus is different from those earlier discussion on change points with time series data.

Many scientific and business inquiries require the modeling of functional data with multiple change points. For example, a phase of a thermodynamic system and the states of matter have uniform physical properties. During a phase transition of a given medium, certain properties of the medium change, often discontinuously, as a result of the change of some external condition such as temperature or pressure. Another example is the financial and economic forecasting problem where two consecutive business cycles may each be described by a smooth process but disconnected at the joining time caused by some unexpected drastic change in economic outlook. Identification of the change-point structure for a period with multiple business cycles may improve the forecasting for production, prices, interest rates and other financial measurements.

We first fix our scope for functional data in this section. In a typical functional-data setting, a sample of n curves are observed at a set of discrete, cross-sectional points; denote by m_i the number of observations for curve i, $i = 1, \cdots, n$. Generally speaking, there could be two different sample settings [42, 160]. Let T_{ij} denote the location of the j-th observation for the i-th curve. In a *common design*, the observations are sampled at the same locations across curves, that is, $T_{1j} = T_{2j} = \cdots = T_{nj} =: T_j$ for all $j = 1, \ldots, m$; mostly likely, the T_j are points on a fixed grid. For example, the stock prices and many other financial indices are usually recorded at the same time for all the companies with equal intervals between two records. A more general setting is when the T_{ij} are independently sampled from a bounded interval $[a, b]$, which we shall refer to as *independent design*. This happens in a typical medical follow-up study where patients may be observed at random follow-up time points during the study period. From another perspective, one may also distinguish between dense and sparse functional data [see, e.g., 168, 285], which roughly correspond to whether each m_i is larger than some power of n or all m_i's are bounded by a finite positive number. Our theoretical results will focus on the independent design setting since doing so allows us to treat the sparse and dense settings in a unified manner. However, we note that when the data are dense, the differences between the common design and independent design settings are small both in theory and computations. We note that when $m_i \to \infty$, it is unnecessary to treat T_{ij} as random. The common design setting, as in the real data examples of this book, can be incorporated into our model setting.

Numerous studies have focused on detecting change points in the mean for univariate series. [264] proposed a kernel-type estimators of jump points in the fixed-design nonparametric regression model. [71] provided an account of early results in change-point detection mainly for data with independent and identically distributed (i.i.d.) error terms. [283] discussed a nonparametric test for change-point detection under dependent and non-stationary errors. [112] proposed a test for shifts in the mean of a heteroscedastic time series which accommodates the possibility of changing variability. See [207], [235], [206] and [108] for additional results on nonparametric change-point detection. For multivariate high dimensional dependent series, [269] considered detecting change points for the variance for blocked time series and dependent panel data. [116] extended the cumulative sum paradigm test statistics for detecting whether the pattern of a spatio-temporal process has changed over a time period. [260] studied the change point in high-dimensional time series of which the mean structure changes in a sparse subset of the coordinates. Recently, [166] proposed a new nonparametric procedure to detect change points in the mean of high-dimensional time series data taking into account both spatial and temporal dependence.

The research on change point detection for functional data only began recently and most of the existing works focus on the change point detection in the mean of functional observations. For example, [26] considered a single break in the mean function assuming that the error terms are i.i.d. curves, and [11] assumed that the change size can be both a constant or approaching zero as the sample size tends to infinity. The results in [26] and [11] were generalized by [9] to deal with an epidemic change model, where after the initial change the mean changes back to the original level after some time. [12] studied the change of regression operator in a functional regression model, and [14] developed a functional test statistics for detecting and dating structural breaks in functional data. [23] introduced several asymptotic methods to test the null hypothesis that the mean structure of a sequence of curves does not change by taking into account their range and other aspects of their shapes. These aforementioned works on change points in the context of functional data all have quite different focuses from ours in this section.

We consider multiple change points for process data and characterize the underlying functional relationship with a statistical model. We may allow the number of change points to diverge with the sample size. Particularly, under common design, when the functions are densely sampled, the number of change points can increase to infinity with the sampling frequency m.

Most of the earlier works on change point detection focused on single change point estimation. The problem of multiple change points is more challenging as the estimation of the unknown number of change points adds an extra layer of difficulty. Recent research has paid more attention to computational and theoretical development for multiple change point. [102] considered a wild binary segmentation method for consistent estimation of the number and locations of multiple change points for independent data. [161] introduced another approach for multiple change point detection where a non-concave penalization approach was adopted. Their framework can be extended to covariate change point in regression analysis. However, these approaches are only suitable for independent data with parametric functionals. For discontinuous nonparametric curve estimation, [266] proposed a jump information criterion to consistently detect the number of jumps through kernel smoothing. Their work is only pertinent to cross-sectional data and cannot be directly applied to functional data.

In this section, we introduce a nonparametric framework proposed by Li et al. (2022) [162] for change-point analysis of functional data. Essentially, we can detect the multiple jump discontinuities in the mean function using a local linear smoother that employs a one-sided kernel. After the jump locations are identified, we will then estimate the jump sizes as well as the smooth component of the mean function and the covariance function. While the within-curve dependency in the functional-data setting poses extra challenges compared with nonparametric regression, it also leads to richer theoretical developments. We will prove a range of convergence rates for the estimates depending on the denseness of the observation points on each curve. In particular, our results for the jump location/size estimation show a "phase transition" of rates as observations per curve changes from sparse to dense. This is akin to the results on mean function estimation in, e.g., [168, 42, 285]. Moreover, the asymptotic normality of the mean estimator is also discussed after the change points are consistently detected, where a pointwise confidence band can be constructed for the estimated mean function. Our results are both general and sharp. Different types of sampling weights are considered and our convergence rates reflect clearly the roles of the smoothing parameters. Furthermore, the functional process is not limited to the Gaussian process and can have heavy-tailed marginal distributions, as demonstrated in the simulation studies in [162].

The remaining sections are organized as follows. In Section 6.1.2, we introduce the FDA model and data structure as well as all the estimation procedures. We establish the theoretical properties of the procedures in Section 6.1.3, where we also discuss the optimal rates of change point estimation and their connections to prominent results in the literature. Some real data applications are offered in Section 6.1.4.

6.1.2 Model Assumptions and Methodology for FDA

6.1.2.1 Model the mean change

Let $\{X(t), t \in [a, b]\}$ be a stochastic process defined on a fixed bounded interval $[a, b]$. Without loss of generality, we assume that $[a, b] = [0, 1]$. Denote the mean and covariance function of $X(t)$ by

$$\mu(t) = \mathbb{E}\{X(t)\}, \qquad R(s,t) = \mathrm{cov}\{X(s), X(t)\},$$

which are assumed to exist. Let

$$U(t) := X(t) - \mu(t)$$

denote the zero-mean stochastic component of $X(t)$.

The goal is to estimate the mean and covariance function based on noisy functional data $\{(Y_{ij}, T_{ij}), i = 1, \dots, n, j = 1, \dots, m_i\}$, where

$$Y_{ij} = X_i(T_{ij}) + e_{ij} = \mu(T_{ij}) + U_i(T_{ij}) + e_{ij}, \tag{6.1}$$

where the X_i's are i.i.d. realizations of X, the T_{ij}'s are i.i.d. random observational points with density function $f_T(\cdot)$, and the e_{ij}'s are i.i.d. random errors with mean zero and finite variance σ^2. The index T may be the temperature variable in a physics research study, the time variable in a financial time series, or the location variable in a spatial or image data. In the developed asymptotic theory, we allow m_i to vary with n and we will address scenarios where m_i's are bounded or tend to ∞.

The mean function μ is a usually assumed to be smooth in the functional data literature [cf. 168, 285]. However, as explained earlier, there are scenarios where it is more realistic to allow for discontinuities in μ. We consider the situation that the mean function $\mu(t)$ has jump discontinuities at M different locations, $\tau_1, \tau_2, \dots, \tau_M \in (0, 1)$, such that

$$\mu(t) = \nu(t) + \sum_{k=1}^{M} d_k I(t \geq \tau_k), \tag{6.2}$$

where $\nu(t)$ is a smooth function with a bounded second-order derivative, and $\{d_k : k = 1, \dots, M\}$ are the jump sizes. For convenience, assume that $|d_1| > |d_2| > \cdots > |d_M| > 0$. Our goal is to conduct inference on the $(\tau_k, d_k), M, \nu$ and R.

In addition to the structural breaks for the overall mean function, we may also allow subject-specific change points for individual functions $U_i(t), i = 1, \cdots, n$. In particular, for the ith curve, suppose there are ι_i thresholds located at $a < \phi_{i1} < \cdots < \phi_{i\iota_i} < b$ such that

$$U_i(t) = \eta_i(t) + \sum_{k=1}^{\iota_i} r_{ik} I(t \geq \phi_{ik}), \tag{6.3}$$

where $\eta_i(t)$ has bounded second-order derivatives and $\{r_{ik} : k = 1, \dots, \iota_i\}$ are the jump sizes at the change points. Searching change points for individual curves may lead to interesting findings when some important subgroups or clusters are present in the data.

6.1.2.2 Estimation of mean function

Since the mean function $\mu(t)$ is smooth except at the change points $\tau_1, \dots, \tau_M$, one may adopt the familiar local linear regression estimator [90] to deal with the smooth part of μ. Let $K(\cdot)$ be a bounded symmetric kernel function with support $[-1, 1]$ and $K_h(t) = (1/h)K(t/h)$ where h is bandwidth. Recall that the local linear estimator $\widehat{\mu}(t; h)$ of $\mu(t)$ in [168] can be obtain as $\widehat{a}_0$, where

$$(\widehat{a}_0, \widehat{a}_1) = \operatorname*{argmin}_{a_0, a_1} \sum_{i=1}^{n} w_i \sum_{j=1}^{m_i} \{Y_{ij} - a_0 - a_1(T_{ij} - t)\}^2 K_h(T_{ij} - t). \tag{6.4}$$

and the weight w_i is attached to each observation for the ith subject such that $\sum_{i=1}^{n} m_i w_i = 1$. It follows that

$$\widehat{\mu}(t;h) = \frac{R_0(t)S_2(t) - R_1 S_1(t)}{S_0(t)S_2(t) - S_1^2(t)}, \tag{6.5}$$

where

$$\begin{aligned} S_r(t) &= \sum_{i=1}^{n} w_i \sum_{j=1}^{m_i} K_h(T_{ij} - t)\{(T_{ij} - t)/h\}^r, \\ R_r(t) &= \sum_{i=1}^{n} w_i \sum_{j=1}^{m_i} K_h(T_{ij} - t)\{(T_{ij} - t)/h\}^r Y_{ij},\ r = 0, 1, 2. \end{aligned} \tag{6.6}$$

This estimator was studied in detail in [168] assuming that $\mu(t)$ has bounded second derivative. Since $\widehat{\mu}(t;h)$ is a local estimator, so long as t is not within a distance of h from any τ_k, the properties of $\widehat{\mu}(t;h)$ are identical to that in [168]. Thus, one approach to estimating the τ_k might be to focus on the locations where $\widehat{\mu}$ changes rapidly. However, this may not be an effective approach.

Let

$$Y_{ij}^* = Y_{ij} - \sum_{k=1}^{M} d_k I(T_{ij} \geq \tau_k) \tag{6.7}$$

for which the mean function is $\nu(t)$ by (6.2). Then the local linear estimator is

$$\widehat{\nu}^*(t;h) = \frac{R_0^*(t)S_2(t) - R_1^*(t)S_1(t)}{S_0(t)S_2(t) - S_1^2(t)},$$

where $R_r^*(t)$ is obtained by replacing Y_{ij} with Y_{ij}^* in $R_r(t)$. We can refer to $\widehat{\nu}^*(t;h)$ the "oracle" estimator of $\nu(t)$ as we assume the jump locations and sizes are known. Clearly,

$$R_r(t) - R_r^*(t) = \sum_{i=1}^{n}\sum_{j=1}^{m_i} K_h(T_{ij} - t)\left(\frac{T_{ij} - t}{h}\right)^r \sum_{k=1}^{M} d_k I(T_{ij} \geq \tau_k). \tag{6.8}$$

Observe that, for t not within a distance of h from any τ_k, we have $\widehat{\mu}(t;h) - \widehat{\nu}^*(t;h) = \sum_k d_k I(t \geq \tau_k)$ and therefore

$$\widehat{\mu}(t;h) - \mu(t) = \widehat{\nu}^*(t;h) - \nu(t). \tag{6.9}$$

For t within a distance of h from some τ_k, the relationship between $\widehat{\mu}(t;h)$ and $\widehat{\nu}^*(t;h)$ is more complicated and we will defer that to Section 6.1.3. In the rest of the book, we focus on the case where the (τ_k, d_k) are unknown.

6.1.2.3 Estimation of discontinuities

In this subsection, we consider estimating the unknown locations and sizes of the discontinuities of μ using a so-called half-kernel approach. Asymmetric, one-sided, kernels were sometimes used to address the boundary issues in the early literature of kernel smoothing [cf. 106, 212]. [207] and [266] investigated a difference kernel estimator based one-sided local linear regression to detect discontinuities in the mean regression curve. [265] proposed a test statistic using the difference between the left and right local averages of observations for identifying the existence of structural breaks in trends. [45] used the half-kernel approach to handle asynchronous longitudinal data.

Let

$$K_-(t) = K(t)\mathbf{1}_{\{t\in[-1,0)\}} \quad \text{and} \quad K_+(t) = K(t)\mathbf{1}_{\{t\in[0,1]\}}$$

be the two one-sided kernel functions corresponding to the symmetric kernel function $K(\cdot)$. Denote

$$K_{h,\ell}(t) = (1/h)K_\ell(t/h), \ell = \pm.$$

for a bandwidth h. Then, for a given interior point $t \in [h, 1-h]$, by using the data points in the two halves $[t-h, t)$ and $[t, t+h]$ of the neighborhood $[t-h, t+h]$ we obtain the two one-sided local linear estimators $\widehat{\mu}_-(t; h)$ and $\widehat{\mu}_+(t; h)$, which are simply obtained in the same way as (6.4) and (6.5) using K_- and K_+ as the kernels. Now, define

$$\Delta_n(t; h) = \widehat{\mu}_+(t; h) - \widehat{\mu}_-(t; h). \tag{6.10}$$

Note that if $[t-h, t)$ does not contain any τ_k, then, as a special case of $\widehat{\mu}(t; h)$, $\widehat{\mu}_-(t; h)$ estimates $\mu(t)$ and its asymptotic properties are no different from the general local linear estimator for continuous $\mu(t)$; the same can be said for $\widehat{\mu}_+(t)$ if $[t, t+h]$ does not contain any τ_k. In particular, if $[t-h, t+h]$ does not contain any τ_k, then both $\widehat{\mu}_-(t; h), \widehat{\mu}_+(t; h)$ estimate $\mu(t)$ under suitable conditions and therefore $\Delta_n(t; h)$ is expected to be small. On the other hand, as we shall see below, if a point t is within a distance h of some τ_k, say $t = \tau_k + \lambda h$ for some $\lambda \in [-1, 1]$ and a suitable bandwidth h, then $\widehat{\mu}_-(t; h)$ and $\widehat{\mu}_+(t; h)$ would be estimating quantities that differ roughly by the jump size. These ideas motivate the approach to be introduced below.

For clarity, formally define the h-neighborhoods of the τ_k as follows. Let

$$D_h = [h, 1-h], \quad D_{h,k} = [\tau_k - h, \tau_k + h], \; k = 1, \ldots, M,$$

and

$$D_{h,M_1\to M_2} = \bigcup_{k=M_1}^{M_2} D_{h,k}, \quad \bar{D}_{h,M_1\to M_2} = D_h \backslash D_{h,M_1\to M_2}, \; 1 \le M_1 \le M_2 \le M.$$

As explained earlier, $\bar{D}_{h,1\to M}$ contains no change points and the detection criterion $|\Delta_n(t; h)|$ is expected to be small in $\bar{D}_{h,1\to M}$. On the other hand, $|\Delta_n(t; h)|$ is expected to reflect the jump sizes in any of the $D_{h,k}$.

We first consider the estimation of the τ_k. To do so we begin by assuming that M is known but will later return to the case of unknown M. Let

$$\widehat{\tau}_1 = \widehat{\tau}_1(h_\tau) := \operatorname*{argmax}_{t\in D_{h_\tau}} |\Delta_n(t; h_\tau)| \tag{6.11}$$

be the first change point candidate, where h_τ is the bandwidth used for this step. Similarly, $\widehat{\tau}_2$ is the maximizer of $|\Delta_n(t; h_\tau)|$ in the set $D_{h_\tau} \backslash \widehat{D}_{\epsilon h_\tau, 1}$, where $\widehat{D}_{\epsilon h_\tau,1} = [\widehat{\tau}_1 - \epsilon h_\tau, \widehat{\tau}_1 + \epsilon h_\tau]$, with $\epsilon > 1$, is the region excluding the ϵh_τ-neighborhood of the first identified change point. There is flexibility in choosing ϵ but for convenience we will let $\epsilon = 2$ in our procedure. The recursive formula of $\widehat{\tau}_k$ is

$$\widehat{\tau}_k = \widehat{\tau}_k(h_\tau) := \operatorname*{argmax}_{t\in D_{h_\tau} \backslash \bigcup_{i=1}^{k-1} \widehat{D}_{\epsilon h_\tau, i}} |\Delta_n(t; h_\tau)|, \quad k = 1, \ldots M, \tag{6.12}$$

where $\widehat{D}_{\epsilon h_\tau, i} = [\widehat{\tau}_i - \epsilon h_\tau, \widehat{\tau}_i + \epsilon h_\tau]$.

For estimating d_k, an obvious candidate is $\Delta_n(\widehat{\tau}_k; h_\tau)$ [cf. 266]. However, this approach does not achieve the optimal rate in some cases for functional data. Thus, we propose the following modified approach. Theorem 6.1.2 shows that $|\widehat{\tau}_k - \tau_k| = O(h_\tau^2 + h_\tau \varsigma_n(h_\tau))$, where

$$\varsigma_n(h) = \left\{ \log n \sum_{i=1}^{n} m_i((m_i - 1) + 1/h)w_i^2 \right\}^{1/2}. \tag{6.13}$$

Thus, if we pick ρ_d to be of a bigger order of magnitude than $h_\tau^2 + h_\tau \varsigma_n(h_\tau)$, then $\widehat{\tau}_k - \rho_d < \tau_k$ and $\widehat{\tau}_k + \rho_d > \tau_k$ for large n. This motivates the estimator

$$\widehat{d}_k(h_\tau, h_d, \rho_d) = \widehat{\mu}_+(\widehat{\tau}_k + \rho_d; h_d) - \widehat{\mu}_-(\widehat{\tau}_k - \rho_d; h_d). \tag{6.14}$$

The considerations for selecting h_τ and h_d are not entirely the same, as will be seen in Theorem 6.1.2 and the discussions in Section 6.1.3.3.

After obtaining the estimates $\widehat{\tau}_k, \widehat{d}_k$, we can define the following estimate for the mean function

$$\widehat{\mu}(t; h_\tau, h_d, \rho_d) = \widehat{\nu}(t; h_\tau) + \sum_{k=1}^{M} \widehat{d}_k(h_\tau, h_d, \rho_d) I(t \geq \widehat{\tau}_k(h_\tau)) \tag{6.15}$$

where $\widehat{\nu}(t; h_\tau)$ is the local linear estimator using bandwidth h_τ, with Y_{ij} replaced by

$$\widetilde{Y}_{ij}(h_\tau, h_d, \rho_d) := Y_{ij} - \sum_{k=1}^{M} \widehat{d}_k(h_\tau, h_d, \rho_d) I(T_{ij} \geq \widehat{\tau}_k(h_\tau)).$$

Note that the same notation $\widehat{\mu}$ is used in (6.5) and (6.14) to streamline notation. However, we distinguish the two by the different tuning parameters.

In applications, M is generally unknown and has to be estimated. A simple and effective way to determine M is to leverage the idea that the uniform rate of $\Delta(t; h_\tau)$ at continuity points is $h_\tau^2 + \varsigma_n(h_\tau)$ (cf. Theorem 6.1.1) under rather general conditions, with ς_n defined by (6.13). Thus, one can proceed in the same way as before in estimating τ_k but would stop whenever

$$\sup_{t \in D_{h_\tau} \setminus \bigcup_{i=1}^{k-1} \widehat{D}_{\epsilon h_\tau, i}} |\Delta_n(t; h_\tau)| < \text{some threshold } \zeta_n \tag{6.16}$$

and declare $\widehat{M} = k - 1$, where ζ_n is of a bigger order of magnitude than $h_\tau^2 + \varsigma_n(h_\tau)$. We will establish that $\widehat{M} \to M$ with probability one under suitable conditions (cf. Theorem 6.1.4).

Remark 6.1. *After obtaining the mean estimate $\widehat{\mu}(t; h_\tau, h_d, \rho_d)$ by (6.15), the corresponding covariance estimator $\widehat{R}(s, t; h_\tau, h_d, \rho_d, h_R)$ can be obtained following [168] and [285] by a local linear smoother applied to the raw residual covariances, $C_{ijl} = [Y_{ij} - \widehat{\mu}(T_{ij})][Y_{il} - \widehat{\mu}(T_{il})]$ with another bandwidth h_R. Briefly, choose some appropriate weight w_i for each C_{ijl} such that $\sum_{i=1}^n m_i^2 \mathsf{w}_i = 1$ and let $\widehat{R}(s,t) = \widehat{b}_0$, where*

$$\begin{aligned}(\widehat{b}_0, \widehat{b}_1, \widehat{b}_2) = \underset{b_0, b_1, b_2}{\operatorname{argmin}} \sum_{i=1}^{n} \Bigg[\mathsf{w}_i \sum_{l \neq j} \{C_{ijl} - b_0 - b_1(T_{ij} - s) - b_2(T_{il} - t)\}^2 \\ \times K_{h_R}(T_{ij} - s) K_{h_R}(T_{il} - t) \Bigg], \end{aligned} \tag{6.17}$$

with $\sum_{l \neq j}$ denoting sum over all $l, j = 1, \ldots, m_i$ such that $l \neq j$. We also note that for dense functional data, after obtaining $\widehat{\mu}(t)$, it is possible to estimate the individual U_i, the stochastic component of X_i, by smoothing the residuals $\widehat{\varepsilon}_{ij} = Y_{ij} - \widehat{\mu}(T_{ij}), i = 1, \ldots, m_i$.

6.1.3 Theoretical Properties

In this section, we conduct a theoretical investigation of the properties of the procedures introduced in Section 6.1.2. Standard notations for asymptotic relations between two real sequences $\{a_n\}$ and $\{b_n\}$ will be used. For instance, $a_n \lesssim b_n$ means $a_n = O(b_n)$; $a_n \prec b_n$ and $a_n \succ b_n$ mean $a_n = o(b_n)$ and $b_n = o(a_n)$, respectively; $a_n \asymp b_n$ stands for $a_n \lesssim b_n$ and $b_n \lesssim a_n$. Also, $a \vee b := \max(a, b)$ and $a \wedge b := \min(a, b)$ for any two real numbers a, b.

Throughout the rest of the section we always assume that $h = h_n \to 0$.

6.1.3.1 Technical assumptions

For clarity, we collect most of the technical conditions in this subsection.

Let $\psi(x)$ be a function on $(0, \infty)$ such that $\psi(x)/x$ is monotone increasing for large enough x. For a kernel function K, define

$$g(\lambda) = g_K(\lambda) := \int_{|\lambda|}^{1} K_+^*(u)du, \ \lambda \in [-1, 1], \tag{6.18}$$

where

$$K_\ell^*(u) = K_\ell(u)\frac{\upsilon_{\ell,2} - \upsilon_{\ell,1}u}{\upsilon_{\ell,0}\upsilon_{\ell,2} - \upsilon_{\ell,1}^2} \quad \text{and} \quad \upsilon_{\ell,r} = \int_{-1}^{1} t^r K_\ell(t)dt, \ \ell = +, -. \tag{6.19}$$

Also, define

$$\mathcal{G}_n(t; h) = \sum_{k=1}^{M} d_k g\left(\frac{t - \tau_k}{h}\right) I(t \in D_{h,k}).$$

The following conditions are needed to facilitate our theoretical results.

1. $K(\cdot)$ is a symmetric probability density function on $[-1, 1]$ which is Lipschitz continuous and $\upsilon_2 = \int t^2 K(t)dt < \infty$, and $\|K\|^2 = \int K(t)^2 dt < \infty$.
2. The kernel K is such that $|g(\lambda)|$ has a unique maximum of 1 at 0.
3. For some constants $m_T > 0$ and $M_T < \infty$, the density function f_T of T_{ij} satisfies $m_T \leq f_T(t) \leq M_T$ for all $t \in [0, 1]$ and f_T is differentiable with a bounded derivative.
4. The smooth component $\nu(t)$ of the mean function is twice differentiable and the second derivative $\nu^{(2)}(t)$ is uniformly bounded.
5. $\mathbf{E}(\psi(|e_{ij}|)) < \infty$ and $\mathbf{E}\left(\sup_{t\in[0,1]} \psi(|U(t)|)\right) < \infty$.
6. $\varsigma_n(h) \to 0$ and
$$\psi\left(\frac{n}{\log n} h\varsigma_n(h)\right) \succ \frac{n}{\log n}. \tag{6.20}$$
where $\varsigma_n(h)$ is defined in (6.13).
7. $\sup_n\{n \max_i(m_i w_i)\} \leq B < \infty$

We briefly discuss the conditions. 1 is a standard condition for the kernel function in smoothing problems. 2 is less common but we conjecture it holds universally. We need this condition explicitly to ensure the desirable rates in the estimates. An intuitive explanation of this is given in (a) of Remark 6.2. We have examined a large number of kernels, either those commonly used or numerically generated, and have not encountered a single one that does not satisfy 2. However, we do not yet know how to prove it for a general K. The condition is stated as a result in the supplement of [266] but the proof appears to be incorrect. 3–4 are standard assumptions in the context of FDA and local polynomial smoothing. In 4, if ν has bounded high-order derivatives, then the corresponding higher-order local polynomial regression procedure would lead to better rates of convergence. 5 and 6 are more general than the standard conditions in the literature where $\psi(x)$ is taken as x^α for some $\alpha \geq 2$ [cf. 168, 285]. While there is little incentive to entertain other choices of ψ for the smooth mean estimation problem, it is different for the change point estimation problem. For instance, for a process X for which the marginal distributions have exponentially decaying tail probabilities (e.g., Gaussian), the almost sure convergence rate of n^{-1} for the jump point location estimation problem can be achieved by using a ψ that is exponentially increasing. See the discussions in Section 6.1.3.3 for details.

6.1.3.2 Main results

We present the following results by first assuming that one has prior knowledge about the number of jump locations M. The jump sizes and locations $\{(d_k, \tau_k) : k = 1, \ldots, M\}$ can be consistently estimated with $\Delta_n(t; h)$, defined in (6.10), using the approach introduced in Section 6.1.2. The various convergence rates are described by the following results. The results under estimated $\widehat{M}$ can be similarly obtained and is straightforward. It is straightforward to establish $\widehat{M} \to M$ almost surely.

Theorem 6.1.1. *Assume that* 1 and 3–7 *hold. Then, with probability one,*

1. $\sup_{t\in\bar{D}_{h,1\to M}} |\Delta_n(t; h)| = O\left(h^2 + \varsigma_n(h)\right)$,
2. $\sup_{t\in D_{h,1\to M}} |\Delta_n(t; h) - \mathcal{G}_n(t; h)| = O\left(h + \varsigma_n(h)\right)$.

The proof of (i) of Theorem 6.1.1 is based on the idea that both $\widehat{\mu}_+(t)$ and $\widehat{\mu}_-(t)$ estimate $\mu(t)$ at rate $O(h^2 + \varsigma_n(h))$ for $t \in \bar{D}_{h,1\to M}$. This is in keeping with what was obtained in [168] for a smooth mean function. Theorem 6.1.1 shows that the $\Delta_n(t; h)$ behaves like $d_k g\left(\frac{t-\tau_k}{h}\right)$ for t in the h-neighborhood of some τ_k and is of vanishing order if this is not the case. This means that, for large n, our algorithm focuses on these neighborhoods to search for change points.

The following is a key result of this section which gives the rates of convergence of $\widehat{\tau}_k, \widehat{d}_k$ defined in (6.11)–(6.15).

Theorem 6.1.2. *Assume that* 1–7 *hold for $h = h_\tau$ and h_d. Let ρ_d be another tuning parameter such that $h_\tau(h_\tau + \varsigma_n(h_\tau)) \prec \rho_d \lesssim h_\tau + \varsigma(h_\tau)$. Then we have, with probability one,*

1. $\max_{k=1,\ldots,M} |\widehat{\tau}_k(h_\tau) - \tau_k| = O\left(h_\tau^2 + h_\tau\varsigma_n(h_\tau)\right)$,
2. $\max_{k=1,\ldots,M} |\widehat{d}_k(h_\tau, h_d, \rho_d) - d_k| = O\left(h_\tau + \varsigma_n(h_\tau) + h_d^2 + \varsigma_n(h_d)\right)$,
3. $\sup_{t\in\bar{D}_{\rho_d,1\to M}} |\widehat{\mu}(t; h_\tau, h_d, \rho_d) - \mu(t)| = O\left(h_\tau + \varsigma_n(h_\tau) + h_d^2 + \varsigma_n(h_d)\right)$.

A few remarks are in order.

Remark 6.2. *(a) We mentioned in the discussion following Theorem 6.1.1 that, asymptotically, $\Delta_n(t; h)$ behaves like $d_k g\left(\frac{t-\tau_k}{h}\right)$ for t in the h-neighborhood of some τ_k. Condition (C2) then enables us to control where the maximum of $|\Delta_n(t; h)|$ is achieved in this interval.*

(b) The proof of Theorem 6.1.2 shows that (iii) remains true if the set $\bar{D}_{\rho_d,1\to M}$ is replaced by any set E_n such that the minimum distance between points in E_n and the τ_k is of a larger order than $h_\tau^2 + h_\tau\varsigma_n(h_\tau)$. An example of such E_n that only depends on data is $\left(\cup_{k=1}^{M}[\widehat{\tau}_k(h_\tau) - \rho_n, \widehat{\tau}_k(h_\tau) - \rho_n]\right)^c$.

(c) As stated in Theorem 6.1.2, the choice of h_d depends on h_τ. Indeed, one possibility for ρ_d is $h_\tau + \varsigma(h_\tau)$. In order to evaluate the optimal rates in (i)–(iii), one would choose h_τ, h_d to minimize the stated error bounds there while taking into account 6. We will defer the discussions until Section 6.1.3.3.

(d) Applying Theorem 6.1.2, the theoretical properties of the estimator defined for the covariance in Remark 6.1 can be established in the same way as in [168] and are omitted.

Next we establish the asymptotic normality of the mean function estimator in the smooth part.

Theorem 6.1.3. *Assume that the conditions of Theorem 6.1.2 with $\psi(x) = x^3$. Also, assume that*

$$\frac{\sum_{i=1}^n m_i w_i^3 (1/h_\tau^2 + (m_i - 1)/h_\tau + (m_i-1)(m_i-2))}{\left(\sum_{i=1}^n m_i w_i^2 (1/h_\tau + (m_i - 1))\right)^{3/2}} \to 0, \tag{6.21}$$

$$\min\left(h_\tau / \sum_{i=1}^n m_i w_i^2, 1/\sum_{i=1}^n m_i(m_i-1)w_i^2\right)(h_\tau + \varsigma_n(h_\tau) + h_d^2 + \varsigma_n(h_d))^2 \to 0, \tag{6.22}$$

and

$$h_\tau \sum_{i=1}^n m_i(m_i-1)w_i^2 / \sum_{i=1}^n m_i w_i^2 \to C_0 \in [0, \infty]. \tag{6.23}$$

Then for any interior point $t \in \bar{D}_{\rho_d,1\to M}$,

$$\Gamma^{-1/2}\left[\widehat{\mu}(t) - \mu(t) - (1/2)h_\tau^2 \upsilon_2 \nu^{(2)}(t)\right] \xrightarrow{d} \mathcal{N}(0,1),$$

where

$$\Gamma = \frac{\sum_{i=1}^n m_i w_i^2}{h_\tau} \|K\|^2 \frac{R(t,t) + \sigma^2}{f_T(t)} + \left(\sum_{i=1}^n m_i(m_i-1)w_i^2\right) R(t,t).$$

Remark 6.3. *Theorem 6.1.3 is useful for providing the asymptotic pointwise confidence band for the estimated mean function. In the covariance function, the unknown quantities Γ, $R(t,t)$, σ^2 and $f_T(t)$ can be replaced by their consistent estimators. In practice, $R(t,t)$ and σ^2 can be estimated using a local linear smoother as described in Remark 6.1, and the density $f_T(t)$ can be estimated via a kernel density estimation [225]. It is possible to extend our theoretical results further to achieve functional weak convergence by verifying the process tightness conditions. See [77] and [290] for some recent examples. Such development may facilitate the construction of simultaneous confidence band. However, it will involve a high level of technicality and is beyond the scope of this book.*

The following corollary provides the central limit theorem of $\Delta_n(t;h)$ which can be used for testing whether a given point is a change point.

Corollary 6.1. *Assume that the conditions of Theorem 3.2 hold with $\psi(x) = x^3$. Also, assume (6.21), (6.23) and $\min\left(h_\tau/\sum_{i=1}^n m_i w_i^2, 1/\sum_{i=1}^n N_i w_i^2\right)(h_\tau + \varsigma_n(h_\tau))^2 \to 0$ for bandwidth h_τ. Then for any point $t \notin \{\tau_1, \ldots, \tau_M\}$,*

$$\Omega^{-1/2}\Delta_n(t) \xrightarrow{d} \mathcal{N}(0,1)$$

where

$$\Omega = \frac{2\sum_{i=1}^n m_i w_i^2}{h_\tau} \|K_+^*\|^2 \frac{R(t,t) + \sigma^2}{f_T(t)} + 2\left(\sum_{i=1}^n m_i(m_i-1)w_i^2\right) R(t,t).$$

Corollary 6.1 provides the basis for a statistical test; in particular, one can reject the null hypothesis that t is not a change point at level α if

$$|\Delta_n(t)\Omega^{-1/2}| > z_{1-\alpha/2},$$

where $z_{1-\alpha/2}$ is the $(1-\alpha/2)$-th quantile of the standard normal distribution. If Ω contains unknown components, then they can be replaced by consistent estimates as discussed in Remark 6.3.

Following Remark 6.3, the unknown quantities $f_T(t)$, σ^2 and $R(t,t)$ can be replaced by consistent estimators. Specifically, the variance terms $R(t,t)$ and σ^2 can be estimated by plugging in the local linear smoothers (6.17), where $\widehat{\mu}(t)$ is chosen to be the one that has smaller residual mean square error among the one-sided local linear estimates $\widehat{\mu}_+(t;h_\tau)$ and $\widehat{\mu}_-(t;h_\tau)$. [283] also provided the central limit theorem of $\Delta_n(t)$ with convergence rate $(nh)^{1/2}$ which is comparable to our result when $m_i = 1$ and all the measurements are independent.

Note that the above asymptotic test does not adjust for multiple comparisons and cannot be regarded as a global test. In particular, it cannot be used to determine M when it is unknown. However, the procedure described above for estimating M is consistent, as the following result shows.

Theorem 6.1.4. *Assume that conditions in Theorem 6.1.2 hold and the threshold parameter ζ_n in (6.16) satisfies $\zeta_n \to 0$ and $\zeta_n/\{h_\tau^2 + \varsigma_n(h_\tau)\} \to \infty$. Then $\widehat{M} \to M$ with probability one.*

6.1.3.3 Optimal rates for $\widehat{\tau}_k$ and $\widehat{d}_k$

In this subsection we discuss optimal choices of the tuning parameters h_τ, h_d for estimating τ_k, d_k by $\widehat{\tau}_k$ and $\widehat{d}_k$, respectively, based on Theorem 6.1.2. We will focus on two examples of $\psi(x)$, namely, $\psi(x) = x^\alpha$ for some $\alpha > 2$ and $\psi(x) = e^x$.

First consider the optimal convergence rate of $\widehat{\tau}_k$. For simplicity, consider $w_i = 1/(nm_i)$ in [168] let $\bar{m}_n = \left(n^{-1}\sum_{i=1}^n m_i^{-1}\right)^{-1}$, the harmonic mean of the m_i. Then,

$$\varsigma_n(h) = O\left(\left\{\frac{\log(n)}{n}\left(1 + \frac{1}{h\bar{m}_n}\right)\right\}^{1/2}\right). \tag{6.24}$$

By (i) of Theorem 6.1.2, we have that

$$\max_{k=1,\ldots,M} |\widehat{\tau}_k(h_\tau) - \tau_k| = O\left(h_\tau^2 + \left\{\frac{\log n}{n}\left(h_\tau^2 + \frac{h_\tau}{\bar{m}_n}\right)\right\}^{1/2}\right). \tag{6.25}$$

For any h_τ that satisfies 6, we consider the rate calculation for two specific functions $\psi(x)$.

(a) Consider $\psi(x) = x^\alpha$, with $\alpha > 2$. In view of (6.24) and (6.25), the optimal h_τ is obtained by selecting the smallest h_τ that satisfies 6, i.e.,

$$\log n/(n\bar{m}_n) \prec h_\tau, \tag{6.26}$$

$$\left(\log n/n\right)^{1-2/\alpha} \prec h_\tau^2 + h_\tau/\bar{m}_n. \tag{6.27}$$

It suffices to focus on (6.27), which amounts to

$$h_\tau = B_n\left(\left(\frac{\log n}{n}\right)^{1/2-1/\alpha} \wedge \bar{m}_n\left(\frac{\log n}{n}\right)^{1-2/\alpha}\right) \tag{6.28}$$

where $B_n \to \infty$ slowly. After some calculations, the rate in (6.25) can be seen to be

$$\begin{aligned} h_\tau^2 + h_\tau\varsigma(h_\tau) &= B_n^2\bar{m}_n^2(\log n/n)^{2-4/\alpha} + B_n(\log n/n)^{1-1/\alpha} \\ &= B_n^2\left\{\bar{m}_n^2 \wedge (n/\log n)^{1-2/\alpha}\right\}(\log n/n)^{2-4/\alpha} + B_n(\log n/n)^{1-1/\alpha}. \end{aligned}$$

Thus, for $\bar{m}_n$ bounded, the rate dominated by $B_n(\log n/n)^{1-1/\alpha}$ for any $B_n \to \infty$. However, if $\bar{m}_n$ is not bounded, then the rate increases in $\bar{m}_n$, with the worst possible rate $B_n(\log n/n)^{1-2/\alpha}$ achieved when $\bar{m}_n > (n/\log n)^{1/2-1/\alpha}$. This is counter to the intuition that denser data lead to better rates.

(b) Next consider $\psi(x) = e^x$, where 6 is essentially

$$\left(h_\tau^2 + \frac{h_\tau}{\bar{m}_n}\right)^{1/2} \left(\frac{n}{\log n}\right)^{1/2} - \log n + \log\log n \to \infty, \tag{6.29}$$

which can be achieved if

$$\left(h_\tau^2 + \frac{h_\tau}{\bar{m}_n}\right)^{1/2} \left(\frac{n}{\log n}\right)^{1/2} \geq \log n$$

Arguing as in the previous case, let

$$h_\tau = \left(\left(\frac{\log^3 n}{n}\right)^{1/2} \wedge \bar{m}_n \frac{\log^3 n}{n}\right).$$

The rate in (6.25) can be seen to be $\log^2 n/n$.

We next briefly discuss the optimal convergence rate of $\widehat{d}_k$. First, ρ_d depends on h_τ, which we have already addressed above. Thus, it is sufficient to focus on h_d. In view of the error bound $h_d^2 + \varsigma_n(h_d)$, the considerations on that are similar to the bandwidth selection problem when the mean function is continuous; see [168]. If $\bar{m}_n$ is bounded and $\psi(x) = x^\alpha$, $\alpha > 2.5$, then the optimal h_d is $(\log n/n)^{1/5}$ which leads to the rate of $(\log n/n)^{2/5}$. On the other hand, if $\bar{m}_n \gtrsim (n/\log n)^{1/4}$ and $\psi(x) = x^\alpha$, $\alpha > 4$, then any bandwidth h_d satisfying $\bar{m}_n^{-1} \lesssim h_d \lesssim (\log n/n)^{1/4}$ leads to the rate $(\log n/n)^{1/2}$. Unlike for $\widehat{\tau}_k$, the rates will not improve if $\psi(x) = e^x$. The overall rate for $\widehat{d}_k$ can be obtained by combining ρ_d and $h_d^2 + \varsigma_n(h_d)$.

For convenience, a brief summary of the preceding discussion on rates is given in Table 6.1.

Table 6.1: Summary of convergence rates for break locations $\widehat{\tau}_k$ and jump sizes $\widehat{d}_k$.

	$\widehat{\tau}_k$		$\widehat{d}_k$	
$\psi(x)$	$\bar{m}_n \lesssim 1$	$\bar{m}_n \gtrsim (n/\log n)^{1/2-1/\alpha}$	$\bar{m}_n \lesssim 1$	$\bar{m}_n \gtrsim (n/\log n)^{1/4}$
x^α	$B_n(\log n/n)^{1-1/\alpha}$	$B_n(\log n/n)^{1-2/\alpha}$	$(\log n/n)^{2/5}$	$(\log n/n)^{1/2}$
e^x	$\log^2 n/n$			

Remark 6.4. *A number of papers have studied the problem of estimating jump points in the nonparametric regression problem. For example, [190], [115] and [130] all established the rates for estimating the discontinuities to be $O_p(n^{-1})$ based on different approaches under the finite variance assumption for the errors. Recall that $\psi(x) = e^x$ is needed in our approach in order to achieve the almost sure rate of $O(n^{-1})$. [266], on the other hand, showed that the almost sure rate of estimating the jump points to be $B_n(\log n/n)^{2/3}$ for any $B_n \to \infty$. That result can be roughly compared to ours in the sparse functional setting with $\alpha = 3$.*

6.1.3.4 Data-driven tuning parameter selection

The methods introduced in the preceding section depend crucially on the tuning parameter h_τ, h_d, and ζ_n. While the theory provides some guidance, in practice we need to determine these tuning parameters using a data-driven approach. In this section we provide further details on how to choose these tuning parameters.

First, we suggest using the cut-off value from the test constructed from Corollary 6.1 as the threshold ζ_n for determining M in (6.16). Specifically, given h_τ, one can choose ζ_n to be $\max_t z_{1-\alpha/2}\Omega^{1/2}(t)$, where the unknown variance terms $R(t,t)$ and σ^2 can be estimated by plugging in the local linear smoothers after we obtain the residuals from fitting the initial one-sided local linear regression, similar to the calculation steps described in Remark 6.1. Then, we can adopt a cross validation approach to select the optimal bandwidths h_τ and h_d. In the simulation studies in the next two subsections, we consider the K-fold cross validation score

$$\mathrm{CV}(h_\tau, h_d) = \frac{1}{K}\sum_{k=1}^{K}\left[\sum_{i\in\mathcal{I}_k} w_i \sum_{j=1}^{m_i} (Y_{ij} - \widehat{\mu}_{-k}(T_{ij}; h_\tau, h_d))^2\right],$$

where $\mathcal{I}_1, \ldots, \mathcal{I}_K$ form a partition of $\{1, \ldots, n\}$, and $\widehat{\mu}_{-k}(\cdot; h_\tau, h_d)$ is our estimator in (6.15), with bandwidths $h_\tau, \rho_d = h_\tau$, and h_d, based all the data Y_{ij} except the ones for which $i \in \mathcal{I}_k$. Then select h_τ and h_d as the minimizer of $\mathrm{CV}(h_\tau, h_d)$ by searching over a 2-dimensional grid. [162] examined this approach using $K = 5$ and 10 and found satisfactory performance in their numerical studies.

6.1.4 Real Data Analysis

6.1.4.1 Average Value Weighted Returns

We now analyze an average value weighted returns data example using the FDA approach in this section. Value weighted return is a type of weighting methodology to calculate portfolio return that gives a weight to an asset in the portfolio based on the asset's market value. The data set can be freely downloaded from Kenneth French's website: `http://mba.tuck.dartmouth.edu/pages/faculty/ken.french/data_library.html`. The original data set consists of the daily simple returns of $n = 49$ industry portfolios from 1927 to 2020. Let Y_{ij} be the daily return of the ith portfolio at the jth time point, $i = 1, \ldots, 49$, $j = 1, \ldots, m_i$. These data have already motivated some well-known financial factor models [167]. In this section, we choose $m_i = 354$ observations from December 2018 through April 2020 for an illustration of the introduced methods, and investigate the mean return curves across different industries $\mu(t)$ over time.

Figures 6.1 and 6.2 present the results for the estimated mean function $\widehat{\mu}(t)$. Thirteen change points $\widehat{\tau}_k$ are detected by our proposed method at 2018-12-24, 2019-01-22, 2019-02-11, 2019-03-04, 2019-05-13, 2019-06-04, 2019-06-24, 2019-07-15, 2019-08-22, 2019-10-02, 2019-10-31, 2019-12-06, 2020-02-10, 2020-03-16, 2020-04-03. The estimated jump sizes $\widehat{d_k}$ at the 15 time points are 5.659, −1.265, 1.620, −1.346, 2.052, 1.774, −1.032, −1.119, −2.347, 2.184, 1.426, −1.029, 1.157, −0.979 and 6.518, respectively. Other than these 15 breaks, the return curve follows a smooth-changing pattern. We also treat the data as independent samples and apply the nonparametric regression method by [266] to estimate the mean function. The resulting curve (green curve in Figure 6.1) contains eight change points that are all detected by our method except the one at 2020-02-28. This approach ignores the data correlation and may not be appropriate for the daily asset data. Figure 6.1 also displays change points detected by a double CUSUM method for panel data [59], which is implemented by R package `hdbinseg`. Three change points are detected at 2018-12-24, 2020-02-21, and 2020-03-23. We note that the first change point was also detected by our method. This approach, however, only gives piecewise constant description of the data. Our results may provide a better separation of different financial and economical cycles over the time period for all asset types across different industries. The 95% pointwise confidence band is shown in Figure 6.2, where we have a wider confidence band in the period of March 2020 compared to other periods.

To compare with the single series analysis, we also present the estimated individual mean functions for 4 selected industry portfolios in Figure 6.3. We note that the estimated change points from these individual industry are distinct but closely related to the change points detected by our approach for the functional series. In practice investigators could combine information from both sources to provide useful macroeconomic analysis and produce sensible predictions.

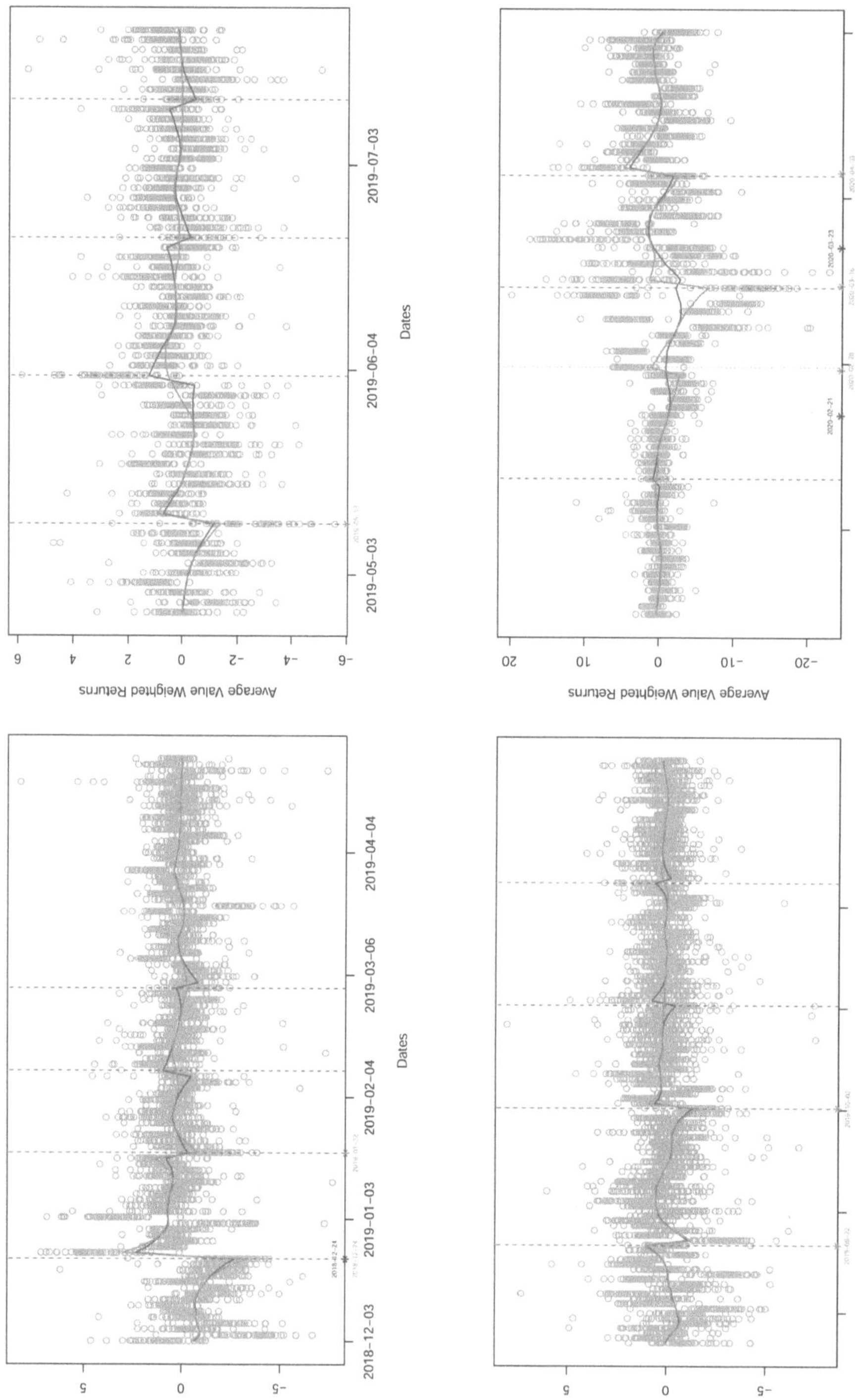

Figure 6.1: Average value weighted returns for 49 industry portfolios. The solid red curve is the mean function estimate using our method, and the vertical dash lines indicate the change point locations. The green curve is the estimate by using [266]. The blue points are change points detected by [59].

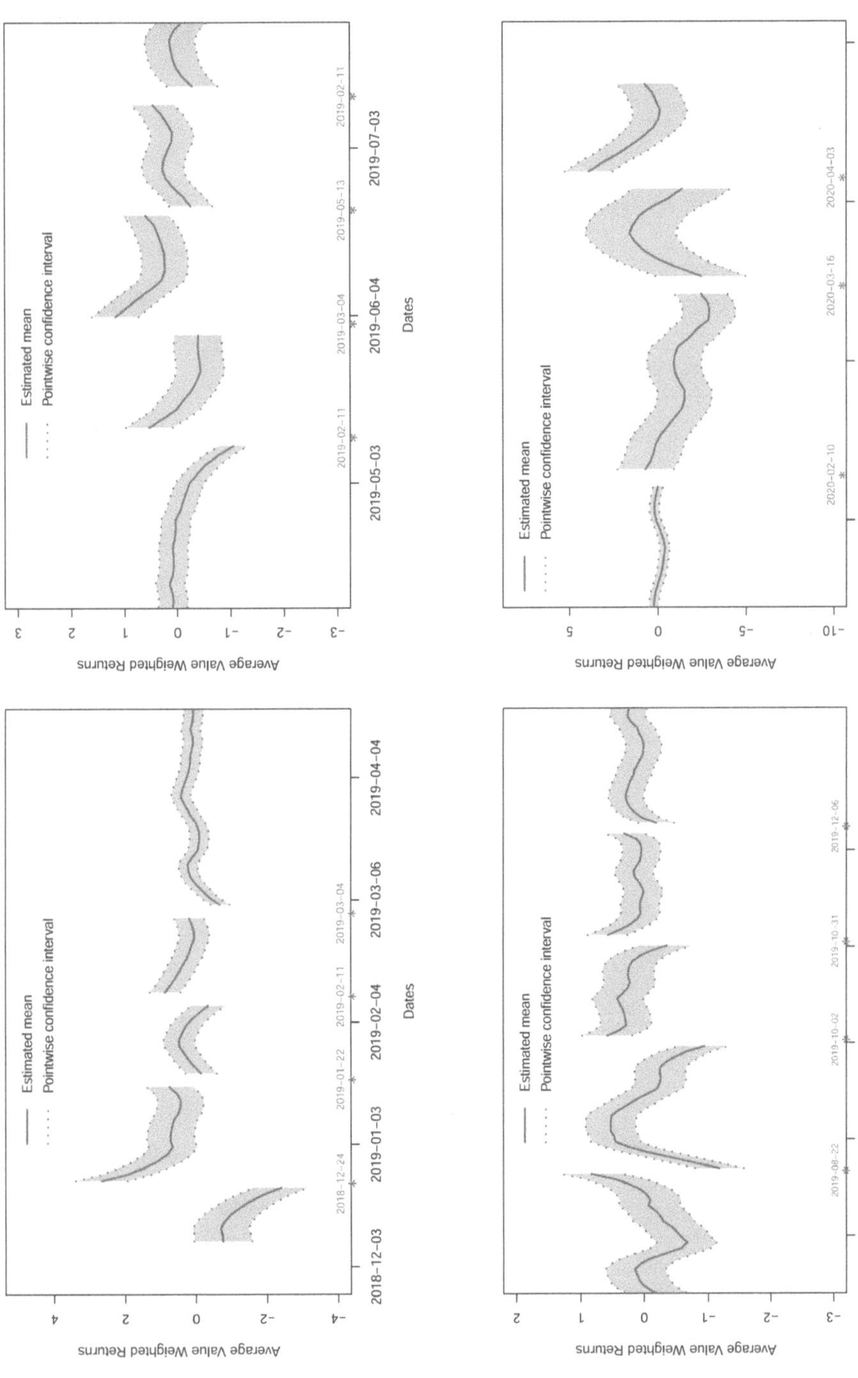

Figure 6.2: Smoothed mean function estimation and 95% pointwise confidence band from asymptotic normal.

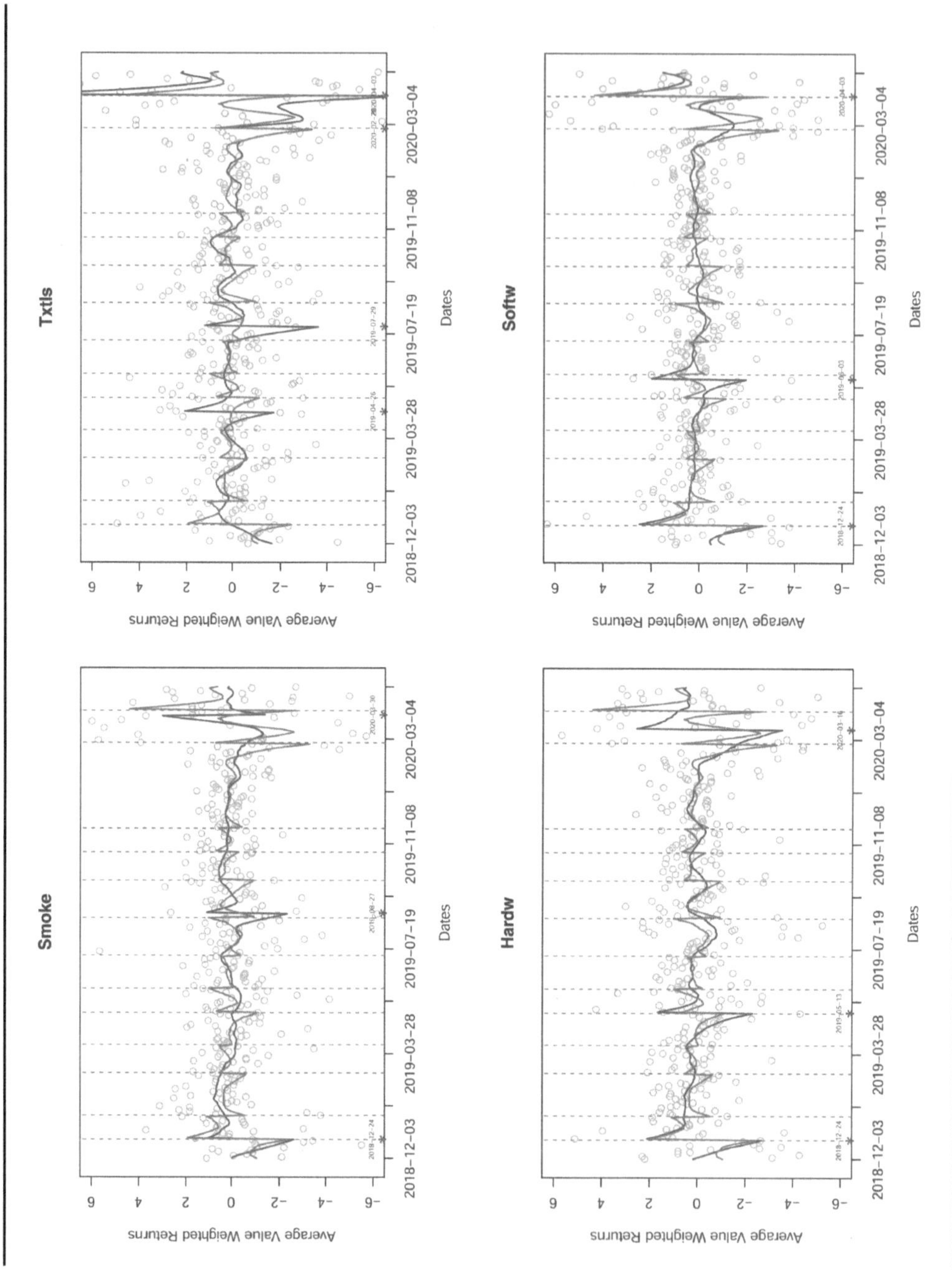

Figure 6.3: Individual curve estimate of average value weighted returns for 4 industry portfolios. The blue solid curve is the individual estimate and the red curve is $\widehat{\mu}(t)$ using our approach.

We next consider monthly maximum average value weighted returns for 49 industry portfolios with a longer time frame from January 1990 to April 2020. This leads to a functional series $\{Y_{ij} : i = 1, \cdots, 49; j = 1 \cdots, 364\}$. Figure 6.4 indicates that 10 jumps $\widehat{\tau}_k$ are detected by our proposed method, at Feb 1992, Oct 1998, Apr 2000, Nov 2001, May 2006, Jun 2008, Apr 2010, Jun 2012, Feb 2015, Jan 2019. These breaks correspond to the recoveries after the financial crises during the periods 1997–1998 and 2008–2012. The estimated jump sizes $\widehat{d}_k$ at the 10 time points are -0.673, -1.009, -2.146, -1.028, 0.996, 2.747, 2.186, 1.135, -1.114 and -2.538, respectively. In contrast, the green curve estimate using the approach of [266] only identifies four change points at Oct 1998, Nov 2001, Apr 2010, Jan 2019 (also all detected by our method) while the change point detection method by [59] identifies two change points at Mar 2000 (also detected by our method) and Jul 2003. Similarly, we have a wide confidence band where the data has a high variation. To compare with the individual estimate, we further present the estimated individual function for 4 selected industry types in Figure 6.5. One can observe that the traditional single curve change point detection method may be inadequate.

6.1.4.2 Monthly U.S. Treasuries

We consider the monthly U.S. treasuries from January 1983 to September 2010. In total, there are $n = 15$ interest rates with maturities of 3, 6, 9, 12, 18, 24, 30, 36, 48, 60, 72, 84, 96, 108, and 120 months. Each maturity has a monthly yield curve of interest rates with $m_i = 333$ observations spanning the observation period. These data have already been studied with different statistical models in [79, 54]. The time period not only covers two financial crises in 1987 and 2008, but also the tenure of the Fed Chairmen Paul Volcker (1979–1987), Alan Greenspan (1987–2006) and Ben Bernanke (2006–2014). Some interesting questions are whether the monetary policies changed with the chairman, a statement which most would probably agree with, and whether there were lasting effects from the two crises on the movement of interest rates. We apply our proposed method to address these questions.

Figure 6.6 presents the mean curve estimation results. Four break points $\widehat{\tau}_k$ are detected at Mar 1987, Feb 1994, Aug 1999, Dec 2007, which includes two aforementioned historical financial crises. The change in Feb 1994 corresponds to the event when the Federal Reserve unexpectedly raised interest rates and caused the bond market crisis or the so-called *Great Bond Massacre*. The change in Aug 1999 reflects a transition to an economic recovery phase. The estimated jump sizes $\widehat{d}_k$ at the 4 change points are 1.554, 2.595, 2.036, and -1.828. The results using [266] and [59] were similarly presented as in the preceding example. We observe again that the method by [266] detected fewer change points. The results from applying [59] also divide the whole period into five segments with 4 change points, located at slightly different time points.

6.1.4.3 Discussion

The change point detection approach for FDA introduced in this section has a broad range of applications. In the real-data examples, we considered two economic data sets for which the functional series are indexed by calendar time. More applications exist in medical and other scientific investigations where the index t could be any continuous variable. For example, in genomic studies researchers may be interested in identifying short intervals of copy number variations in multiple DNA sequences; in environmental monitoring, it may be important to detect anomalies such as the presence of hazardous materials or intruders typically existing only in some small areas. Our proposed change point detection method produces meaningful segmentation of such functional data and suggests important subgroups or sub-regions delineated by discontinuous boundaries.

An immediate extension of our approach is change point detection for multivariate functional data. In such problems, one might be also interested in identifying structural changes of the covariance function as well as the mean function. Indeed, many economic patterns are impacted by the

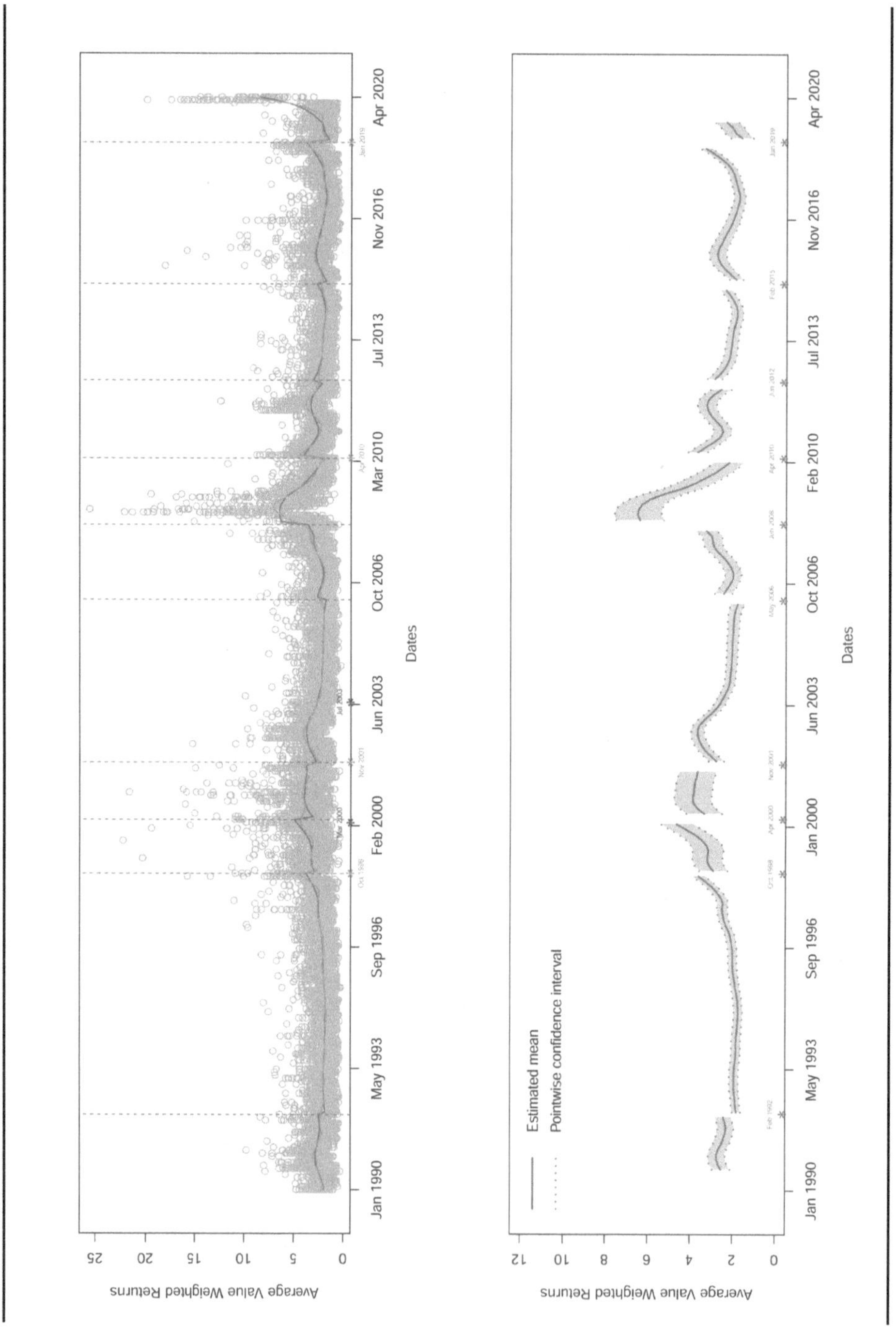

Figure 6.4: Monthly maximum average value weighted returns for 49 industry portfolios. Top panel: The red solid line is the mean function estimate $\widehat{\mu}(t)$, the vertical dash lines show the change point locations, the green curve is the estimate using [266], and the blue points are change detected by [59]. Bottom panel: $\widehat{\mu}(t)$ and 95% pointwise confidence band using our method.

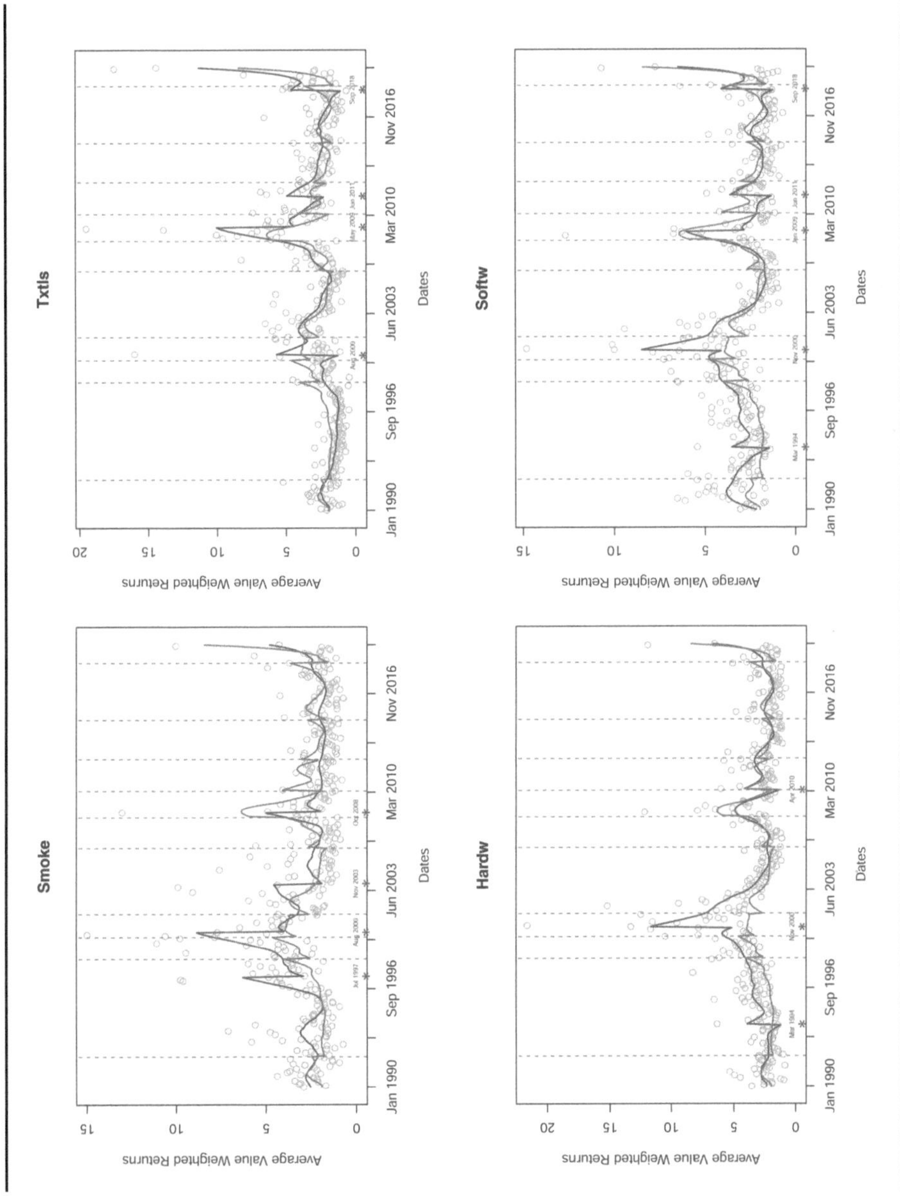

Figure 6.5: Individual curve estimates of monthly maximum average value weighted returns for 4 industry portfolios. The blue curve is the individual estimate and the red curve is $\widehat{\mu}(t)$ using our approach.

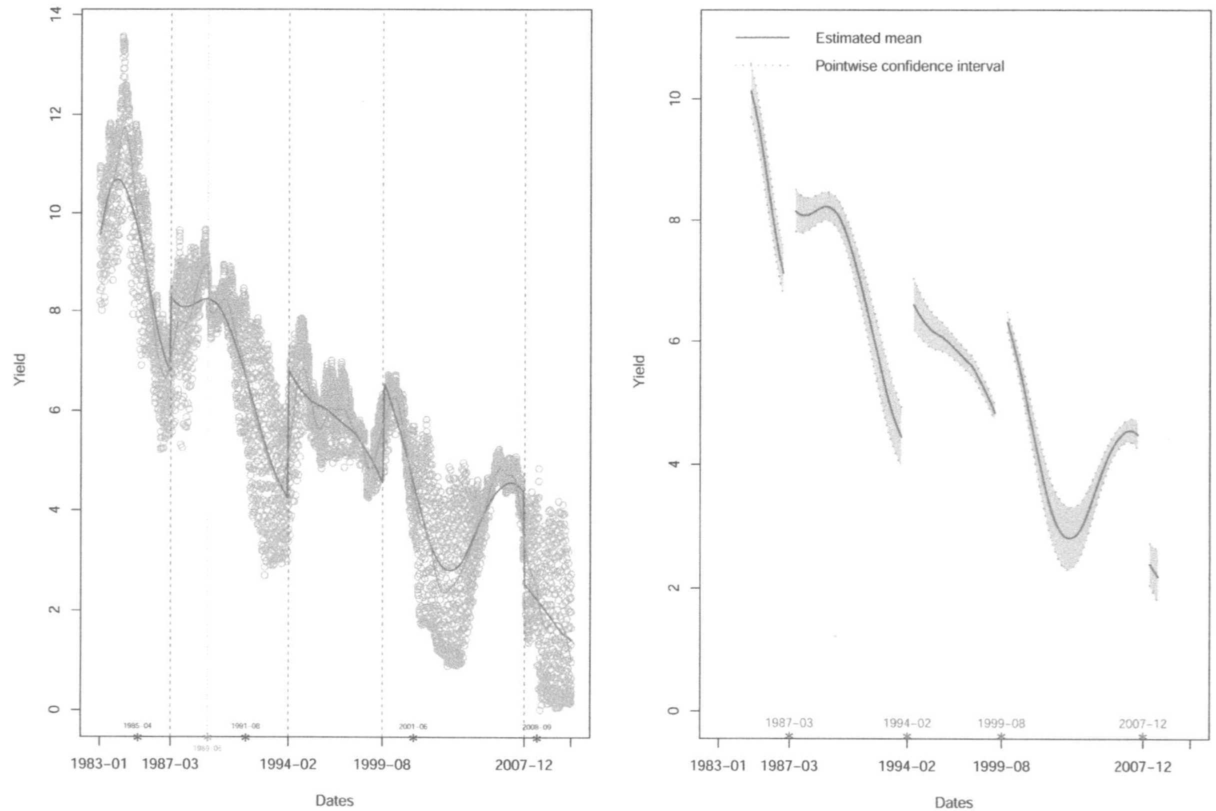

Figure 6.6: End-of-month U.S. Treasury bill and bond yields at maturities ranging from three months to ten years. Left panel: The red solid line is the mean function estimate $\widehat{\mu}(t)$, the vertical dash lines show the change points $\widehat{\tau}_k$, the green curve is obtained by using [266], and the blue points are breaks detected by [59]. Right panel: Mean function estimation and 95% pointwise confidence band.

underlying volatility or correlation process that is subject to sudden breaks triggered by external events. In such situations, both the theoretical development and the computational burden present new challenges.

6.2 Change detection in network data

6.2.1 Dynamic network

Many systems of scientific interest are composed of a large number of interacting elements, examples ranging from interacting within each living cell to people interacting with one another within and across societies. These and many other systems can be conceptualized as *networks*, where the network nodes represent the elements in a given system and network edges represent interactions between the elements. *Dynamic network* is also called temporal, evolutionary, snapshot, longitudinal or time-varying network. It has been very popular in recent decades, with the increasing amount of sequential data collected from structured and evolving systems, such as online communication platforms, daily email networks, monthly transaction networks, traffic networks, swarm behavior monitoring, sensor networks, and biological networks, among others. Following are some specific examples:

- Social networks have garnered significant interest due to their relevance in various social processes, such as information dissemination, distributed search, and the diffusion of social influence. Social scientists have long studied networks not only for their static properties but also as dynamic processes where individuals continuously form and dissolve social ties, thereby reshaping the network structure. The formation of social networks is complex, with individuals striving to achieve their goals while navigating multiple, often conflicting, constraints. For example, individuals tend to interact with others who are similar to them, a phenomenon known as homophily, and they seek to avoid conflict while leveraging overlapping social circles. Yet, the realization of these interactions is shaped by both spatial and social proximity. In cases where cooperative relationships are beneficial, individuals may favor embedded ties within locally dense clusters, such as forming relationships with friends of friends—a process referred to as triadic closure. Conversely, individuals may also seek access to novel resources by forming ties outside their immediate circle, thereby bridging structural holes between unconnected individuals. Social ties may also dissolve for various reasons, including lack of support from other relationships or conflicts with existing ties. Understanding the degree to which these mechanisms influence social dynamics requires longitudinal network data that captures changes over time and includes information on individuals' attributes and affiliations.

- A significant current challenge in neuroscience is the reconstruction of the dynamic interactions between brain regions during both task-based and resting-state functional magnetic resonance imaging (fMRI) studies. Successfully achieving this reconstruction can profoundly enhance our understanding of the brain's functional organization. Since the early 1990s, the introduction of blood oxygen level-dependent (BOLD) contrast has revolutionized fMRI due to its high spatial resolution and non-invasive nature. BOLD contrast is based on the relationship between blood flow and neuronal activation, meaning that when a brain region becomes active, increased blood flow to that region can be detected.

  In fMRI studies, the brain is typically divided into small cubic regions called voxels, with brain activity measured in each voxel over time, generating time-series data. Functional connectivity (FC) analysis explores the relationships between these time-series data, either between voxels or specific brain regions. Simple methods for estimating FC include covariance, correlation, and precision matrices. More advanced methods involve spectral measures such as coherence and partial coherence. The functional connections between brain regions or voxels can be represented as networks.

Functional connectivity patterns can change for various reasons, many of which are of interest to researchers. For example, changes may occur during a task due to habituation, learning, or delays in task state induction. In resting-state studies, changes can result from shifts in self-directed thought or spontaneous cognitive processes. Moreover, it has been demonstrated that the temporal dynamics of brain connectivity vary across individuals, suggesting that capturing intra-experimental connectivity changes can offer valuable insights.

- Correlation networks, often referred to as similarity networks, are constructed by defining edges based on a similarity or correlation measure between pairs of nodes, where nodes typically represent multivariate observations from a sample. For instance, in financial applications, the second moment dynamics of multivariate processes are crucial for understanding relationships between economic and financial variables. There is an extensive body of research on multivariate volatilities in such contexts. Correlation networks appear frequently in both financial and biological fields, such as in stock market price data and gene expression data.

  When constructing correlation networks, the entire dataset is used to estimate correlations between nodes, implying that the network structure may evolve over time. For example, in the stock market, a correlation network might be built using almost two decades of stock prices, as done by Onnela et al. [193]. However, over such a long time period, the relationships between stocks, and thus the underlying network structure, are likely to change. This issue is typically addressed by dividing the data into shorter time windows to capture the dynamic nature of the network. Similarly, in fMRI studies, brain connectivity can vary across tasks or even within the same task, meaning that assuming a static brain activity correlation network across trials with multiple tasks may lead to inaccurate conclusions.

  Correlation networks are more complex than directly observed networks because they are inferred from correlation data. Once a network is inferred from correlations, it can be treated like any other familiar network. However, the process of inferring these networks is not straightforward. A common approach is to threshold the correlations to construct an adjacency matrix, but the resulting networks can be highly sensitive to the chosen threshold. Therefore, applying standard network analysis methods to correlation networks is challenging without first addressing the issue of inferring the network itself. Ongoing research is actively exploring solutions to these challenges.

In general, adding a time component to network data is a richer representation which allows for a more powerful analysis of these systems. We may consider the aforementioned FC analysis as an example. The idea of studying the brain as an FC network is helpful as it can be viewed as a system with various interacting regions which produce complex behaviors. Similar to other biological networks, understanding the complex network organization of the brain can lead to profound clinical implications. Earlier methods usually assume that the network series are stationary over time, i.e. the dependency or the FC between the brain regions remains constant over the time. However, it is only a mathematically convenient assumption and may not hold for a dynamic network study. Modelling the temporal evolution of a network is especially important when the latter is governed by a non-stationary underlying process, whose dynamics undergo abrupt switches or breaks, called change points. For instance, social network interactions display different structures along time, which can be affected by external events such as social unrest or criminal attacks. Moreover, detecting structural breaks and finding stationary phases in dynamic networks have diverse applications, from brain connectivity state segmentation to phase discovery in financial correlation networks. In addition, real-world dynamic networks are often structured around functional groups or densely connected communities. Therefore, the temporal evolution of such networks has been often analyzed through changes in these substructures such as size growths and decays, merges and splits of communities, etc.

To fix idea, we need some graph-theory terminology, which can be found in many standard texts. Mathematically, a static network $\mathbf{X} = (V, E)$ is an undirected simple graph with vertex set V and edge set $E \subset V \times V$. The vertices in V correspond to the objects and may also be referred to as nodes in the literature. In the social network, each node may be one individual human subject; in the neural image network, each node may be a voxel or region of interest. The edges in E indicate some relation between pairs of these objects and are also called ties, links or connections. For example, the existence of an edge in a social network corresponds to a friendship between two subjects; an edge in neural image network may correspond to a strong correlation between two voxels. In an undirected network, an edge is an unordered pair $(u, v) \in E$ indicating that vertices u and v are connected. That is, we regard (u, v) and (v, u) as being identical. The network size is defined as the cardinality of V, i.e., the number of vertices in V. A path in the network $\mathscr{G}$ is a subgraph with a sequence of successive edges which joins a sequence of distinct vertices. For example, let vertex set $\{v_1, \cdots, v_m\}$ be a nonempty subset of V, such that the pair of vertices (v_i, v_{i+1}) is connected by an edge in E, for any $i = 1, \cdots, m-1$. Then the m vertices together with these edges form a path in $\mathscr{G}$. We say that either vertex v_1 or v_m is an endpoint, and the path is from v_1 to v_m (or, reversely, from v_m to v_1). Since a path of size m contains exactly $m-1$ edges, its length is said to be $m-1$. In many applications, vertices in a network are associated with individual information. If such information of a vertex can be represented as a real-valued vector, we call it a covariate. Specially, if a covariate is degenerated to a real number, we simply call it attribute.

We now introduce the discrete time representation of temporal networks and denote a dynamic network $\mathscr{G}_I = \{\mathbf{X}_t, t \in I\}$ to be a sequence of graph snapshots, where I is an ordered set such as the set of natural numbers, and for each $t \in I$, $\mathbf{X}_t$ is a static graph, which could be directed, have edge weights or node attributes. In order to formalize the concept of change point in this context, we assume that each snapshot $\mathbf{X}_t$ of the dynamic network is independently generated from a random graph distribution $\mathscr{P}_t$. For example, the unknown generative distribution $\mathscr{P}_t$ could be an inhomogeneous Bernoulli network distribution, a stochastic block model, or a graphon model. A value $\tau \in I$ is a change point if $\mathscr{P}_\tau \neq \mathscr{P}_{\tau+1}$. As in chapter 3, a dynamic network sequence may contain multiple change points $\tau_1 < \tau_2 < \cdots$, corresponding to multiple distribution changes. We note that a change of the distribution will be a shift in structures of the underlying parameters, which is fundamentally different from a simple mean-shift considered in earlier chapters.

The change of distribution could be understood in the following way by invoking the notion of network community. A community correspond to a subset of nodes with much higher connectivity within the group than across groups. A great many methods for graph partitioning are essentially variations on the more general concept of hierarchical clustering. There are numerous techniques that have been proposed for the general clustering problem, differing primarily in how they evaluate the quality of proposed clustering and the algorithms by which they seek to optimize that quality. Hierarchical methods are classified as either agglomerative, being based on the successive coarsening of partitions through the process of merging, or divisive, being based on the successive refinement of partitions through the process of splitting. At each stage, the current candidate partition is modified in a way that minimizes a specified measure of cost. In agglomerative methods, the least costly merge of two previously existing partition elements is executed, while in divisive methods, it is the least costly split of a single existing partition element into two that is executed. These approaches can be regarded as unsupervised subgrouping methods. In this chapter, we consider that these subgroups or communities may change over time. Real-world communication structure changes can be complicated. In various settings, the change may correspond to the emergence of a new community, switching community memberships, change in the number of communities, among others. In practice there are many ways to identify the communities. A very popular method is the spectral clustering method which has a solid theoretical foundation, usually associate the connectivity of a graph with the eigen-analysis of certain matrices, such as the adjacency matrix or the Laplacian.

In addition, there are many practical algorithms for community detection such as some greedy algorithm, iterative removal of edges, the modularity method, block modeling or a community score. Furthermore there are also online detection methods proposed in recent years, mainly in the machine learning and computer science literature.

6.2.2 Autoregressive network models

Understanding and being able to model the network changes over time is of immense importance for, e.g., monitoring anomalies in internet traffic networks, predicting demand and setting prices in electricity supply networks, managing natural resources in environmental readings in sensor networks, and understanding how news and opinion propagates in online social networks. In spite of the existence of a large body of literature on dynamic networks, the development of the foundation for dynamic network models is still in its infancy [145]. As for dealing with dynamic changes of networks, early attempts are based on the evolution analysis of network snapshots over time without really modeling the changes directly [2, 83]. A related work is [100] which applied a latent space model for bipartite networks to study interlocking boards in leading Irish company directories over 11 years. It is not uncommon that the networks observed at different times are assumed to be independent [197, 27]. or independent conditionally on some latent processes [85, 183]. Although this reflects the fact that most networks change slowly over time, it provides little insight on the dynamics underlying the changes and is almost powerless for future prediction. Speedily increasing availability of large network data recorded over time also calls for more tools to reveal underlying dynamic structures more explicitly. The popular approaches for modeling dynamic changes include, among others, Markov process models [233, 179], the exponential random graph models [121, 150], and latent process based models [100, 85, 183]. The estimation for those models is compute-intensive, relying on various MCMC or EM algorithms.

In this section we consider a simple first-order autoregressive (i.e. AR(1)) model for dynamic network processes of which the edges changes over time while the nodes are unchanged [134]. The autoregressive equation depicts the changes over time explicitly. Though this setting is a special case of the Markov chain network models (see [277], and also [233] and [179]), a simple AR(1) structure makes it possible to measure explicitly the underlying dynamic properties such as auto-correlation coefficients, and the Hamming distance. It facilitates the maximum likelihood estimation (MLE) for the parameters in the model in a simple and direct manner with uniform error rates. Furthermore diagnostic checking for the fitted network models can be performed in terms of an easy permutation test, which is impossible under a merely Markovian structure.

The setting in this section can be applied to any network processes with various underlying structures as long as the edges are independent with each other, which we illustrate through an AR(1) stochastic block model. The latent communities in our setting are characterized by the transition probabilities over time, instead of the (static) connection probabilities—the approach often adopted from static stochastic block models; see [197] and the references therein. This new structure also paves the way for a new spectral clustering algorithm which identifies the latent communities more effectively—a phenomenon corroborated by both the asymptotic theory and the simulation results. To cater for possible structure changes of underlying processes, we incorporate a change point detection mechanism in the AR(1) stochastic block modeling. Again the change point is estimated by the maximum likelihood method. The AR(1) continuous time stochastic block model of [179] is based on a sophisticated construction. Its estimation is based on a reversible jump MCMC, though a discrete-time version of their model admits the same Markov Chain representation (6.32) below.

Theoretical developments for dynamic stochastic block models in the literature were typically based on the assumption that networks observed at different times are independent; see [197, 27] and references therein. The autoregressive structure considered in this section brings the extra complexity due to serial dependence. By establishing the α-mixing property with exponentially decaying coefficients for the AR(1) network processes, we are able to show that the proposed spectral clustering algorithm leads to a consistent recovery of the latent community structure. On the other hand, an extra challenge in detecting a change point in the dynamic stochastic block network process is that the estimation for latent community structures before and after a possible change point is typically not consistent during the search for the change point. Comparing with [277] where the membership structure is not considered, the additional uncertainty brought by spectral clustering makes our problem unconventional, especially in the case where the membership structures before and after the change point are different. To overcome this obstacle, we introduce a truncation technique which breaks the searching interval into two parts such that the error bounds for the estimated change point can be established.

The computational methods in this section only apply to the dynamic networks observed on discrete times. Even so the relevant literature is large, across mathematics, computer science, engineer, statistics, biology, genetics and social sciences. We can only list a small selection of more statistics-oriented papers in addition to the aforementioned references. [105] proposed a state space mixed membership stochastic block model (with a logistic normal prior). [121] proposed an exponential random-graph model which specifies the conditional distribution of a network given its lagged values as a separable exponential function containing an implicit normalized constant. The inference for the model was conducted by an MCMC method. [150] further developed some separable exponential models. [100] applied a latent space model for modeling dynamic structures of bipartite networks. to study interlocking boards in leading Irish company directories over 11 years. [85] assumed that the elements of adjacency matrix at each time are conditionally independent Bernoulli random variables given two latent processes, and the conditional probabilities are the functions of those two processes. They further developed two so-called locally adaptive dynamic inference methods based on MCMC. [68] studied the limit properties of Markovian, exchangeable and càdlàg (i.e. every edge remains in each state which it visits for a positive amount of time) dynamic network. [183] proposed a variational EM-algorithm for a dynamic stochastic block network model. [197] studied the theoretical properties (such as the minimax lower bounds for the risk) of a dynamic stochastic block model, assuming "smooth" connectivity probabilities. The literature on change point detection in dynamic networks includes [271, 256, 262, 289, 27, 291]. [144, 293, 294, 49, 292] adopted autoregressive models for modeling continuous responses observed from the nodes of a network process. [138] used dynamic network as a tool to model non-stationary vector autoregressive processes. For the development on continuous-time dynamic networks, we refer readers to [233], [184], [179] and [66]. We acknowledge that there are published works which treat time as continuous variable. See for example [184], [66].

6.2.2.1 AR(1) models

Let $\{\mathbf{X}_t, t = 0, 1, 2, \cdots\}$ be a dynamic network process defined on the p fixed nodes, denoted by $\{1, \cdots, p\}$, where $\mathbf{X}_t \equiv (X_{i,j}^t)$ denotes the $p \times p$ adjacency matrix at time t. We also assume that all networks are Erdös-Renyi in the sense that $X_{i,j}^t$, $(i, j) \in \mathcal{J}$, are independent and take values either 1 or 0, where $\mathcal{J} = \{(i, j) : 1 \leq i \leq j \leq p\}$ for undirected networks, $\mathcal{J} = \{(i, j) : 1 \leq i < j \leq p\}$ for undirected networks without self-loops, $\mathcal{J} = \{(i, j) : 1 \leq i, j \leq p\}$ for directed networks, and $\mathcal{J} = \{(i, j) : 1 \leq i \neq j \leq p\}$ for directed networks without self-loops. Note that an edge from node i to j is indicated by $X_{i,j} = 1$, and no edge is denoted by $X_{i,j} = 0$. For undirected networks, $X_{i,j}^t = X_{j,i}^t$.

Definition 6.1. *An AR(1) network process is defined as*

$$X_{i,j}^t \;=\; X_{i,j}^{t-1}\, I(\varepsilon_{i,j}^t = 0) \;+\; I(\varepsilon_{i,j}^t = 1), \quad t \geq 1, \tag{6.30}$$

where $I(\cdot)$ denotes the indicator function, the innovations $\varepsilon_{i,j}^t$, $(i, j) \in \mathcal{J}$, are independent, and

$$P(\varepsilon_{i,j}^t = 1) = \alpha_{i,j}^t, \quad P(\varepsilon_{i,j}^t = -1) = \beta_{i,j}^t, \quad P(\varepsilon_{i,j}^t = 0) = 1 - \alpha_{i,j}^t - \beta_{i,j}^t. \tag{6.31}$$

In the above expression, $\alpha_{i,j}^t, \beta_{i,j}^t$ are non-negative constants, and $\alpha_{i,j}^t + \beta_{i,j}^t \leq 1$.

Equation (6.30) is an analogue of the noisy network model of [48]. The innovation (or noise) $\varepsilon_{i,j}^t$ is added via the two indicator functions to ensure that $X_{i,j}^t$ is still binary. Obviously, $\{\mathbf{X}_t, t = 0, 1, 2, \cdots\}$ is a Markov chain, and

$$P(X_{i,j}^t = 1 | X_{i,j}^{t-1} = 0) = \alpha_{i,j}^t, \quad P(X_{i,j}^t = 0 | X_{i,j}^{t-1} = 1) = \beta_{i,j}^t, \tag{6.32}$$

or collectively,

$$P(\mathbf{X}_t | \mathbf{X}_{t-1}, \cdots, \mathbf{X}_0) = P(\mathbf{X}_t | \mathbf{X}_{t-1}) = \prod_{(i,j)\in\mathcal{J}} P(X_{i,j}^t | X_{i,j}^{t-1}) \tag{6.33}$$

$$= \prod_{(i,j)\in\mathcal{J}} (\alpha_{i,j}^t)^{X_{i,j}^t(1-X_{i,j}^{t-1})}(1-\alpha_{i,j}^t)^{(1-X_{i,j}^t)(1-X_{i,j}^{t-1})}(\beta_{i,j}^t)^{(1-X_{i,j}^t)X_{i,j}^{t-1}}(1-\beta_{i,j}^t)^{X_{i,j}^t X_{i,j}^{t-1}}.$$

It is clear that the smaller $\alpha_{i,j}^t$ is, the more likely the no-edge status at time $t-1$ (i.e. $X_{i,j}^{t-1} = 0$) will be retained at time t (i.e. $X_{i,j}^t = 0$); and the smaller $\beta_{i,j}^t$ is, the more likely an edge at time $t-1$ (i.e. $X_{i,j}^{t-1} = 1$) will be retained at time t (i.e. $X_{i,j}^t = 1$). For most slowly changing networks (such as social networks), we expect $\alpha_{i,j}^t$ and $\beta_{i,j}^t$ to be small.

It is natural to model dynamic networks by a Markov chain. See, e.g. [121, 150, 267, 277, 100, 68, 183, 211, 179]. For example, the Markovian transition probabilities under a discrete version of the stationary independent arcs network model of Snijders (2005, Section 5) [233] can be written equivalently as (6.32) with $\alpha_{i,j}^t \equiv \alpha$ and $\beta_{i,j}^t \equiv \beta$. In this section we build the Markovian structure based on the explicit AR(1) model (6.30), which enables us to study the theoretical properties of the network processes, and to develop simple and efficient inference methods with appropriate theoretical guarantee.

6.2.2.2 Stationarity

Note that $\{\mathbf{X}_t\}$ is a homogeneous Markov chain if

$$\alpha_{i,j}^t \equiv \alpha_{i,j} \quad \text{and} \quad \beta_{i,j}^t \equiv \beta_{i,j} \quad \text{for all } t \geq 1 \text{ and } (i,j) \in \mathcal{J}. \tag{6.34}$$

Specify the distribution of the initial network $\mathbf{X}_0 = (X_{i,j}^0)$ as follows:

$$P(X_{i,j}^0 = 1) = \pi_{i,j} = 1 - P(X_{i,j}^0 = 0), \tag{6.35}$$

where $\pi_{i,j} \in (0,1)$, $(i,j) \in \mathcal{J}$, are constants.

Proposition 6.1. *Let the homogeneity condition (6.34) hold with* $\alpha_{i,j} + \beta_{i,j} \in (0,1]$, *and*

$$\pi_{i,j} = \alpha_{i,j}/(\alpha_{i,j} + \beta_{i,j}), \quad (i,j) \in \mathcal{J}. \tag{6.36}$$

Then $\{\mathbf{X}_t, t = 0, 1, 2, \cdots\}$ *is a strictly stationary process. Furthermore for any* $(i,j), (\ell, m) \in \mathcal{J}$ *and* $t, s \geq 0$,

$$E(X_{i,j}^t) = \frac{\alpha_{i,j}}{\alpha_{i,j} + \beta_{i,j}}, \qquad var(X_{i,j}^t) = \frac{\alpha_{i,j}\beta_{i,j}}{(\alpha_{i,j} + \beta_{i,j})^2}, \tag{6.37}$$

$$\rho_{i,j}(|t-s|) \equiv \text{Corr}(X_{i,j}^t, X_{\ell m}^s) = \begin{cases} (1 - \alpha_{i,j} - \beta_{i,j})^{|t-s|} & \text{if } (i,j) = (\ell, m), \\ 0 & \text{otherwise.} \end{cases} \tag{6.38}$$

The Hamming distance counts the number of different edges in the two networks, and is a measure the closeness of two networks [83].

Definition 6.2. *For any two matrices* $\mathbf{A} = (A_{i,j})$ *and* $\mathbf{B} = (B_{i,j})$ *of the same size, the Hamming distance is defined as* $D_H(\mathbf{A}, \mathbf{B}) = \sum_{i,j} I(A_{i,j} \neq B_{i,j})$.

Proposition 6.2. *Let* $\{\mathbf{X}_t, t = 0, 1, \cdots\}$ *be a stationary network process satisfying the condition of Proposition 6.1. Let* $d_H(|t-s|) = E\{D_H(\mathbf{X}_t, \mathbf{X}_s)\}$ *for any* $t, s \geq 0$*. Then* $d_H(0) = 0$*, and it holds for any* $k \geq 1$ *that*

$$d_H(k) = d_H(k-1) + \sum_{(i,j)\in\mathcal{J}} \frac{2\alpha_{i,j}\beta_{i,j}}{\alpha_{i,j}+\beta_{i,j}}(1-\alpha_{i,j}-\beta_{i,j})^{k-1} \tag{6.39}$$

$$= \sum_{(i,j)\in\mathcal{J}} \frac{2\alpha_{i,j}\beta_{i,j}}{(\alpha_{i,j}+\beta_{i,j})^2}\{1-(1-\alpha_{i,j}-\beta_{i,j})^{k}\}. \tag{6.40}$$

Proposition 6.2 indicates that the expected Hamming distance $d_H(d) = E\{D_H(\mathbf{X}_t, \mathbf{X}_{t+k})\}$ increases strictly, as k increases, initially from $d_H(1) = \sum \frac{2\alpha_{i,j}\beta_{i,j}}{\alpha_{i,j}+\beta_{i,j}}$ towards the limit $d_H(\infty) = \sum \frac{2\alpha_{i,j}\beta_{i,j}}{(\alpha_{i,j}+\beta_{i,j})^2}$ which is also the expected Hamming distance of the two independent networks sharing the same marginal distribution of $\mathbf{X}_t$.

Proposition 6.3 below shows that $\{\mathbf{X}_t,\ t = 0, 1, \cdots\}$ is α-mixing with exponentially decaying coefficients. Note that the conventional mixing results for ARMA processes do not apply here, as they typically require that the innovation distribution is continuous; see, e.g., Section 2.6.1 of [93]. Let $\mathcal{F}_a^b$ be the σ-algebra generated by $\{X_{i,j}^k, a \leq k \leq b\}$. The α-mixing coefficient of process $\{X_{i,j}^t, t = 0, 1, \cdots\}$ is defined as

$$\alpha^{i,j}(\tau) = \sup_{k\in\mathbb{N}} \sup_{A\in\mathcal{F}_0^k, B\in\mathcal{F}_{k+\tau}^{\infty},} |P(A\cap B) - P(A)P(B)|.$$

Proposition 6.3. *Let condition (6.34) hold,* $\alpha_{i,j}, \beta_{i,j} > 0$*, and* $\alpha_{i,j} + \beta_{i,j} \leq 1$*. Then* $\alpha^{i,j}(\tau) \leq \rho_{i,j}(\tau) = (1-\alpha_{i,j}-\beta_{i,j})^{\tau}$ *for any* $\tau \geq 1$.

6.2.2.3 Estimation

To simplify the notation, we assume the availability of the observations $\mathbf{X}_0, \mathbf{X}_1, \cdots, \mathbf{X}_n$ from a stationary network process which satisfies the condition of Proposition 6.1. Without imposing any further structure on the model, the parameters $(\alpha_{i,j}, \beta_{i,j})$, for different (i, j), can be estimated separately. Conditionally on $\mathbf{X}_0$, the joint probability function of $\mathbf{X}_1, \cdots, \mathbf{X}_n$ is $\prod_{1\leq t\leq n} P(\mathbf{X}_t|\mathbf{X}_{t-1})$. It then follows from (6.33) that the maximum likelihood estimators are

$$\widehat{\alpha}_{i,j} = \frac{\sum_{t=1}^n X_{i,j}^t(1-X_{i,j}^{t-1})}{\sum_{t=1}^n(1-X_{i,j}^{t-1})}, \qquad \widehat{\beta}_{i,j} = \frac{\sum_{t=1}^n(1-X_{i,j}^t)X_{i,j}^{t-1}}{\sum_{t=1}^n X_{i,j}^{t-1}}. \tag{6.41}$$

For definiteness we shall set $0/0 = 1$. To state the asymptotic properties, we list some regularity conditions first.

C1. There exists a constant $l > 0$ such that $\alpha_{i,j}, \beta_{i,j} \geq l$ and $\alpha_{i,j} + \beta_{i,j} \leq 1$ for all $(i, j) \in \mathcal{J}$.

C2. $n, p \to \infty$, and $(\log n)(\log\log n)\sqrt{\frac{\log p}{n}} \to 0$.

Condition C1 defines the parameter space, and condition C2 indicates that the number of nodes is allowed to diverge in a smaller order than $\exp\left\{\frac{n}{(\log n)^2(\log\log n)^2}\right\}$.

Proposition 6.4. *Let conditions (6.34), C1 and C2 hold. For any constant* $c > 2$*, there exists a large enough constant* $C > 0$ *such that,*

$$P\left(\max_{(i,j)\in\mathcal{J}} |\widehat{\alpha}_{i,j} - \alpha_{i,j}| \geq l^{-1}C\sqrt{\frac{\log p}{n}}\right) \geq 2p^{2-c},$$
$$P\left(\max_{(i,j)\in\mathcal{J}} |\widehat{\beta}_{i,j} - \beta_{i,j}| \geq l^{-1}C\sqrt{\frac{\log p}{n}}\right) \geq 2p^{2-c}.$$

Consequently, as $n, p \to \infty$*, we have:*

$$\max_{(i,j)\in\mathcal{J}} |\widehat{\alpha}_{i,j} - \alpha_{i,j}| = O_p\left(\sqrt{\frac{\log p}{n}}\right) \quad \text{and} \quad \max_{(i,j)\in\mathcal{J}} |\widehat{\beta}_{i,j} - \beta_{i,j}| = O_p\left(\sqrt{\frac{\log p}{n}}\right).$$

Proposition 6.4 provides a uniform convergence rate for the MLEs in (6.41). Proposition 6.5 below implies that any fixed number of estimators $\{\widehat{\alpha}_{i,j}, \widehat{\beta}_{i,j}\}$ are jointly asymptotically normal; noting that p diverges together with n. To state this joint asymptotic normality, we introduce some notation first: let $J_1 = \{(i_1, j_1), \dots, (i_{m_1}, j_{m_1})\}$, $J_2 = \{(k_1, \ell_1), \dots, (k_{m_2}, \ell_{m_2})\}$ be two arbitrary subsets of $\mathcal{J}$ with $m_1, m_2 \geq 1$ fixed. Denote $\mathbf{\Theta}_{J_1,J_2} = (\alpha_{i_1,j_1}, \dots, \alpha_{i_{m_1},j_{m_1}}, \beta_{k_1,\ell_1}, \dots, \beta_{k_{m_2},\ell_{m_2}})^\top$, and correspondingly denote the MLEs as $\widehat{\mathbf{\Theta}}_{J_1,J_2} = (\widehat{\alpha}_{i_1,j_1}, \dots, \widehat{\alpha}_{i_{m_1},j_{m_1}}, \widehat{\beta}_{k_1,\ell_1}, \dots, \widehat{\beta}_{k_{m_2},\ell_{m_2}})^\top$.

Proposition 6.5. *Let conditions (6.34), C1 and C2 hold. Then* $\sqrt{n}(\widehat{\mathbf{\Theta}}_{J_1,J_2} - \mathbf{\Theta}_{J_1,J_2}) \to N(\mathbf{0}, \mathbf{\Sigma}_{J_1,J_2})$*, where* $\mathbf{\Sigma}_{J_1,J_2} = \mathrm{diag}(\sigma_{11}, \dots, \sigma_{m_1+m_2,m_1+m_2})$ *is a diagonal matrix with*

$$\sigma_{rr} = \frac{\alpha_{i_r,j_r}(1-\alpha_{i_r,j_r})(\alpha_{i_r,j_r}+\beta_{i_r,j_r})}{\beta_{i_r,j_r}}, \quad 1 \leq r \leq m_1,$$
$$\sigma_{rr} = \frac{\beta_{k_r,\ell_r}(1-\beta_{k_r,\ell_r})(\alpha_{k_r,\ell_r}+\beta_{k_r,\ell_r})}{\alpha_{k_r,\ell_r}}, \quad m_1+1 \leq r \leq m_1+m_2.$$

6.2.2.4 Model diagnostic check

As with any other regression models we should consider checking the goodness-of-fit after fitting this AR(1) model. Based on estimators in (6.41), we define the residual $\widehat{\varepsilon}_{i,j}^t$, resulted from fitting model (6.30) to the data, as the estimated value of $E(\varepsilon_{i,j}^t | X_{i,j}^t, X_{i,j}^{t-1})$, i.e.

$$\begin{aligned}\widehat{\varepsilon}_{i,j}^t =& \frac{\widehat{\alpha}_{i,j}}{1-\widehat{\beta}_{i,j}} I(X_{i,j}^t = 1, X_{i,j}^{t-1} = 1) - \frac{\widehat{\beta}_{i,j}}{1-\widehat{\alpha}_{i,j}} I(X_{i,j}^t = 0, X_{i,j}^{t-1} = 0)\\ &+ I(X_{i,j}^t = 1, X_{i,j}^{t-1} = 0) - I(X_{i,j}^t = 0, X_{i,j}^{t-1} = 1), \quad (i,j) \in \mathcal{J},\ t = 1, \cdots, n.\end{aligned}$$

One way to check the adequacy of the model is to test for the independence of $\widehat{\mathcal{E}}_t \equiv (\widehat{\varepsilon}_{i,j}^t)$ for $t = 1, \cdots, n$. Since $\widehat{\varepsilon}_{i,j}^t$, $t = 1, \cdots, n$, only take 4 different values for each $(i,j) \in \mathcal{J}$, we adopt the two-way, or three-way contingency table to test the independence of $\widehat{\mathcal{E}}_t$ and $\widehat{\mathcal{E}}_{t-1}$, or $\widehat{\mathcal{E}}_t$, $\widehat{\mathcal{E}}_{t-1}$ and $\widehat{\mathcal{E}}_{t-2}$. For example the test statistic for the two-way contingency table is

$$T = \frac{1}{n|\mathcal{J}|} \sum_{(i,j)\in\mathcal{J}} \sum_{k,\ell=1}^{4} \{n_{i,j}(k,\ell) - n_{i,j}(k,\cdot)n_{i,j}(\cdot,\ell)/(n-1)\}^2 / \{n_{i,j}(k,\cdot)n_{i,j}(\cdot,\ell)/(n-1)\}, \tag{6.42}$$

where $|\mathcal{J}|$ denotes the cardinality of $\mathcal{J}$, and for $1 \le k, \ell \le 4$,

$$n_{i,j}(k,\ell) = \sum_{t=2}^{n} I\{\widehat{\varepsilon}_{i,j}^{t} = u_{i,j}(k),\ \widehat{\varepsilon}_{i,j}^{t-1} = u_{i,j}(\ell)\},$$

$$n_{i,j}(k,\cdot) = \sum_{t=2}^{n} I\{\widehat{\varepsilon}_{i,j}^{t} = u_{i,j}(k)\}, \quad n_{i,j}(\cdot,\ell) = \sum_{t=2}^{n} I\{\widehat{\varepsilon}_{i,j}^{t-1} = u_{i,j}(\ell)\}.$$

In the above expressions, $u_{i,j}(1) = -1, u_{i,j}(2) = -\frac{\widehat{\beta}_{i,j}}{1-\widehat{\alpha}_{i,j}}, u_{i,j}(3) = \frac{\widehat{\alpha}_{i,j}}{1-\widehat{\beta}_{i,j}}$ and $u_{i,j}(4) = 1$. We calculate the P-values of the test T based on the following permutation algorithm:

1. Permute $\widehat{\mathcal{E}}_1, \cdots, \widehat{\mathcal{E}}_n$ to obtain a new sequence $\mathcal{E}_1^\star, \cdots, \mathcal{E}_n^\star$. Calculate the test statistic $T^\star$ in the same manner as T with $\{\widehat{\mathcal{E}}_t\}$ replaced by $\{\mathcal{E}_t^\star\}$.
2. Repeat 1 above M times, obtaining permutation test statistics $T_j^\star,\ j = 1, \cdots, M$, where $M > 0$ is a large integer. The P-value of the test (for rejecting the stationary AR(1) model) is then

$$\frac{1}{M}\sum_{j=1}^{M} I(T < T_j^\star).$$

6.2.3 Autoregressive stochastic block models

The general setting in Section 6.2.2 may apply to various network processes with some specific underlying structures as long as the edges are independent with each other. In this section we illustrate the idea with a new dynamic stochastic block (DSB) model.

6.2.3.1 Models for community

The DSB networks are undirected (i.e. $X_{i,j}^t \equiv X_{j,i}^t$) with no self-loops (i.e. $X_{i,i}^t \equiv 0$). Most available DSB models assume that the networks observed at different times are independent [197, 27] or conditionally independent [268, 85, 183] as connection probabilities and node memberships evolve over time. We take a radically different approach as we impose autoregressive structure (6.30) in the network process. Furthermore, instead of assuming that the members in the same communities share the same (unconditional) connection probabilities, we entertain the idea that the transition probabilities (6.32) for the members in the same communities are the same. This reflects more directly the dynamic behavior of the process, and implies the former assumption on the unconditional connection probabilities under the stationarity. See (6.36). Furthermore, it is only natural under the AR(1) DSB model to identify the latent community structure using the information on both $\alpha_{i,j}$ and $\beta_{i,j}$, instead of that on $\pi_{i,j} = \alpha_{i,j}/(\alpha_{i,j} + \beta_{i,j})$ only. Both our theory (Theorem 6.2.1 and also remarks below) and the numerical experiments in [134] confirm that this spectral clustering algorithm based on $\alpha_{i,j}$ and $\beta_{i,j}$ provides more accurate estimation than that based on $\pi_{i,j}$ only.

Let ν_t be the membership function at time t, i.e. for any $1 \le i \le p$, $\nu_t(i)$ takes an integer value between 1 and q ($\le p$); indicating that node i belongs to the $\nu_t(i)$-th community at time t, where q is a fixed integer. This effectively assumes that the p nodes are divided into the q communities. We assume that q is fixed though some communities may contain no nodes at some times.

Definition 6.3. *An AR(1) stochastic block network process $\{\mathbf{X}_t = (X_{i,j}^t), t = 0, 1, 2, \cdots\}$ is defined by (6.30), where for $1 \le i < j \le p$,*

$$P(\varepsilon_{i,j}^t = 1) = \alpha_{i,j}^t = \theta_{\nu_t(i),\nu_t(j)}^t, \quad P(\varepsilon_{i,j}^t = -1) = \beta_{i,j}^t = \eta_{\nu_t(i),\nu_t(j)}^t, \tag{6.43}$$
$$P(\varepsilon_{i,j}^t = 0) = 1 - \alpha_{i,j}^t - \beta_{i,j}^t = 1 - \theta_{\nu_t(i),\nu_t(j)}^t - \eta_{\nu_t(i),\nu_t(j)}^t.$$

In the above expressions, $\theta_{k,\ell}^t, \eta_{k,\ell}^t$ *are non-negative constants, and* $\theta_{k,\ell}^t + \eta_{k,\ell}^t \leq 1$ *for all* $1 \leq k \leq \ell \leq q$.

The evolution of membership process ν_t and/or the connection probabilities was often assumed to be driven by some latent Markov processes. The statistical inference for those models is carried out using computational methods such as MCMC, EM algorithms or extended Kalman filters. See, for example, [271, 268, 267, 85, 183, 211]. [27] adopted a change point approach: assuming both the membership and the connection probabilities remain constants either before or after a change point. See also [179, 262]. This reflects the fact that many networks (e.g. social networks) hardly change, and a sudden change is typically triggered by some external events.

We adopt a change point approach in this book. Sections 6.2.3.2 to 6.2.3.5 introduce the estimation procedure for both the community membership and transition probabilities when there are no change points in the process. This will serve as a building block for the inference for the processes with a change point in Section 6.2.3.6. Note that detecting change points in dynamic networks is a surging research area. In addition to the aforementioned references, more recent developments include [256, 291, 257]. Also note that the method of [289] can be applied to detect multiple change points for any dynamic networks.

6.2.3.2 Estimation without change points: settings

We first consider a simple scenario of no change points in the observed period, i.e.

$$\nu_t(\cdot) \equiv \nu(\cdot) \quad \text{and} \quad (\theta_{k,\ell}^t, \eta_{k,\ell}^t) \equiv (\theta_{k,\ell}, \eta_{k,\ell}), \quad t = 1, \cdots, n, \quad 1 \leq k \leq \ell \leq q. \tag{6.44}$$

Then fitting the DSB model consists of two steps: (i) estimating $\nu(\cdot)$ to cluster the p nodes into q communities, and (ii) estimating transition probabilities $\theta_{k,\ell}$ and $\eta_{k,\ell}$ for $1 \leq k \leq \ell \leq q$. To simplify the presentation, q is assumed to be known, which is the assumption taken by most papers on change point detection for DSB networks. In practice, when the number of cluster q is unspecified, one can determine a proper q by, for example, the jittering method of [47], or a Bayesian information criterion (BIC). Below we first introduce a Laplacian eigendecomposition which provides the underpinning for the proposed spectral clustering algorithm for estimating network memberships. With the estimated membership, the MLEs for $\theta_{k,\ell}$ and $\eta_{k,\ell}$ can be derived in the similar manner as in (6.41).

6.2.3.3 Laplacian eigendecomposition

The stochastic block model with p nodes and q communities can be parameterized by a pair of matrices $(\mathbf{Z}, \mathbf{\Omega})$, where $\mathbf{Z} = (z_{i,j}) \in \{0,1\}^{p \times q}$ is the membership matrix such that it has exactly one 1 in each row and at least one 1 in each column, and $\mathbf{\Omega} = (\omega_{k,\ell})_{q \times q} \in [0,1]^{q \times q}$ is a symmetric and full rank connectivity matrix, with $\omega_{k,\ell} = \frac{\theta_{k,\ell}}{\theta_{k,\ell}+\eta_{k,\ell}}$. Then $z_{i,j} = 1$ if and only if the i-th node belongs to the j-th community. On the other hand, $\omega_{k,\ell}$ is the connection probability between the nodes in community k and the nodes in community ℓ, and $s_k \equiv \sum_{i=1}^p z_{i,k}$ is the size of community $k \in \{1, \ldots, q\}$. Clearly, matrix $\mathbf{Z}$ and function $\nu(\cdot)$ are the two equivalent representations for the community membership of the network nodes.

Under model (6.44), the edge connection probability matrix is given as $E(\mathbf{X}_t) = \mathbf{W} - \text{diag}(\mathbf{W})$, where $\mathbf{W} = \mathbf{Z}\mathbf{\Omega}\mathbf{Z}^\top = (\omega_{\nu(i),\nu(j)})$. Then the standard spectral clustering algorithm for estimating the community memberships are based on Laplacian eigen-decomposition of $\boldsymbol{W}$ or its normalized version [215], for which connection probability $\omega_{\nu(i),\nu(j)}$ is taken as a similarity measure between nodes i and j. To take the advantage of the dynamic structure of dynamic networks, we take the transition probabilities $\theta_{\nu(i),\nu(j)}$ and $(1 - \eta_{\nu(i),\nu(j)})$, instead of connection probability $\omega_{\nu(i),\nu(j)}$, as the similarity measures (see Remark 1(i) below). To this end, define

$$\mathbf{\Omega}_1 = (\theta_{k,\ell})_{q\times q}, \qquad \mathbf{\Omega}_2 = (1-\eta_{k,\ell})_{q\times q},$$
$$\mathbf{W}_1 = \mathbf{Z}\mathbf{\Omega}_1\mathbf{Z}^\top = (\alpha_{i,j})_{p\times p}, \qquad \mathbf{W}_2 = \mathbf{Z}\mathbf{\Omega}_2\mathbf{Z}^\top = (1-\beta_{i,j})_{p\times p},$$

where $\alpha_{i,j} = \theta_{\nu(i),\nu(j)}$, $\beta_{i,j} = \eta_{\nu(i),\nu(j)}$. Let $\mathbf{D}_1$ and $\mathbf{D}_2$ be two $p \times p$ diagonal matrices with, respectively, $d_{i,1}, d_{i,2}$ as their (i,i)-th elements, where

$$d_{i,1} = \sum_{j=1}^{p} \alpha_{i,j}, \qquad d_{i,2} = \sum_{j=1}^{p} (1-\beta_{i,j}).$$

The normalized Laplacian matrices based on $\mathbf{W}_1$ and $\mathbf{W}_2$ are then defined as:

$$\mathbf{L}_1 = \mathbf{D}_1^{-1/2}\mathbf{W}_1\mathbf{D}_1^{-1/2}, \qquad \mathbf{L}_2 = \mathbf{D}_2^{-1/2}\mathbf{W}_2\mathbf{D}_2^{-1/2}, \qquad \mathbf{L} = \mathbf{L}_1 + \mathbf{L}_2. \tag{6.45}$$

The following lemma shows that the block structure in the membership matrix $\mathbf{Z}$ can be recovered by the leading eigenvectors of $\mathbf{L}$.

Proposition 6.6. *Let $rank(\mathbf{L}) = q$, and $\mathbf{\Gamma}_q \mathbf{\Lambda} \mathbf{\Gamma}_q^\top$ be the eigen-decomposition of $\mathbf{L}$, where $\mathbf{\Lambda} = diag\{\lambda_1, \ldots, \lambda_q\}$ is the diagonal matrix consisting of the nonzero eigenvalues of $\mathbf{L}$ arranged in the order $|\lambda_1| \geq \cdots \geq |\lambda_q| > 0$. There exists a matrix $\mathbf{U} \in \mathcal{R}^{q\times q}$ such that $\mathbf{\Gamma}_q = \mathbf{Z}\mathbf{U}$. Furthermore, for any $1 \leq i, j \leq p$, $\mathbf{z}_i\mathbf{U} = \mathbf{z}_j\mathbf{U}$ if and only if $\mathbf{z}_i = \mathbf{z}_j$, where $\mathbf{z}_i$ denotes the i-th row of $\mathbf{Z}$.*

Remark 6.5. *Both $\theta_{\nu(i),\nu(j)}$ and $(1-\eta_{\nu(i),\nu(j)})$ can be regarded as a similarity measure between nodes i and j. This is due to the fact that the communities in a network are often formed in the way that the members within the same community are more likely to be connected with each other, and the members belong to different communities are unlikely or less likely to be connected. Hence when nodes i and j belong to the same community, $\alpha_{i,j}$ tends to be large and $\beta_{i,j}$ tends to be small (see (6.32)). The converse is true when the two nodes belong to two different communities.*

Remark 6.6. *As $\mathbf{Z}$ is $p \times q$ with rank q, it is reasonable to assume rank$(\mathbf{L}) = q$ in Proposition 6.6. The q columns of $\mathbf{\Gamma}_q$ are the orthonormal eigenvectors of $\mathbf{L}$ corresponding to the q non-zero eigenvalues. Proposition 6.6 implies that there are only q distinct rows in the $p \times q$ matrix $\mathbf{\Gamma}_q$, and two nodes belong to the same community if and only if the corresponding rows in $\mathbf{\Gamma}_q$ are the same. Intuitively the discriminant power of $\mathbf{\Gamma}_q$ can be understood as follows. For any unit vector $\gamma = (\gamma_1, \cdots, \gamma_p)^\top$, it holds that*

$$\gamma^\top \mathbf{L}\gamma = 2 - \sum_{1\leq i<j\leq p} \alpha_{i,j}\Big(\frac{\gamma_i}{\sqrt{d_{i,1}}} - \frac{\gamma_j}{\sqrt{d_{j,1}}}\Big)^2 - \sum_{1\leq i<j\leq p} (1-\beta_{i,j})\Big(\frac{\gamma_i}{\sqrt{d_{i,2}}} - \frac{\gamma_j}{\sqrt{d_{j,2}}}\Big)^2. \tag{6.46}$$

For γ being an eigenvector corresponding to the largest (positive) eigenvalue of $\mathbf{L}$, the sum of the 2nd and the 3rd terms on the RHS (6.46) is minimized. Thus $|\gamma_i - \gamma_j|$ is minimized when $\alpha_{i,j}$ and/or $(1-\beta_{i,j})$ are large; noting that $d_{i,k} = d_{j,k}$ for $k = 1, 2$ when nodes i and j belong to the same community. The eigenvectors corresponding to negative eigenvalues are capable to identify the so-called heterophilic communities, see pp.1892-3 of [215].

6.2.3.4 Estimation of membership $\nu(\cdot)$

It follows from Proposition 6.1, (6.43) and (6.44) that

$$P(X_{ij}^t = 1) = \theta_{\nu(i),\nu(j)}/(\theta_{\nu(i),\nu(j)} + \eta_{\nu(i),\nu(j)}) \equiv \omega_{\nu(i),\nu(j)}, \quad 1 \leq i < j \leq p,$$

provided that X_{ij}^0 is initiated with the same marginal distribution. A simple approach adopted in literature is to apply a community detection method for static stochastic block models using the averaged data $\bar{\mathbf{X}} = \sum_{1\le t\le n} \mathbf{X}_t/n$ to detect the latent communities characterized by the connection probabilities $\{\omega_{k,\ell}, 1 \le k \le \ell \le q\}$. We take a different approach based on estimators $\{(\widehat{\alpha}_{i,j}, \widehat{\beta}_{i,j}), 1 \le i < j \le p\}$ defined in (6.41) to identify the clusters determined by the transition probabilities $\{(\theta_{k,\ell}, \eta_{k,\ell}), 1 \le k \le \ell \le q\}$ instead. More precisely, we propose a new spectral clustering algorithm to estimate $\mathbf{\Gamma}_q$ based on Proposition 6.6 above.

Let $\widehat{\boldsymbol{W}}_1, \widehat{\boldsymbol{W}}_2$ be two $p \times p$ matrices with, respectively, $\widehat{\alpha}_{i,j}, (1-\widehat{\beta}_{i,j})$ as their (i,j)-th elements for $i \neq j$, and 0 on the main diagonals. Let $\widehat{\mathbf{D}}_1, \widehat{\mathbf{D}}_2$ be two $p \times p$ diagonal matrices with, respectively, $\widehat{d}_{i,1}, \widehat{d}_{i,2}$ as their (i,i)-th elements, where

$$\widehat{d}_{i,1} = \sum_{j=1}^{p} \widehat{\alpha}_{i,j}, \qquad \widehat{d}_{i,2} = \sum_{j=1}^{p} (1-\widehat{\beta}_{i,j}).$$

Define two normalized Laplacian matrices

$$\widehat{\mathbf{L}}_1 = \widehat{\mathbf{D}}_1^{-1/2} \widehat{\boldsymbol{W}}_1 \widehat{\mathbf{D}}_1^{-1/2}, \qquad \widehat{\mathbf{L}}_2 = \widehat{\mathbf{D}}_2^{-1/2} \widehat{\boldsymbol{W}}_2 \widehat{\mathbf{D}}_2^{-1/2}. \tag{6.47}$$

Perform the eigen-decomposition for the sum of $\mathbf{L}_1$ and $\mathbf{L}_2$:

$$\widehat{\mathbf{L}} \equiv \widehat{\mathbf{L}}_1 + \widehat{\mathbf{L}}_2 = \widehat{\mathbf{\Gamma}}\, \text{diag}(\widehat{\lambda}_1, \cdots, \widehat{\lambda}_p) \widehat{\mathbf{\Gamma}}^\top, \tag{6.48}$$

where the eigenvalues are arranged in the order $\widehat{\lambda}_1^2 \ge \ldots \ge \widehat{\lambda}_p^2$, and the columns of the $p \times p$ orthogonal matrix $\widehat{\mathbf{\Gamma}}$ are the corresponding eigenvectors. We call $\widehat{\lambda}_1, \ldots, \widehat{\lambda}_q$ the q leading eigenvalues of $\widehat{\mathbf{L}}$. Denote by $\widehat{\mathbf{\Gamma}}_q$ the $p \times q$ matrix consisting of the first q columns of $\widehat{\mathbf{\Gamma}}$, which are called the leading eigenvectors of $\widehat{\mathbf{L}}$. The spectral clustering applies the k-means clustering algorithm to the p rows of $\widehat{\mathbf{\Gamma}}_q$ to obtain the community assignments for the p nodes $\widehat{\nu}(i) \in \{1, \cdots, q\}$ for $i = 1, \cdots, p$.

Remark 6.7. *Proposition 6.6 implies that the true memberships can be recovered by the q distinct rows of $\mathbf{\Gamma}_q$. Note that*

$$\widehat{\mathbf{L}} = \widehat{\mathbf{L}}_1 + \widehat{\mathbf{L}}_2 \approx \mathbf{L}_1 - \textit{diag}(\mathbf{L}_1) + \mathbf{L}_2 - \textit{diag}(\mathbf{L}_2) = \mathbf{L} - \textit{diag}(\mathbf{L}).$$

We shall see that the effect of the term diag$(\mathbf{L})$ on the eigenvectors $\mathbf{\Gamma}_q$ is negligible when p is large ([134]), and hence the rows of $\widehat{\mathbf{\Gamma}}_q$ should be slightly perturbed versions of the q distinct rows in $\mathbf{\Gamma}_q$.

The following theorem justified the validity of using $\widehat{\mathbf{L}}$ for spectral clustering. Note that $\|\cdot\|_2$ and $\|\cdot\|_F$ denote, respectively, the L_2 and the Frobenius norm of matrices.

Theorem 6.2.1. *Let conditions (2.5), C1 and C2 hold, and $\lambda_q^{-2}\left(\sqrt{\frac{\log(pn)}{np}} + \frac{1}{n} + \frac{1}{p}\right) \to 0$, as $n, p \to \infty$. Then it holds that*

$$\max_{i=1,\ldots,p} |\lambda_i^2 - \widehat{\lambda}_i^2| \le \|\widehat{\mathbf{L}}\widehat{\mathbf{L}} - \mathbf{L}\mathbf{L}\|_2 \le \|\widehat{\mathbf{L}}\widehat{\mathbf{L}} - \mathbf{L}\mathbf{L}\|_F = O_p\left(\sqrt{\frac{\log(pn)}{np}} + \frac{1}{n} + \frac{1}{p}\right). \tag{6.49}$$

Moreover, for any constant $B > 0$, there exists a constant $C > 0$ such that the inequality

$$\|\widehat{\mathbf{\Gamma}}_q - \mathbf{\Gamma}_q \mathbf{O}_q\|_F \le 4\lambda_q^{-2} C \left(\sqrt{\frac{\log(pn)}{np}} + \frac{1}{n} + \frac{1}{p}\right) \tag{6.50}$$

holds with probability greater than $1 - 16p\left[(pn)^{-(1+B)} + \exp\{-B\sqrt{p}\}\right]$, where $\mathbf{O}_q$ is a $q \times q$ orthogonal matrix.

It follows from (6.49) that the leading eigenvalues of $\mathbf{L}$ can be consistently recovered by the leading eigenvalues of $\widehat{\mathbf{L}}$. By (6.50), the leading eigenvectors of $\mathbf{L}$ can also be consistently estimated, subject to a rotation (due to the possible multiplicity of some leading eigenvalues $\mathbf{L}$). Proposition 6.6 indicates that there are only q distinct rows in $\mathbf{\Gamma}_q$, and, therefore, also q distinct rows in $\mathbf{\Gamma}_q\mathbf{O}_q$, corresponding to the q latent communities for the p nodes. This paves the way for the k-means algorithm stated below. Put

$$\mathcal{M}_{p,q} = \{\mathbf{M} \in \mathcal{R}^{p\times q} : \mathbf{M} \text{ has } q \text{ distinct rows}\}.$$

The k-means clustering algorithm: Let

$$(\widehat{\mathbf{c}}_1, \cdots, \widehat{\mathbf{c}}_p)^\top = \arg\min_{\mathbf{M}\in\mathcal{M}_{p,q}} \|\widehat{\mathbf{\Gamma}}_q - \mathbf{M}\|_F^2.$$

There are only q distinct vectors among $\widehat{\mathbf{c}}_1, \cdots, \widehat{\mathbf{c}}_p$, forming the q communities. Theorem 6.2.2 below shows that they are identical to the latent communities of the p nodes under (6.50) and (6.51). The latter holds if $\sqrt{s_{\max}}\lambda_q^{-2}C\left(\sqrt{\frac{\log(pn)}{np}} + \frac{1}{n} + \frac{1}{p}\right) \to 0$, where $s_{\max} = \max\{s_1, \ldots, s_q\}$ is the size of the largest community.

Theorem 6.2.2. *Let (6.50) hold and*

$$\sqrt{\frac{1}{s_{\max}}} > 4\sqrt{2}\lambda_q^{-2}C\left(\sqrt{\frac{\log(pn)}{np}} + \frac{1}{n} + \frac{1}{p}\right). \tag{6.51}$$

Then $\widehat{\mathbf{c}}_i = \widehat{\mathbf{c}}_j$ if and only if $\nu(i) = \nu(j)$, $1 \le i, j \le p$.

Remark 6.8. *By Lemma A.1 of [215], the error bound for the standard spectral clustering algorithm (with $n = 1$) is $O_p\left(\frac{\log p}{\sqrt{p}} + \frac{1}{p}\right)$, where the term $\frac{1}{p}$ reflects the bias caused by the inconsistent estimation of diagonal terms (see equation (A.5) and subsequent derivations in [215]). This bias comes directly from the removal of the diagonal elements of $\mathbf{L}$. Although the algorithm was designed for static networks, it has often been applied to dynamic networks using $\frac{1}{n}\sum_t \mathbf{X}_t$ in the place of a single observed network; see, e.g. [27]. With some simple modification to the proof of Lemma A.1 of [215], it can be shown that the error bound is then reduced to*

$$O_p\left(\frac{\log(pn)}{\sqrt{np}} + \frac{1}{p}\right), \tag{6.52}$$

provided that the observed networks are i.i.d. The error would only increase when the observations are not independent. On the other hand, our proposed spectral clustering algorithm for (dependent) dynamic networks entails the error rate specified in (6.49) and (6.50) which is smaller than (6.52) as long as n is sufficiently large (i.e. $(p/n)^{\frac{1}{2}} / \log(np) \to 0$). Note that we need n to be large enough in relation to p in order to capture the dynamic dependence of the networks.

6.2.3.5 Estimation of $\theta_{k,\ell}$ and $\eta_{k,\ell}$

For any $1 \le k \le \ell \le q$, we define

$$S_{k,l} = \begin{cases} \{(i,j) : 1 \le i \ne j \le p, \nu(i) = k, \nu(j) = \ell\} & \text{if } k \ne l, \\ \{(i,j) : 1 \le i < j \le p, \nu(i) = k = \nu(j) = \ell\} & \text{if } k = l, \end{cases} \tag{6.53}$$

Clearly the cardinality of $S_{k,\ell}$ is $n_{k,\ell} = s_k s_\ell$ when $k \ne \ell$ and $n_{k,\ell} = s_k(s_k - 1)/2$ when $k = \ell$.

Based on the procedure presented in Section 6.2.3.4, we obtain an estimated membership function $\widehat{\nu}(\cdot)$. Consequently, the MLEs for $(\theta_{k,\ell}, \eta_{k,\ell})$, $1 \le k \le \ell \le q$, admit the form

$$\widehat{\theta}_{k,\ell} = \sum_{(i,j)\in\widehat{S}_{k,\ell}} \sum_{t=1}^{n} X_{i,j}^{t}(1 - X_{i,j}^{t-1}) \Big/ \sum_{(i,j)\in\widehat{S}_{k,\ell}} \sum_{t=1}^{n} (1 - X_{i,j}^{t-1}), \tag{6.54}$$

$$\widehat{\eta}_{k,\ell} = \sum_{(i,j)\in\widehat{S}_{k,\ell}} \sum_{t=1}^{n} (1 - X_{i,j}^{t})X_{i,j}^{t-1} \Big/ \sum_{(i,j)\in\widehat{S}_{k,\ell}} \sum_{t=1}^{n} X_{i,j}^{t-1}, \tag{6.55}$$

where

$$\widehat{S}_{k,\ell} = \begin{cases} \{(i,j) : 1 \le i \ne j \le p,\ \widehat{\nu}(i) = k,\ \widehat{\nu}(j) = \ell\} & \text{if } k \ne \ell, \\ \{(i,j) : 1 \le i < j \le p,\ \widehat{\nu}(i) = \widehat{\nu}(j) = k\} & \text{if } k = \ell. \end{cases}$$

See (6.41) and also (6.43).

Theorem 6.2.2 implies that the memberships of the nodes can be consistently recovered. Consequently, the consistency and the asymptotic normality of the MLEs $\widehat{\theta}_{k,\ell}$ and $\widehat{\eta}_{k,\ell}$ can be established in the same manner as for Propositions 6.4 and 6.5. We state the results below.

Let $\mathcal{K}_1 = \{(i_1, j_1), \ldots, (i_{m_1}, j_{m_1})\}$ and $\mathcal{K}_2 = \{(k_1, \ell_1), \ldots, (k_{m_2}, \ell_{m_2})\}$ be two arbitrary subsets of $\{(k, \ell) : 1 \le k \le \ell \le q\}$ with $m_1, m_2 \ge 1$ fixed. Let

$$\mathbf{\Psi}_{\mathcal{K}_1,\mathcal{K}_2} = (\theta_{i_1,j_1}, \ldots, \theta_{i_{m_1},j_{m_1}}, \eta_{k_1,\ell_1}, \ldots, \eta_{k_{m_2},\ell_{m_2}})',$$

and let $\widehat{\mathbf{\Psi}}_{\mathcal{K}_1,\mathcal{K}_2}$ denote its MLE. Put $\mathbf{N}_{\mathcal{K}_1,\mathcal{K}_2} = \text{diag}(n_{i_1,j_1}, \ldots, n_{i_{m_1},j_{m_1}}, n_{k_1,\ell_1}, \ldots, n_{k_{m_2},\ell_{m_2}})$ where $n_{k,\ell}$ is the cardinality of $S_{k,\ell}$ defined as in (6.53).

Theorem 6.2.3. *Let conditions (2.5), C1 and C2 hold, and* $\frac{\sqrt{s_{\max}}}{\lambda_q^2}\left(\sqrt{\frac{\log(pn)}{np}} + \frac{1}{n} + \frac{1}{p}\right) \to 0$. *Then it holds that*

$$\max_{1\le k,\ell\le q} |\widehat{\theta}_{k,\ell} - \theta_{k,\ell}| = O_p\left(\sqrt{\frac{\log q}{ns_{\min}^2}}\right) \quad \text{and} \quad \max_{1\le k,\ell\le q} |\widehat{\eta}_{k,\ell} - \eta_{k,\ell}| = O_p\left(\sqrt{\frac{\log q}{ns_{\min}^2}}\right),$$

where $s_{\min} = \min\{s_1, \ldots, s_q\}$.

Theorem 6.2.4. *Let the condition of Theorem 6.2.3 hold. Then*

$$\sqrt{n}\mathbf{N}_{\mathcal{K}_1,\mathcal{K}_2}^{\frac{1}{2}}(\widehat{\mathbf{\Psi}}_{\mathcal{K}_1,\mathcal{K}_2} - \mathbf{\Psi}_{\mathcal{K}_1,\mathcal{K}_2}) \to N(\mathbf{0}, \widetilde{\mathbf{\Sigma}}_{\mathcal{K}_1,\mathcal{K}_2}),$$

where $\widetilde{\mathbf{\Sigma}}_{\mathcal{K}_1,\mathcal{K}_2} = \text{diag}(\widetilde{\sigma}_{11}, \ldots, \widetilde{\sigma}_{m_1+m_2,m_1+m_2})$ *with*

$$\widetilde{\sigma}_{rr} = \frac{\theta_{i_r,j_r}(1 - \theta_{i_r,j_r})(\theta_{i_r,j_r} + \eta_{i_r,j_r})}{\eta_{i_r,j_r}}, \quad 1 \le r \le m_1,$$

$$\widetilde{\sigma}_{rr} = \frac{\eta_{k_r,\ell_r}(1 - \eta_{k_r,\ell_r})(\theta_{k_r,\ell_r} + \eta_{k_r,\ell_r})}{\theta_{k_r,\ell_r}}, \quad m_1 + 1 \le r \le m_1 + m_2.$$

Finally to prepare for the inference in Section 6.2.3.6 below, we introduce some additional notations. First we denote $\widehat{\nu}$ by $\widehat{\nu}^{1,n}$, to reflect the fact that the community clustering was carried out using the data $\mathbf{X}_1, \cdots, \mathbf{X}_n$ (conditionally on $\mathbf{X}_0$). See Section 6.2.3.4 above. Further we denote the maximum log likelihood by

$$\widehat{l}(1, n;\ \widehat{\nu}^{1,n}) = l(\{\widehat{\theta}_{k,\ell}, \widehat{\eta}_{k,\ell}\};\ \widehat{\nu}^{1,n}) \tag{6.56}$$

to highlight the fact that both the node clustering and the estimation for transition probabilities are based on the data $\mathbf{X}_1, \cdots, \mathbf{X}_n$.

6.2.3.6 Inference with a change point

For dynamic network it is reasonable to assume a change point structure, similar to traditional time series data. Many change point detection methods are proposed recently along this direction. We cannot present all of them. Instead we choose a simple strategy in this section to showcase the method. Other more complicated methods should be used for deeper or more complete analysis in practice.

Now we assume that there is a change point τ_0 at which both the membership of nodes and the transition probabilities $\{\theta_{k,\ell}, \eta_{k,\ell}\}$ change. It is necessary to assume $n_0 \leq \tau_0 \leq n - n_0$, where n_0 is an integer and $n_0/n \equiv c_0 > 0$ is a small constant, as we need enough information before and after the change in order to detect τ_0. We assume that within the time period $[0, \tau_0]$, the network follows a stationary model (6.44) with parameters $\{(\theta_{1,k,\ell}, \eta_{1,k,\ell}) : 1 \leq k, l \leq q\}$ and a membership map $\nu^{1,\tau_0}(\cdot)$. Within the time period $[\tau_0 + 1, n]$ the network follows a stationary model (6.44) with parameters $\{(\theta_{2,k,\ell}, \eta_{2,k,\ell}) : 1 \leq k, l \leq q\}$ and a membership map $\nu^{\tau_0+1,n}(\cdot)$. Though we assume that the number of communities is unchanged after the change point, our results can be easily extended to the case that the number of communities also changes.

We estimate the change point τ_0 by the maximum likelihood method:

$$\widehat{\tau} = \arg \max_{n_0 \leq \tau \leq n - n_0} \{\widehat{l}(1, \tau; \widehat{\nu}^{1,\tau}) + \widehat{l}(\tau + 1, n; \widehat{\nu}^{\tau+1,n})\}, \tag{6.57}$$

where $\widehat{l}(\cdot)$ is given in (6.56).

To measure the difference between the two sets of transition probabilities before and after the change, we put

$$\Delta_F^2 = \frac{1}{p^2}(\|\boldsymbol{W}_{1,1} - \boldsymbol{W}_{2,1}\|_F^2 + \|\boldsymbol{W}_{1,2} - \boldsymbol{W}_{2,2}\|_F^2),$$

where the four $p \times p$ matrices are defined as

$$\boldsymbol{W}_{1,1} = (\theta_{1,\nu^{1,\tau_0}(i),\nu^{1,\tau_0}(j)}), \quad \boldsymbol{W}_{1,2} = (1 - \eta_{1,\nu^{1,\tau_0}(i),\nu^{1,\tau_0}(j)}),$$

$$\boldsymbol{W}_{2,1} = (\theta_{2,\nu^{\tau_0+1,n}(i),\nu^{\tau_0}(j)+1,n}), \quad \boldsymbol{W}_{2,2} = (1 - \eta_{1,\nu^{\tau_0+1,n}(i),\nu^{\tau_0+1,n}(j)}).$$

Note that Δ_F can be viewed as the signal strength for detecting the change point τ_0. Let $s_{\max}$, $s_{\min}$ denote, respectively, the largest, and the smallest community size among all the communities before and after the change. Similar to (6.45), we denote the normalized Laplacian matrices corresponding to $\boldsymbol{W}_{i,j}$ as $\mathbf{L}_{i,j}$ for $i, j = 1, 2$. Let $|\lambda_{i,1}| \geq |\lambda_{i,2}| \geq \ldots \geq |\lambda_{i,q}|$ be the absolute nonzero eigenvalues of $\mathbf{L}_{i,1} + \mathbf{L}_{i,2}$ for $i = 1, 2$, and we denote $\lambda_{\min} = \min\{|\lambda_{1,q}|, |\lambda_{2,q}|\}$. Now some regularity conditions are in order.

C3. For some constant $l > 0$, $\theta_{i,k,\ell}, \eta_{i,k,\ell} > l$, and $\theta_{i,k,\ell} + \eta_{i,k,\ell} \leq 1$ for all $i = 1, 2$ and $1 \leq k \leq \ell \leq q$.

C4. $\log(np)/\sqrt{p} \to 0$, and $\sqrt{s_{\max}}\lambda_{\min}^{-2}\left(\sqrt{\log(pn)/np} + \frac{1}{n} + \frac{1}{p} + \frac{\log(np)/n + \sqrt{\log(np)/(np^2)}}{\Delta_F^2}\right) \to 0$.

C5. $\frac{\Delta_F^2}{\log(np)/n + \sqrt{\log(np)/(np^2)}} \to \infty$.

Condition C3 is similar to C1. The condition $\log(np)/\sqrt{p} \to 0$ in C4 controls the misclassification rate of the k-means algorithm. Recall that there is a bias term $O(p^{-1})$ in spectral clustering caused by the removal of the diagonal of the Laplacian matrix (see Remark 2 above). Intuitively, as p increases, the effect of this bias term on the misclassification rate of the k-means algorithm becomes negligible. On the other hand, note that the length of the time interval for searching for the change point in (6.57) is of order $O(n)$; the $\log(n)$ term here in some sense reflects the effect of the difficulty in detecting the true change point when the searching interval is extended as n increases. The second condition in C4 is similar to (6.51), which ensures that the true communities can be recovered. Condition C5 requires that the average signal strength $\Delta_F^2 = p^{-2}[\|\boldsymbol{W}_{1,1} - \boldsymbol{W}_{2,1}\|_F^2 + \|\boldsymbol{W}_{1,2} - \boldsymbol{W}_{2,2}\|_F^2]$ is of higher order than $\frac{\log(np)}{n} + \sqrt{\frac{\log(np)}{np^2}}$ for change point detection.

[277] deals with the MLE for a change-point in a network Markov chain but without a latent community structure. Hence it does not have the complication to estimate the community memberships in addition. Allowing the membership change in our setting leads to an extra challenge: in the process of searching for the location of the change-point, the estimation for the latent communities before or after a specified location may not be consistent. To overcome this obstacle, we introduce a truncation which breaks the searching interval into two parts such that the error in the estimated change point can be bounded. [27] also allows the membership change. But it assumes that the networks observed at different times are independent with each other.

Theorem 6.2.5. *Let conditions C2–C5 hold. Then the following assertions hold.*

(i) When $\nu^{1,\tau_0} \equiv \nu^{\tau_0+1,n}$,

$$\frac{|\tau_0 - \widehat{\tau}|}{n} = O_p\left(\frac{\frac{\log(np)}{n} + \sqrt{\frac{\log(np)}{np^2}}}{\Delta_F^2} \times \min\left\{1, \frac{\min\left\{1, (n^{-1}p^2\log(np))^{\frac{1}{4}}\right\}}{\Delta_F s_{\min}}\right\}\right).$$

(ii) When $\nu^{1,\tau_0} \neq \nu^{\tau_0+1,n}$,

$$\frac{|\tau_0 - \widehat{\tau}|}{n} = O_p\left(\frac{\frac{\log(np)}{n} + \sqrt{\frac{\log(np)}{np^2}}}{\Delta_F^2} \times \min\left\{1, \frac{\min\left\{1, (n^{-1}p^2\log(np))^{\frac{1}{4}}\right\}}{\Delta_F s_{\min}} + \frac{1}{\Delta_F^2}\right\}\right).$$

Notice that for $\tau < \tau_0$, the observations in the time interval $[\tau + 1, n]$ are a mixture of the two different network processes if $\nu^{1,\tau_0} \neq \nu^{\tau_0+1,n}$. In the worst case scenario then, all q communities can be changed after the change point τ_0. This causes the extra estimation error term $\frac{1}{\Delta_F^2}$ in Theorem 6.2.5(ii).

Proofs for all the propositions and theorems in this section can be found in [134].

6.2.4 Asian trade data

We now present a real data illustration. This example concerns the annual international trades among $p = 50$ Asian countries between 1950 and 2014 (i.e. $n = 65$). We define an edge between two countries to be 1 as long as there exist trades between the two countries in that year (regardless the direction), and 0 otherwise. We take this simplistic approach to illustrate the AR(1) stochastic block model with change point. The data used are a subset of the openly available global trade data for 205 countries in 1870–2014 [22, 21].

Applying the maximum likelihood estimator (6.57), we obtained an estimated change point at year 1991. In Figure 6.7, we compare the averaged adjacency matrices of Asian trade clusters before and after 1991. In each matrix, a cold blue hue signifies smaller values, while a warm red hue indicates larger values. Before 1991, the major cluster was centered around the Gulf States in the Middle East, along with two major economic powers—China and Japan. The intense red in the matrix's lower left corner reflects robust trade activities among these nations. After 1991, the two clusters merged into one cluster.

It should be noted that the year 1990 marked a significant turning point in the process of globalization. The disintegration of the Soviet Union in 1989, coupled with the fall of the Berlin Wall and the cessation of the Cold War by 1991, catalyzed a new era of global inter-connectivity. Previously isolated nations, which had been part of the communist bloc, began their integration into the global market economy. This period saw a notable increase in trade and investment, accompanied by a reduction in barriers to migration and cultural exchange.

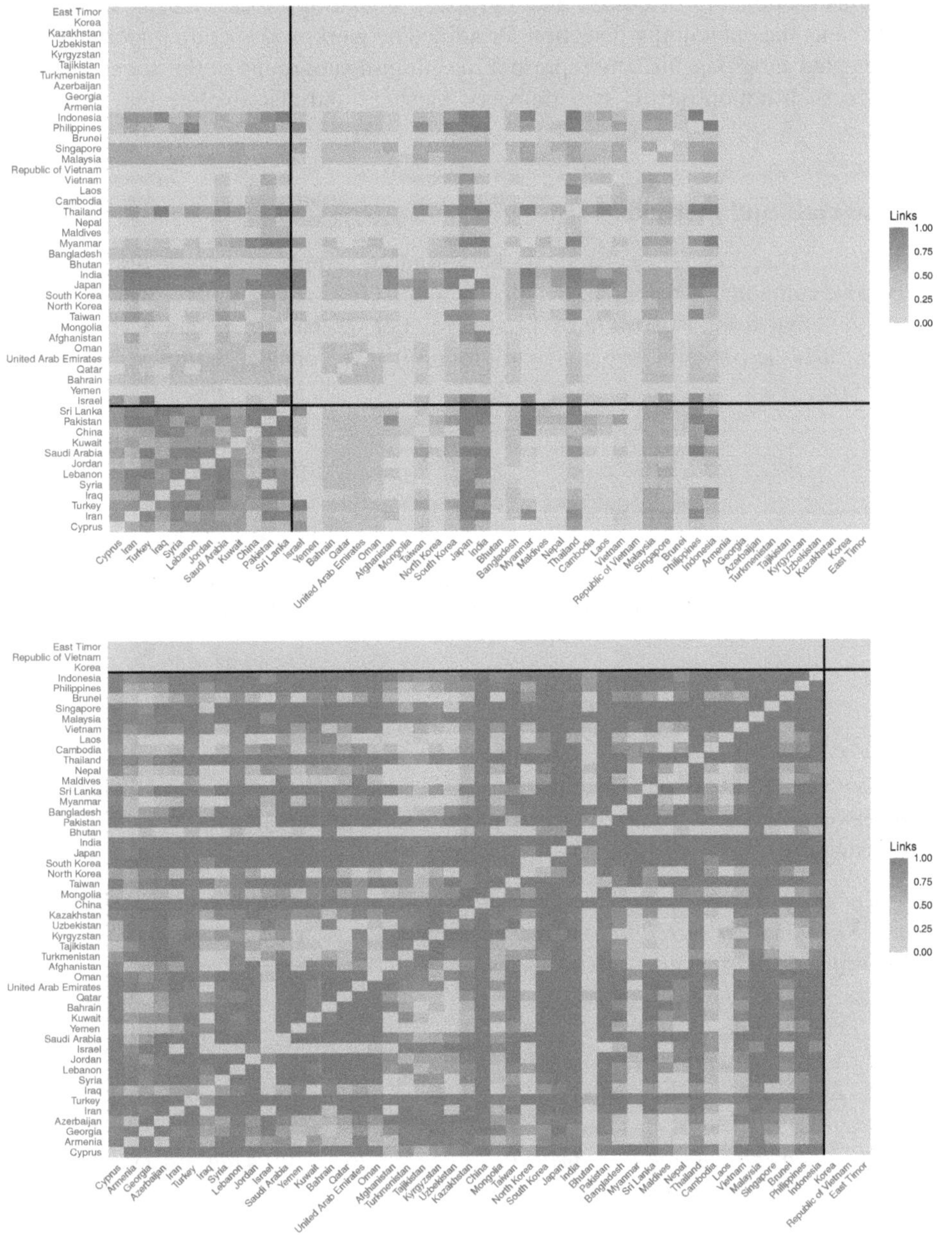

Figure 6.7: Comparison of Asian Trade Clusters Before (top) and After (bottom) 1991, based on the averaged adjacency matrices. The heatmap color scheme is low = "lightblue", high = "red", mid = "orange", midpoint = 0.5, limit = c(0, 1).

6.3 Learning block structure in U-statistic based matrices

In the previous section, we discussed autoregressive stochastic block models. In addition to the change over time, the community detection for a fixed network is also quite important in the literature. The detected subgroups usually represent meaningful subgroups of the nodes. In this section, we generalize such notion into U-statistic-based matrices and also explore the underlying block structure.

6.3.1 Materials and methods

Let $\{X_i = (X_{i1}, \dots, X_{ip})^{\mathrm{T}}, i = 1, \dots, n\}$ be a random sample of some p-dimensional random variable, whose components belong to one of K groups $G_1, \dots, G_K$ with corresponding cardinalities $p_1, \dots, p_K$ that form a partition of $\{1, \dots, p\}$. For the ease of notation, let $i \in G_k$ represent the event that the ith variable belongs to group G_k. We consider the following U-statistic for $i \in G_k, j \in G_l$,

$$\hat{\rho}_{ij} = \hat{\rho}_{ij}(\varphi; m) = \binom{n}{m}^{-1} \sum_{1 \le l_1 < \dots < l_m \le n} \varphi_{ij}(X_{l_1}, \dots, X_{l_m}), \tag{6.58}$$

where $\varphi_{ij}(X_{l_1}, \dots, X_{l_m}) = \varphi\{(X_{l_1 i}, X_{l_1 j}), \dots, (X_{l_m i}, X_{l_m j})\}$ is the kernel function of order m, which is a symmetric and uniformly bounded Borel measurable function. To account for the group structure and variable specific characteristics when $i \in G_k, j \in G_l$, let the expectation of (6.58) satisfy

$$E(\hat{\rho}_{ij}) = \begin{cases} \tau_{kl}\sigma_i\sigma_j, & i \neq j, \\ \tau_k \kappa_i, & i = j, \end{cases} \tag{6.59}$$

where $\tau_{kl} = \tau_{lk}, \tau_k \ge 0, \sigma_i > 0, \sigma_j > 0$ and $\kappa_i = \sigma_i^2$ or 1 are well-defined unknown parameters to avoid non-identifiability issue. The equations (6.58) and (6.59) define a broad class of proximity measures between variables. Many empirical quantities describing global properties of a population fall in this category.

Example 6.1. *The U-statistic kernels and parameters in (6.59) can be specified easily for the sample covariance, Kendall's tau correlations and the energy distance when $i \in G_k, j \in G_l$, respectively.*

(1) The sample covariance: $\varphi_{ij}(X_\alpha, X_\beta) = (X_{\alpha i} - X_{\beta i})(X_{\alpha j} - X_{\beta j})/2$ with $\sigma_i^2 = var(X_{\alpha i})$, and $-1 \le \tau_{kl} \le 1, \tau_k = 1, \kappa_i = \sigma_i^2$.

(2) The Kendall's Tau-a coefficient: $\varphi_{ij}(X_\alpha, X_\beta) = sign(X_{\alpha i} - X_{\beta i})sign(X_{\alpha j} - X_{\beta j})$, with $\sigma_i = 1$ and $-1 \le \tau_{kl} \le 1, \tau_k = 1, \kappa_i = 1$.

(3) The Kendall's Tau-b coefficient: $\varphi_{ij}(X_\alpha, X_\beta) = \{n_0/(n_0 - n_i)\}^{1/2} sign(X_{\alpha i} - X_{\beta i})\{n_0/(n_0 - n_j)\}^{1/2} sign(X_{\alpha j} - X_{\beta j})$, with $\sigma_i^2 = 1/pr\{(X_{\alpha i} - X_{\beta i})^2 > 0\}$ and $-1 \le \tau_{kl} \le 1, \tau_k = 1, \kappa_i = 1$, where $n_0 = n(n-1)/2$, $n_t = \sum_s u_{ts}(u_{ts} - 1)/2$ and u_{ts} is the number of tied values in the sth different values for the tth quantity.

(4) The energy distance: $\varphi_{ij}(X_\alpha, X_\beta, X_\gamma) = 2|X_{\alpha i} - X_{\beta j}| - |X_{\alpha i} - X_{\gamma i}| - |X_{\beta j} - X_{\gamma j}|$ with $\sigma_i = 1$ for all i and $\tau_k = 0$.

Let $M_n = (\hat{\rho}_{ij})_{i,j=1}^p$ be the resulting U-statistic based $p \times p$ matrix defined by (6.58). By (6.59), the expectation $E(M_n)$ possesses a block structure which includes not only the case with equal correlations for the variables within each group and for the pair of variables between two different groups ($\sigma_i = \sigma_j$), but the cases where block structures exist while the within/between associations are unequal ($\sigma_i \neq \sigma_j$). Our primary goal is to estimate the unknown group number K and identify the group structure based on M_n. When the dimensionality p is fixed but the sample size n diverges, classical limit theorems guarantee the ease of identifying the group structure. For example, if the

p variables of interest can be divided into K distinct groups with equal correlations for variables within each group, and equal correlations between any given pair of variables from different groups, with the diagonals $\tau_k \kappa_i^2$ for $i \in G_k$ or all $\sigma_i = 1 (i = 1, \ldots, p)$, then under some mild conditions, we have $\hat{\rho}_{ij}/\sqrt{(\hat{\rho}_{ii}\hat{\rho}_{jj})} \rightarrow_{a.s.} \tau_{kl}/\sqrt{(\tau_k \tau_l)}$ for $i \in G_k$, $j \in G_l$ and $i \neq j$ [see Section 5.4 of 223], then we can directly use classical clustering algorithms to reveal the block structure based on these estimates $\{\hat{\rho}_{ij}/\sqrt{(\hat{\rho}_{ii}\hat{\rho}_{jj})}, i = 1, \ldots, p-1, j = i+1, \ldots, p\}$.

Since the curse of dimensionality imposes many obstacles for variables clustering [236], directly clustering the elements of M_n usually fails as $p > n$. New approaches are needed for this scenario. In order to reveal, understand and utilize the properties regarding the eigendecomposition in a block-structured matrix, we rewrite the equation (6.58) as summation of a linear and degenerate part by the Hoeffding decomposition when $i \in G_k, j \in G_l$, i.e.,

$$\hat{\rho}_{ij} = E(\hat{\rho}_{ij}) + \sum_{s=0}^{m} \binom{m}{s} \hat{\rho}_{ij}(\varphi_s), \tag{6.60}$$

where

$$\hat{\rho}_{ij}(\varphi_s) = \binom{n}{s}^{-1} \sum_{1 \le i_1 < \ldots < i_s \le n} [\varphi_{ij,s}(X_{i_1}, \ldots, X_{i_s}) - E\{\varphi_{ij,s}(X_{i_1}, \ldots, X_{i_s})\}],$$

with $\varphi_{ij,s}(x_1, \ldots, x_s) = \sum_{r=0}^{s} (-1)^{s-r} \sum_{1 \le l_1 < \ldots < l_r \le s} \tilde{\varphi}_{ij,r}(x_{l_1}, \ldots, x_{l_r})$ and $\tilde{\varphi}_{ij,r}(x_1, \ldots, x_r) = E\{\varphi_{ij}(x_1, \ldots, x_r, X_{r+1}, \ldots, X_m)\}$.

Let $S = (s_{ij})_{i,j=1}^{p}$ be a $p \times p$ matrix with $s_{ij} = \sum_{s=1}^{m} \binom{m}{s} \hat{\rho}_{ij}(\varphi_s)$, then the matrix form of (6.60) is

$$M_n = M_p^* + S, \tag{6.61}$$

where

$$M_p^* = E(M_n) = J + \sum_{1 \le k,l \le K} \tau_{kl} f_k f_l^\top = J + \sum_{1 \le k,l \le K} \|f_k\| \|f_l\| \tau_{kl} \tilde{f}_k \tilde{f}_l^\top. \tag{6.62}$$

Here f_k is a $p \times 1$ vector with the jth element being 0 if $j \notin G_k$ for $j = 1, \ldots, p$, and its normalized version $\tilde{f}_k = f_k/\|f_k\|$. J is a $p \times p$ diagonal matrix with the jth diagonal element being $\tau_k \kappa_j - \tau_{kk}\sigma_j^2$ if $j \in G_k$. Denote $K \times K$ matrix by A whose (k, l)th element is $a_{kl} = \|f_k\| \|f_l\| \tau_{kl}/p$. Then applying the orthogonal decomposition to A, we have

$$a_{kl} = e_k^\top \Big(\sum_{r=1}^{K} \lambda_r \phi_r \phi_r^\top \Big) e_l, \tag{6.63}$$

where e_k is the kth canonical basis of $\mathbb{R}^K$, (λ_r, ϕ_r) is the pair of eigenvalue and corresponding eigenvector of A $(r = 1, \ldots, K)$ with $|\lambda_1| \ge \ldots \ge |\lambda_K|$. Notice that $\{\lambda_r\}_{r=1}^{K}$ depend on p and have limits because the entries of A has. For simplification, we abuse the notation for both the ones depending on p, or their limits. Plugging (6.63) into (6.61) we get

$$\frac{1}{p} M_n = \sum_{r=1}^{K} \lambda_r v_r v_r^\top + \frac{1}{p} J + \frac{1}{p} S, \tag{6.64}$$

where $v_r = \sum_{k=1}^{K} \tilde{f}_k e_k^\top \phi_r$ satisfies

$$v_r^\top v_s = \begin{cases} 0, & r \neq s \\ 1, & r = s \end{cases}. \tag{6.65}$$

Because $\|\tilde{f}_k\| = 1$ and $\phi_r, r = 1, \ldots, K$ are eigenvectors, the elements of the matrix $J/p + S/p$ tend to zero as p goes infinity, and the first term in the right side of equation (6.64) can be seen as the approximate singular value decomposition of $\frac{1}{p}M_n$.

Let δ_i's be the eigenvalues of M_n/p satisfying $|\delta_1| \geq |\delta_2| \geq \ldots \geq |\delta_p|$, and u_i $(i = 1, \ldots, p)$ be the corresponding eigenvectors. Clearly, the smallest $p - K$ eigenvalues are expected to be negligible as the dimensionality p goes infinity. According to the asymptotical behavior of M_n/p shown in Theorem 6.3.1, there will be a gap between the largest K and smallest $p-K$ eigenvalues asymptotically at rate $O\{\log(n)\}$, we therefore define the following empirical eigenvalue-threshold estimator for the unknown group size K,

$$\hat{K} = \max\left(k : |\delta_k| \geq c\left[\frac{1}{\log(n)} + \frac{\log\{\log(p)\}}{p}\right]\right), \tag{6.66}$$

where c is a constant. The selection of c is discussed in more detail in Remark 6.9 and simulation studies in the next section. Obviously, the estimator (6.66) is easy to implement in practice. It is intuitively based on the properties of eigenvalues.

Theorem 6.3.1. *For the U-statistic based symmetric matrix defined by (6.58), assume $|\lambda_K| > c_0$ for some positive constant c_0 and $E\varphi_{ij}^4(X_{l_1}, \ldots, X_{l_m}) < \infty$. Assume $\sigma_i > \sigma_0$ for all $i = 1, \ldots, p$, and $\lim_{p\to\infty} \|f_j\|^2/p = \varsigma_j > \varsigma_0$, for all $j = 1, \ldots, K$, where σ_0 and ς_0 are positive constants. Let $\hat{K}$ be the estimator of K defined by (6.66), then*

$$\hat{K} \to K, \ a.s. \ n, p \to \infty.$$

Remark 6.9. *From the proof of Theorem 6.3.1 in [284], we note that the constant c in (6.66) asymptotically satisfies that $c \leq |\delta_K| p\log(n)/[\log(n)\log\{\log(p)\} + p]$ and $c \geq |\delta_{K+1}| p\log(n)/[\log(n)\log\{\log(p)\} + p] \geq (\sum_{j=K+1}^{p} \delta_j^2/p)^{1/2} p\log(n)/[\log(n)\log\{\log(p)\} + p]$. Then, we have*

$$\sum_{j=K+1}^{p} \delta_j^2/p \geq 2log\{\log(n)\}\sum_{i=1}^{p}\sum_{j=1}^{p}\sigma_{k_1,k_2}^2(i,j)^2/(np^3) + \sum_{j=1}^{p}(\tau_k(j) - \tau_{kk}\sigma_j^2)^2/p^3 \tag{6.67}$$

with probability 1, where $\sigma_{k_1,k_2}^2(i,j) = var(\varphi_{ij,1}(X_l))$ with $i \in G_{k_1}$ and $j \in G_{k_2}$. For correlation matrix M_n whose diagonals are ones, it is valid that $\sigma_{k_1,k_2}^2(i,j) \leq 1$ and $(\tau_k(j) - \tau_{kk}\sigma_j^2)^2 \leq \{1 + 1/(p-1)\}^2$ for all i, j, and therefore the constant c satisfies that

$$\left[\frac{2\log\{\log(n)\}}{np} + \frac{1}{(p-1)^2}\right]^{1/2} \frac{p\log(n)}{\log(n)\log\{\log(p)\} + p} \leq c \leq |\delta_1|\frac{p\log(n)}{\log(n)\log\{\log(p)\} + p}. \tag{6.68}$$

Notice that the lower limit tends to zero and the upper limit goes to infinity as $n, p \to \infty$, thus c can be an arbitrary positive constant as long as n, p are sufficiently large. However, the lower limit is recommended to c for the case of small sample and week correlation. For example, in the case where $n = 50$ and $p = 100$, the lower limit of the constant c by (6.68) is 0.094.

If $n/\{p\log(n)\} \to 0$, (6.66) can be simplified as

$$\hat{K} = \max\left(k : |\delta_k| \geq c\left[\frac{1}{\log(n)}\right]\right),$$

and the constant c satisfies that

$$\left[\frac{2\log\{\log(n)\}\sum_{i=1}^{p}\sum_{j=1}^{p}\sigma_{k_1,k_2}^2(i,j)}{np^3}\right]^{1/2}\log(n) \leq c \leq |\delta_1|\log(n).$$

where $\sigma^2_{k_1,k_2}(i,j) = (1+\rho^2_{k_1,k_2})\sigma_i^2\sigma_j^2 \leq 2\sigma_i^2\sigma_j^2$ *for the covariance matrix and* $\sigma^2_{k_1,k_2}(i,j) = 2(1-\rho_{k_1,k_2})(1-2/\pi) \leq 2(1-\rho_{k_1,k_2})2/\pi = E(\varphi_{ij,1}(X_l))^2$ *for the Gini under the normal distribution. Empirical studies in the next section show that* $c = 0.1$ *usually works satisfactorily in various combinations of sample size and correlation levels. A distinctive gap between the eigenvalues can also provide reference in case of conservative bound. One can also consider to estimate the right side of (6.67) to get a more precise lower bound for* c *by using the their sample versions.*

The problem to determinate the number of groups by (6.66) is similar to the problem of estimating the number of significant components in high-dimensional principal component analysis [17] or the problem of estimating the number of signals in signal processing [210]. The most common approach to solve this problem is to use the information theoretic criteria, and in particular the minimum description length (MDL) [213], the Bayesian information criterion (BIC) and the Akaike information criterion (AIC). The information theoretic criterion for a specific application is generically a function of the log-likelihood of the maximum-likelihood estimator of the parameters in the assumed model and a term which depends on the number of parameters of the model that penalizes overfitting of the model order. These methods, even though applicable to the group size detection problem, are usually computationally intensive. On the contrary, the proposed estimator (6.66) is computationally fast and theoretically consistent.

Theorem 6.3.1 and the decomposition (6.64) together imply that the top $\hat{K}$ eigenvectors, i.e., the eigenvectors associated with the largest $\hat{K}$ eigenvalues, provide an insight into the natural grouping of variables. Noticing that the matrix A is full rank, and each eigenvector will have identical values in entries corresponding to a single block or cluster, the resulting top $\hat{K}$ eigenvectors of M_p^*/p are therefore linearly independent and can be used to identify the block structure by classical clustering techniques. This approach is commonly accepted for community detection in network data studies. The following theorem assures that the empirical eigenvectors are approaching the corresponding population eigenvectors consistently.

Theorem 6.3.2. *Under the conditions of Theorem 6.3.1, let all the distinct eigenvalues of* A *be* $\tilde{\lambda}_1 > \ldots > \tilde{\lambda}_s$ *satisfying* $\min|\tilde{\lambda}_i| \geq c_0 > 0$, *with algebraic multiplicities* $k_1, k_2, \ldots, k_s$, *i.e.,* $\lambda_{k_{i-1}+1} = \ldots = \lambda_{k_i} = \tilde{\lambda}_i, i = 1, \ldots, s$ *where* $k_0 = 0, k_s = K$. *Denote the corresponding eigenvectors of* $\tilde{\lambda}_i$ *by* $\tilde{V}_i = (v_{k_{i-1}+1}, \ldots, v_{k_{i-1}+k_i})$, *and the corresponding eigenvectors of* M_n/p *by* $\tilde{U}_i = (u_{k_{i-1}+1}, \ldots, u_{k_{i-1}+k_i})$, *respectively. Write* $C_i = \tilde{V}_i^\top \tilde{U}_i$, *then the following properties hold for* $i = 1, \ldots, s$,

(1) C_i *satisfies* $I_{k_i} - C_i^\top C_i \geq 0$ *and* $\|I_{k_i} - C_i^\top C_i\|_F = O_{a.s.}([1/p+\log\{\log(n)\}]/n)$. *Furthermore,* $\|\tilde{U}_i - \tilde{V}_i C_i\|_F = O_{a.s.}([1/p + \log\{\log(n)\}/n]^{1/2})$.

(2) $\|\tilde{U}_i - \tilde{V}_i C_i\|_\infty = O_{a.s.}[1/p + \log\{\log(n)\}/n]$.

Remark 6.10. *If* $n/\{p\log(n)\} \to 0$, *Theorem 6.3.2 implies that* $\|I_{k_i} - C_i^\top C_i\|_F = o_{a.s.}\{\log(n)/n\}$, $\|\tilde{U}_i - \tilde{V}_i C_i\|_F = o_{a.s.}[\{\log(n)/n\}^{1/2}]$ *and* $\|\tilde{U}_i - \tilde{V}_i C_i\|_\infty = o_{a.s.}\{\log(n)/n\}$ *for* $i = 1, \ldots, s$. *If* $k_i = 1$, *then* C_i *is a scalar* c_i *which satisfies that* $1 - c_i^2 \geq 0$ *and* $c_i^2 = 1 + o_{a.s.}\{\log(n)/n\}$. *Let* $c_i > 0$. *Theorem 6.3.2 states that* $\|u_i - v_i\|_F = o_{a.s.}[\{\log(n)/n\}^{1/2}]$ *and* $\|u_i - v_i\|_\infty = o_{a.s.}(\log(n)/n)$. *This means that the top* K *eigenvectors of* M_n/p *are asymptotically equivalent to the corresponding* K *eigenvectors of* M_p^*/p *almost surely as* n, p *goes to infinity. Therefore, the variables can be well separated to different groups by any eigenvectors. For the case of repeated eigenvalues, the eigenvectors for the same eigenvalue are linearly independent and constitute a subspace with the dimension of the eigenvalue's multiplicity. Theorem 6.3.2 asserts that the corresponding sample eigenvectors converges to their projection in the subspace spanned by the eigenvectors of repeated eigenvalues. If all* $\sigma_i = 1$, $i = 1, \ldots, p$, *the singular value decomposition of* M_n/p *can be exactly obtained by [40], such that applying the same derivation process of Theorem 6.3.2 causes the conclusions have a little difference for example* $\|I_{k_i} - C_i^\top C_i\|_F = o_{a.s.}\{\log(n)/n\}$,

$\|\tilde{U}_i - \tilde{V}_i C_i\|_F = o_{a.s.}[\{\log(n)/n\}^{1/2}]$ and $\|\tilde{U}_i - \tilde{V}_i C_i\|_\infty = o_{a.s.}[\{\log(n)/(np)\}^{1/2} + \log(n)/n]$, but if we assume $n/\{p\log(n)\} \to 0$ then we will also have $\|\tilde{U}_i - \tilde{V}_i C_i\|_\infty = o_{a.s.}\{\log(n)/n\}$ which is same as that in Theorem 6.3.2.

To identify the block structure of the correlation matrix, we can cluster the variables of interest based on the selected top $\hat{K}$ eigenvectors. To see this point more clearly, let us denote $V_K = (\tilde{V}_1 C_1, \tilde{V}_2 C_2, \ldots, \tilde{V}_s C_s)$. Denote $W = (w_1^\top, w_2^\top, \ldots, w_p^\top)^\top$ where w_i is the i-th row of V_K divided by the Euclidean norm of the i-th row such that $\|w_i\| = 1$.

Notice that C_i is an approximate orthogonal matrix by Assertion (1) of Theorem 6.3.2, from (6.65), the Euclidean norm of the i-th row of V_K is close to $f_i/\|f_k\|$ with $i \in G_k$, and then

$$w_i \to w_k^* \equiv (\phi_{k1}, \ldots, \phi_{kk})^\top, \forall\, i \in G_k$$

with probability 1 by Theorem 6.3.2, where $\phi_k = (\phi_{1k}, \ldots, \phi_{kk})^\top, k = 1, \ldots, K$ are the eigenvectors of A.

Similarly, let $U_K = (u_1, u_2, \ldots, u_K)$. Denote $Z = (z_1^\top, z_2^\top, \ldots, z_p^\top)^\top$ where z_i is the ith row of U_K divided by the Euclidean norm of the ith row such that $\|z_i\| = 1$. By Theorem 6.3.2, recalling that $\sigma_i > \sigma_0$ for $i = 1, \ldots, p$ and $\lim_{p\to\infty} \|f_j\|^2/p = \varsigma_j > \varsigma_0$ for $j = 1, \ldots, K$, we have with probability 1, there exist two positive constants c and c^* satisfying

$$\begin{aligned} \|z_{j_1} - z_{j_2}\| &\le \|z_{j_1} - w_k^*\| + \|z_{j_2} - w_k^*\| \le \frac{c\|f_k\|}{f_0}\|\tilde{U}_i - \tilde{V}_i C_i\|_\infty \\ &\le c^*\sqrt{p}\log\{\log(n)\}/n + c^*/\sqrt{p}, \quad \text{if } j_1, j_2 \in G_k, \end{aligned} \tag{6.69}$$

and if $j_1 \in G_{k_1}, j_2 \in G_{k_2}, k_1 \ne k_2$,

$$\begin{aligned} \big|\|z_{j_1} - z_{j_2}\| - \|w_{k_1}^* - w_{k_2}^*\|\big| &\le \|z_{j_1} - w_{k_1}^* - z_{j_2} + w_{k_2}^*\| \\ &\le c^*\sqrt{p}\log\{\log(n)\}/n + c^*/\sqrt{p}. \end{aligned} \tag{6.70}$$

The vector $(\phi_{k1}, \ldots, \phi_{kk})$ is the kth row of the orthogonal matrix $(\phi_1, \cdots, \phi_K)$. By the properties of orthogonal matrix, we have $w_k^*, k = 1, \ldots, K$ are orthogonal, and then $\|w_{k_1}^* - w_{k_2}^*\| = \sqrt{2}$. This, combined with Theorem 6.3.2, (6.69) and (6.70), will lead to the following corollary.

Corollary 6.2. *Under the conditions of Theorem 6.3.2 and $\sqrt{p}\log\{\log(n)\}/n \to 0$ as $n \to \infty$, then the following properties hold with probability 1,*

(1) $\max_{j_1, j_2 \in G_k, k=1,\ldots,K} \|z_{j_1} - z_{j_2}\| \to 0$,

(2) $\min_{j_1 \in G_{k_1}, j_2 \in G_{k_2}, k_1 \ne k_2 \in \{1,\ldots,K\}} \|z_{j_1} - z_{j_2}\| > \sqrt{2}$.

In summary, we learn the block structure in a U-statistic-based matrix through the following clustering block structure learning algorithm.

The following theorem affirms the algorithms are consistent.

Theorem 6.3.3. *Under the conditions of Theorem 1 and Corollary 6.2, the proposed clustering algorithm (10) is consistent, that is, it places two variables into the same cluster if and only if they belong to the same group as n goes to infinity.*

Proofs for this section's theoretical results can be found in [284].

6.3.2 Simulations

We demonstrate the performance of the block structure learning approach in a block-wise Kendall's tau correlation matrix through Simulation 1 (group size detection), Simulation 2 (cluster structure

Algorithm 10 Block structure learning (BSL) algorithms

1. Compute the U-statistic based matrix and its eigenvalues $\delta_i, \ldots, \delta_p$ and eigenvectors $u_1, \ldots, u_p$.
2. *Hierarclust*: Use the EET estimator (6.66) to estimate the number of groups $\hat{K}$. Let $U_{\hat{K}} = (u_1, u_2, \ldots, u_{\hat{K}})$. Denote $\hat{Z} = (\hat{z}_1^\top, \hat{z}_2^\top, \ldots, \hat{z}_p^\top)^\top$ where $\hat{z}_i$ is the i-th row of $U_{\hat{K}}$ divided by the Euclidean norm of the i-th row such that $\|\hat{z}_i\| = 1$. To identify the group structures in the variables $\hat{z}_1, \ldots, \hat{z}_p$, use the complete linkage hierarchical clustering algorithm under Euclidean distance to build a cluster tree (Dendrogram):
   (1). Cut the dendrogram at estimated number of groups $\hat{K}$;
   (2). (Optional) Cut the dendrogram at a given height $\sqrt{2}$ and obtain $\tilde{K}$ groups, pick out the clusters with minimum number of groups.

identification) and Simulation 3 (Selection of constant c). Then we compare this approach with other methods (Simulation 4). Finally, we show in Simulation 5 that the proposed approach can serve as an improved estimator for the linear correlation coefficients when the variables are elliptically distributed. In the simulation studies, we that the Kendall's tau coefficient ρ^K and Pearson's correlation coefficient ρ are related in the bivariate normal distribution by the equation [113, p. 290]

$$\rho^K = \frac{2}{\pi} arcsin(\rho). \tag{6.71}$$

[95] proved that the above relation (6.71) still holds for a class of distributions called meta-elliptical distributions, which includes various distributions, such as elliptically contoured distributions, the meta-Gaussian distributions and various asymmetric distributions.

Simulation 1: Group size detection. This simulation is designed to investigate the performance of estimator (6.66) under different levels of correlation, group size imbalance and sample size. We set the number of variables p to be 200, the sample size n to be 50 and 200, respectively. Let $rB_0 = (\tau_{kl})$ be the $K \times K$ matrix of correlation coefficients between K blocks of variables, where the diagonal entries of B_0 are 3 and off-diagonal entries are 1. The correlation levels are chosen at $r \in \{0.05, 0.1, 0.15, 0.2, 0.25\}$. The population block-wise Kendall's tau correlation matrix M_p^* can then be obtained by rB_0. For group size, we set the size of the smallest group to be p_1, and the size of each of the remaining $K-1$ groups to be $(p - p_1)/(K-1)$. For each combination of (r, K, p_1, n), we generate 100 data sets of size n from a p-variate Gaussian distribution with mean zero and covariance matrix whose entries are transformed from the corresponding Kendall's tau correlation coefficients using relationship (6.71), where the p variables are randomly permuted in each replication. The proposed eigenvalue threshold estimation approach in (6.66) is then carried out for these 100 data sets, where c takes value 0.1. Figure 6.8 shows the proportion of correct group size detection among these 100 repetitions as functions of r for different p_1 and $K \in \{2, 3, 4\}$. As expected, the performance is better as r and p_1 increase. Particularly, the proposed estimation approach can perfectly choose the true number of groups in case of the moderate and larger correlation or larger sample size. When the correlation is weak, the proposed approach can correctly pick out the true group size in a high rate even for case with small sample size, and perfectly estimate the group size for cases with large sample sizes.

Simulation 2: Cluster structure identification. This simulation is designed to demonstrate the performance of proposed clustering method in block structure identification with simultaneous selection K. The population block-wise Kendall's tau correlation matrix is the same as simulation 1. We consider $K \in \{2, 3, 4\}$, $n \in \{100, 200, 400\}$ and $p \in \{300, 600\}$ with equal number of variables in each block. For each combination, we generate 100 data sets from a p-variate Gaussian distribution with mean zero and covariance matrix whose entries are transformed from the corresponding Kendall's tau correlation coefficients using relationship (6.71), where the p variables are permuted in each case. Table 6.2 reports the accuracy rate of correct group size K estimation (ACC), the average predictive precision value (PPV, the proportion of variables rightly grouped together in

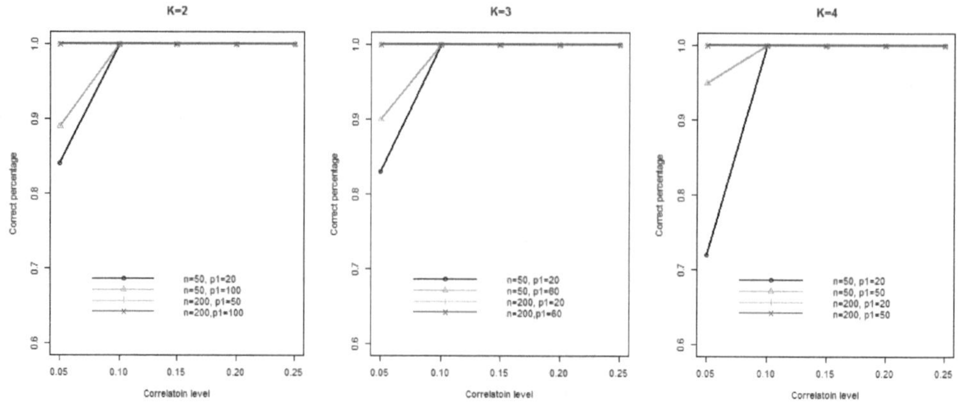

Figure 6.8: Results for Simulation 1. The profanations of 100 simulated datasets for which K is correctly estimated.

the estimated structures) and the average recall values (TPR, a.k.a, true positive rate, the proportion of rightly grouped together variables in the true structures). As expected, the group size detection approach (6.66) can perfectly choose the true number of blocks for all the cases of p and n. Compare to the k-means clustering method, the proposed clustering method captured the supposed cluster structure extremely well, especially for the case of $n = 200$, where it perfectly identifies the correct block structure.

Table 6.2: Results for simulation 2. Reported are averages of ACC, PPV and TPR.

		$K=2$			$K=3$			$K=4$		
n	p	ACC	PPV	TPR	ACC	PPV	TPR	ACC	PPV	TPR
100	300	100	100	100	100	97.69	93.75	100	96.64	89.83
	600	100	100	100	100	97.85	94.21	100	97.05	91.25
200	300	100	100	100	100	98.08	94.66	100	96.80	90.48
	600	100	100	100	100	98.00	94.65	100	96.65	90.15
400	300	100	100	100	100	97.55	93.31	100	95.82	87.55
	600	100	100	100	100	97.83	94.21	100	97.12	91.33

Simulation 3: Selection of constant c. This simulation is designed to demonstrate the effect of different constant c in the EET estimator (6.66). The block-wise Kendall's tau correlation matrix is defined by rB_0, where B_0 is given in simulation 1. We consider the correlation level $r \in \{0.1, 0.15, 0.2\}$, the group size $K \in \{2, 5, 10\}$, and $n, p \in \{100, 150, 200\}$. For each combination, we generate 100 data sets from a p-variate Gaussian distribution with mean zero and covariance matrix whose entries are transformed from the corresponding Kendall's tau correlation coefficients using relationship (6.71), where the p variables are permuted in each case. Figure 6.9 shows the proportion of correct group size detection among the 100 repetitions as functions of c for nine orthogonal design of p, r and K. It can be seen that optimal c ranges from 0.07 to 0.13. $c = 0.1$ works well almost for all cases except the case with large group size and weak correlation, where smaller c, say, 0.07 is preferred.

Simulation 4: Comparison to other methods. This simulation is designed to compare the proposed group size detection approach (6.66) and the block structure identification approach to competing approaches. The block-wise Kendall's tau correlation matrix is defined by rB_0, where B_0 is given in simulation 1. We consider the correlation level $r \in \{0.1, 0.15, 0.2\}$, the group size $K \in \{2, 5, 10\}$, $n \in \{100, 200\}$ and $p \in \{100, 200\}$. For each combination, we generate 100 data

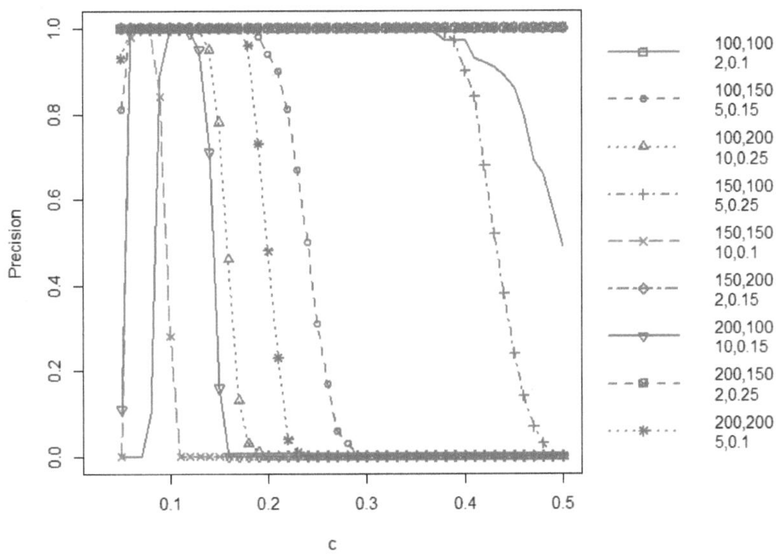

Figure 6.9: Results for Simulation 3. The precision of 100 simulated datasets for which K is correctly estimated.

sets from a p-variate Gaussian distribution with mean zero and covariance matrix whose entries are transformed from the corresponding Kendall's tau correlation coefficients using relationship (6.71), where the p variables are permuted in each case. [17] investigated the strong consistency of the estimation criteria AIC and BIC under a high-dimensional asymptotic framework. We compare our proposed EET estimator (6.66) to the AIC and BIC criteria for group size detection. Given estimated group size, the proposed hierarchical clustering (Hierarclust) approach is compared to K-means approach for block structure identification. Table 6.3 reports the results for all the cases. Overall, the EET estimator, AIC and BIC are competitive in cases of small block size and moderate and large correlation levels. The EET estimator outperforms the other two approaches in the cases of large block size and weak correlation levels. At the same time, the proposed Hierarclust algorithm both with specified number and groups or minimum distance outperforms the K-means algorithm almost in all the cases.

Simulation 5: Improved estimator for block-structured linear correlation matrix. As pointed by formula (6.71), there is a one-to-one relationship between linear correlation coefficient and Kendall's tau correlation for meta-elliptical distribution. Thus, we can improve the standard linear correlation coefficient estimator by averaging out the entries block-wise in Kendall's tau correlation matrix and then inverting the relationship to get the estimate of corresponding linear correlation coefficient. We set the population linear correlation matrix as $0.1B_0$, where B_0 is given in Simulation 1. We consider the group size $K \in \{2, 5\}$, $n \in \{100, 200\}$ and $p \in \{100, 200\}$ with equal number of variables in each block. For each combination of p and n, we generate 100 data sets of size n from a p-variate Gaussian distribution with mean zero and specified correlation matrix whose entries are transformed from the corresponding Kendall's tau correlation coefficients using relationship (6.71), where the p variables are permuted in each case. Table 6.4 reports the Frobenius norms of the biases of standard sample correlation and the Kendall's tau transform estimate, respectively. Clearly, the simulation results suggest that this simple method performs better than commonly used sample correlation estimation.

Table 6.3: Results for simulation 4. Reported are averages of the accuracy rate of correct group size estimation, PPV and TPR.

				Group size detection			Kmeans		Hierarclust			
n	p	K	r	EET	AIC	BIC	PPV	TPR	PPV	TPR	PPV*	TPR*
100	100	2	0.1	100.00	100.00	100.00	100.00	100.00	99.98	99.98	99.98	99.98
			0.15	100.00	100.00	100.00	100.00	100.00	100.00	100.00	100.00	100.00
			0.2	100.00	100.00	100.00	100.00	100.00	100.00	100.00	100.00	100.00
		5	0.1	100.00	99.00	0.00	96.88	90.10	99.92	99.98	99.94	99.94
			0.15	100.00	100.00	98.00	96.28	87.61	100.00	100.00	100.00	100.00
			0.2	100.00	100.00	100.00	96.15	86.99	100.00	100.00	100.00	100.00
		10	0.1	93.00	0.00	0.00	96.13	88.00	91.51	99.80	99.02	99.16
			0.15	100.00	78.00	0.00	95.35	82.20	100.00	100.00	100.00	100.00
			0.2	100.00	100.00	100.00	95.53	82.25	100.00	100.00	100.00	100.00
	200	2	0.1	100.00	100.00	100.00	100.00	100.00	100.00	100.00	100.00	100.00
			0.15	100.00	100.00	100.00	100.00	100.00	100.00	100.00	100.00	100.00
			0.2	100.00	100.00	100.00	100.00	100.00	100.00	100.00	100.00	100.00
		5	0.1	100.00	100.00	100.00	96.41	88.48	99.75	100.00	99.94	99.94
			0.15	100.00	100.00	100.00	96.06	87.22	100.00	100.00	100.00	100.00
			0.2	100.00	100.00	100.00	95.80	86.13	100.00	100.00	100.00	100.00
		10	0.1	99.00	93.00	93.00	96.36	86.68	94.44	99.96	99.81	99.63
			0.15	100.00	100.00	100.00	95.62	83.74	100.00	100.00	100.00	100.00
			0.2	100.00	100.00	100.00	95.33	82.23	100.00	100.00	100.00	100.00
200	100	2	0.1	100.00	100.00	100.00	100.00	100.00	100.00	100.00	100.00	100.00
			0.15	100.00	100.00	100.00	100.00	100.00	100.00	100.00	100.00	100.00
			0.2	100.00	100.00	100.00	100.00	100.00	100.00	100.00	100.00	100.00
		5	0.1	100.00	100.00	96.00	96.30	87.64	100.00	100.00	100.00	100.00
			0.15	100.00	100.00	100.00	96.31	87.64	100.00	100.00	100.00	100.00
			0.2	100.00	100.00	100.00	95.86	86.78	100.00	100.00	100.00	100.00
		10	0.1	100.00	88.00	0.00	96.14	85.20	100.00	100.00	100.00	100.00
			0.15	100.00	100.00	31.00	95.93	84.03	100.00	100.00	100.00	100.00
			0.2	100.00	100.00	100.00	95.77	83.55	100.00	100.00	100.00	100.00
	200	2	0.1	100.00	100.00	100.00	100.00	100.00	100.00	100.00	100.00	100.00
			0.15	100.00	100.00	100.00	100.00	100.00	100.00	100.00	100.00	100.00
			0.2	100.00	100.00	100.00	100.00	100.00	100.00	100.00	100.00	100.00
		5	0.1	100.00	100.00	100.00	96.05	87.32	100.00	100.00	100.00	100.00
			0.15	100.00	100.00	100.00	96.11	87.36	100.00	100.00	100.00	100.00
			0.2	100.00	100.00	100.00	96.13	87.46	100.00	100.00	100.00	100.00
		10	0.1	100.00	100.00	0.00	95.06	81.38	100.00	100.00	100.00	100.00
			0.15	100.00	100.00	95.00	95.01	81.11	100.00	100.00	100.00	100.00
			0.2	100.00	100.00	100.00	94.90	80.10	100.00	100.00	100.00	100.00

Column 10–11 represent the results under Hierarclust with specified number of groups($\hat{K}$); Column 12–13 represent the results under Hierarclust with minimum distance($\frac{1}{\sqrt{p}}\sqrt{\frac{2}{1-(\hat{K}-1)/(\hat{K}+1)}} = \sqrt{(\hat{K}+1)/p}$).

Table 6.4: Results for simulation 5. Reported are Frobenius norms of the bias of standard sample correlation (F_S) and the Kendall's tau transform estimate (F_K).

		$p=100$		$p=200$				$p=100$		$p=200$	
n	K	F_S	F_K	F_S	F_K	n	K	F_S	F_K	F_S	F_K
100	2	9.53	3.06	19.16	6.18	200	2	6.68	1.87	13.52	4.22
	5	9.80	3.73	19.61	6.85		5	6.89	2.31	13.83	4.51

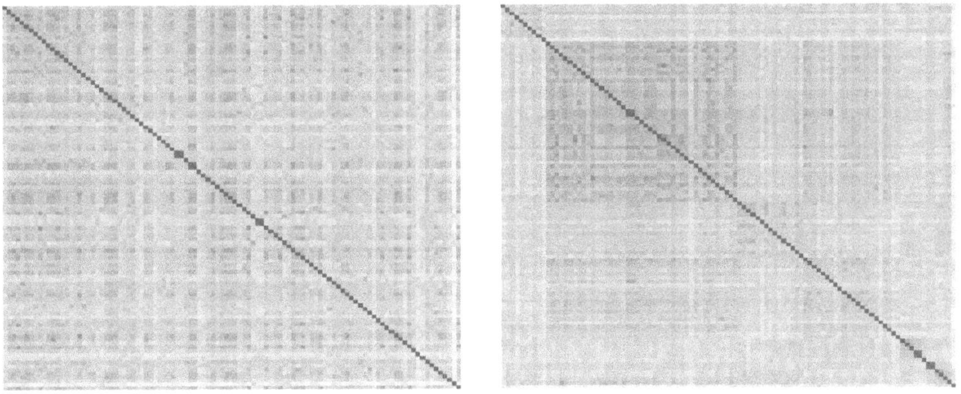

Figure 6.10: Plots of correlation matrix. (a) Original empirical block-wise Kendall's tau correlation matrix (b) Re-grouped empirical block-wise Kendall's tau correlation matrix by estimated block structure.

6.3.3 Real data analysis

Data example 1: analysis of the NASDAQ-100 index. The data consist of daily closing prices of all stocks that belonged to the NASDAQ-100 index from January 2, 2018 to November 28, 2018. This yields a total of $p = 103$ stocks with $n = 230$ observations from eight industry sectors: Capital Goods (4), Consumer Non-Durables (5), Consumer Services (23), Health Care (18), Miscellaneous (6), Public Utilities (3), Technology (41), Transportation (3). Data were obtained from the Yahoo Finance database. The information on the components of the index (company name, sectors and industries) was taken from the official website of NASDAQ. To study the dependence structure among all these stocks, each individual stock returns is fitted by a GARCH(1,1) model with parameters estimated by Quasi-maximum likelihood estimation. Then we apply our approach to the series of residuals. Figure 6.10 (a) presents the empirical Kendall's tau correlation matrix to the original labeling of the $p = 103$ stocks. The pairwise Kendall's correlation coefficients range from –0.16 to 0.92. Clearly, it is not so easy to directly tell any obvious pattern from this matrix. The estimated number of clusters is $\hat{K} = 7$ by using the proposed group size detection method (6.66), Figure 6.10 (b) reports the re-grouped empirical Kendall's tau correlation matrix. Now, a block structure is revealed by grouping the variables according to the estimated clusters.

Figure 6.11 illustrates the estimated clusters for the 103 stock returns. The proposed hierarchical clustering approach produces seven clusters as indicated by the same leaf symbol. The December NASDAQ classification was represented by different colors on the leaf label. The precision and recall are 0.617 and 0.595, respectively. Clusters have intuitive interpretations in terms of business sectors or known affiliations. Cluster 1 (circle) is almost exclusively a collection of stocks corresponding to technology; Cluster 2 (triangle) is mainly composed of stocks corresponding to healthcare; Cluster 3 (square) corresponds to communication services and transportation; Cluster 4 (diamond) contains stocks from the biopharmaceutical industry and miscellaneous sector; Cluster 5 (solid circle) includes retail and other specialty stores; cluster 6 (inverted triangle) is a collection of stocks corresponding to television services; cluster 7 (solid square) is purely composed of the FOX Group (NASDAQ: FOX and FOXA). Furthermore, some strongly correlated stocks are also clustered together: GOOG and GOOGL (0.923); FOX and FOXA (0.875); LBTYA and LBTYK (0.875).

Data example 2: analysis of gene expression profiles. We analyze the publicly available expression dataset of breast tumor tissues from patients $n = 200$ (GSE11121, Gene Expression Omnibus) [217](Schmidt et al., 2008). We downloaded the raw data and performed normalization using

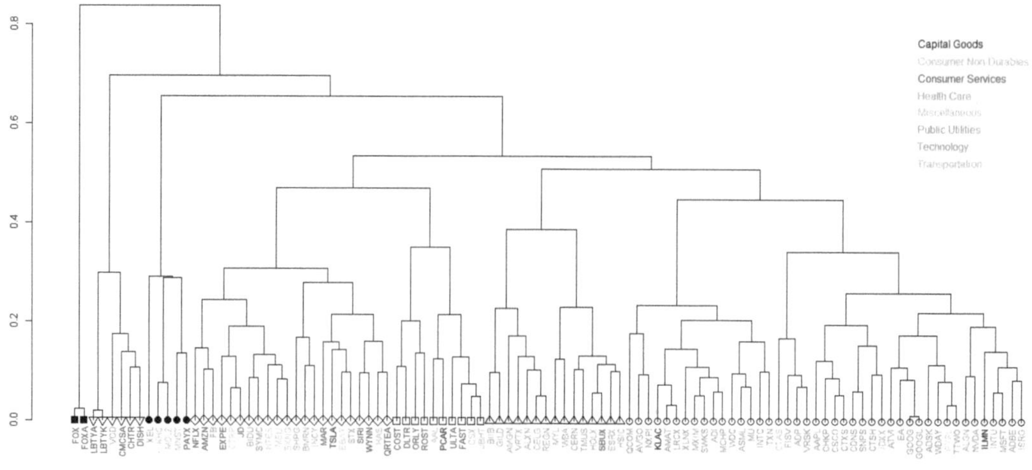

Figure 6.11: Stock returns clustering.

the R package `affy` [107], and then selected genes with variance greater than 0.2. This results in $p = 4506$ differentially expressed genes (DEGs). Since the underlying module (cluster) structure is not known in real data application, the key to evaluating a detected module is biological interpretability. Pathway enrichment analysis is a common method to assess the quality and interpretability of a detected cluster since in gene co-expression networks, modules are comprised of highly interconnected genes and are often highly enriched in genes with shared functional annotations such as pathways. We computed the Spearman correlation matrix and then identified $K = 9$ gene clusters from these DEGs. Next, we applied the R package `clusterProfiler` [274] to compare and visualize functional profiles between gene groups. As illustrated in Figure 6.12, the detected DEG groups are indeed related to different biological processes: for example, cluster 1 is related to the cell cycle and progression of oocytes; cluster 2 is related to several diseases; cluster 5 is related to the phagosome maturation and virus infection; and cluster 8 is related to protein interactions.

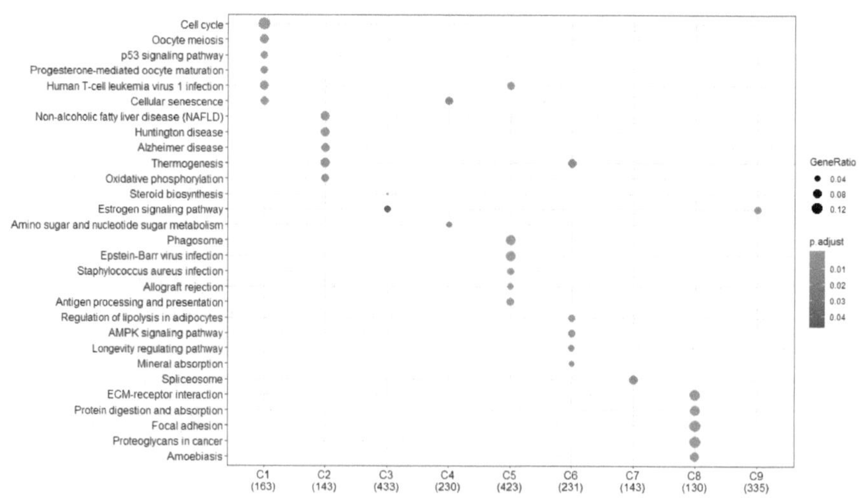

Figure 6.12: Comparison of gene co-expression enrichment of gene clusters.

6.4 Other problems and applications

Before we conclude this book, we point out that numerous emerging data types could potentially benefit from a change point analysis. Beyond the financial and medical examples we have discussed, there has been increasing attention to various business and scientific applications, such as streaming data, online decision-making, and multi-source (multi-modal) data. In particular, when analyzing rapidly generated data like online data [82], distributional computing becomes essential, often leading to the creation of heterogeneous samples for analysis. Each of these samples can be viewed as a subgroup. By pooling similar groups to form relatively homogeneous clusters, we can apply the change point detection procedures introduced in this book to better analyze and interpret these complex datasets.

Consider the problem of combining data from smaller subproblems that have some shared structure. Suppose there are multiple supervised learning datasets that share the same predictors but have sparse true coefficients that vary between subproblems. This variation might stem from differences in measurement methods or other underlying disparities, such as the application of different treatments across datasets. We may conjecture that, due to these similarities, a combined analysis could yield greater insights across all datasets by sharing data in a meaningful way. This approach is particularly relevant in fields like biostatistics, econometrics, and computer sciences. If the true regression coefficients are identical across all datasets, pooling the data and fitting a single model would typically provide the most efficient solution. On the other hand, if the coefficients differ significantly between groups, fitting separate models for each dataset would be more appropriate. Current methodologies generally focus on cases between these extremes, where some coefficients are shared across models while others are unique to each dataset. These methods often assume the group structures are known in advance. However, in scenarios where the group structures are unknown, subgroup analysis methods—like those discussed in this book—can be applied to enhance the analysis by identifying the appropriate groupings and shared structures.

A related problem is data integration. Many medical studies face the challenge of limited resources, making it difficult to perform efficient statistical analysis. Typically, researchers collect individual-level data from their own investigations, referred to as the internal study in this context. However, internal studies often suffer from small sample sizes due to funding constraints and recruitment challenges, leading to inefficient and potentially unreliable inference results. In some cases, investigators may have access to external studies—related earlier studies that have explored similar medical questions. Incorporating information from these external studies could significantly improve the statistical efficiency of the analysis. However, individual-level data from external studies are rarely available to the public due to ethical, legal, or administrative barriers. Instead, researchers often only have access to summary statistics, which may be extracted from published articles, public databases, or related projects. To address this, appropriate statistical methods are required to leverage the aggregated auxiliary information from external studies to enhance the analysis of the internal study [80, 81]. A key challenge arises when heterogeneity exists between the internal and external studies, requiring careful handling of multiple data groups to ensure valid and efficient statistical inference. This integration of data from multiple sources can substantially improve the power of statistical analyses, especially when internal studies are limited in size.

Multimodal data are new data structure for medical study. It is also called multi-view or multi-source data. Examples include multi-omics data in genomics, multimodal neuroimaging data in neuroscience, multi-modal electronic health records data in health care administration. Combining diverse but usually complementary information from different types of data, an integrative analysis of multimodal data is often beneficial. Challenges for this research problem include high-dimensionality and high correlation. For instance, in genetics, DNA methylation and gene expression are two distinct yet related types of data, each reflecting different biological processes. In neuroimaging, fMRI (functional Magnetic Resonance Imaging) and DTI (Diffusion Tensor Imaging) provide complementary views of brain function and structure, respectively. Merging these diverse datasets in a single statistical framework requires sophisticated methods, including those discussed in this book for clustering and subgroup identification.

We acknowledge that more examples from environmental research and social studies are available but due to the page limit we cannot add further details. We have particularly paid attention to various theoretical issues in this book. There are more focused studies in different directions in the literature, such as convergence speed of algorithms and minimax properties.

Bibliography

[1] Adelfio, G., 2012. Change-point detection for variance piecewise constant models. Communications in Statistics-Simulation and Computation 41, 437–448.

[2] Aggarwal, C., Subbian, K., 2014. Evolutionary network analysis: A survey. ACM Computing Surveys (CSUR) 47, 1–36.

[3] Akaike, H., 1974. A new look at the statistical model identification. IEEE transactions on automatic control 19, 716–723.

[4] Altzerinakou, M.A., Paoletti, X., 2021. Change-point joint model for identification of plateau of activity in early phase trials. Statistics in Medicine 40, 2113–2138.

[5] Anastasiou, A., Chen, Y., Cho, H., Fryzlewicz, P., 2020. Breakfast: methods for fast multiple change-point detection and estimation. R package version 2.

[6] Anastasiou, A., Fryzlewicz, P., 2022. Detecting multiple generalized change-points by isolating single ones. Metrika 85, 141–174.

[7] Andrews, D.W., 1988. Laws of large numbers for dependent non-identically distributed random variables. Econometric theory 4, 458–467.

[8] Areosa, W.D., McAleer, M., Medeiros, M.C., 2011. Moment-based estimation of smooth transition regression models with endogenous variables. Journal of Econometrics 165, 100–111.

[9] Aston, J.A., Kirch, C., 2012. Detecting and estimating changes in dependent functional data. Journal of Multivariate Analysis 109, 204–220.

[10] Atchade, Y., Bybee, L., 2017. A scalable algorithm for gaussian graphical models with change-points. arXiv preprint arXiv:1707.04306 .

[11] Aue, A., Gabrys, R., Horváth, L., Kokoszka, P., 2009. Estimation of a change-point in the mean function of functional data. Journal of Multivariate Analysis 100, 2254–2269.

[12] Aue, A., Hörmann, S., Horváth, L., Hu*v*sková, M., 2014. Dependent functional linear models with applications to monitoring structural change. Statistica Sinica , 1043–1073.

[13] Aue, A., Kirch, C., 2024. The state of cumulative sum sequential changepoint testing 70 years after page. Biometrika 111, 367–391.

[14] Aue, A., Rice, G., Sönmez, O., 2018. Detecting and dating structural breaks in functional data without dimension reduction. Journal of the Royal Statistical Society: Series B (Statistical Methodology) 80, 509–529.

[15] Bai, J., 1995. Least absolute deviation estimation of a shift. Econometric Theory 11, 403–436.

[16] Bai, J., 1997. Estimation of a change point in multiple regression models. Review of Economics and Statistics 79, 551–563.

[17] Bai, Z.D., Choi, K.P., Fujikoshi, Y., 2018. Consistency of aic and bic in estimating the number of significant components in high-dimensional principal component analysis. The Annals of Statistics 46, 1050–1076.

[18] Baiocchi, M., Cheng, J., Small, D.S., 2014. Instrumental variable methods for causal inference. Statistics in Medicine 33, 2297–2340.

[19] Banerjee, M., McKeague, I.W., 2007. Confidence sets for split points in decision trees. Annals of Statistics 35, 543–574.

[20] Baranowski, R., Chen, Y., Fryzlewicz, P., 2019. Narrowest-over-threshold detection of multiple change points and change-point-like features. Journal of the Royal Statistical Society Series B: Statistical Methodology 81, 649–672.

[21] Barbieri, K., Keshk, O.M.G., 2016. Correlates of War Project Trade Data Set Codebook, Version 4.0. Online: http://correlatesofwar.org.

[22] Barbieri, K., Keshk, O.M.G., Pollins, B., 2009. Trading data: Evaluating our assumptions and coding rules. Conflict Management and Peace Science 26, 471–491.

[23] Bardsley, P., Horváth, L., Kokoszka, P., Young, G., 2017. Change point tests in functional factor models with application to yield curves. The Econometrics Journal 20, 86–117.

[24] Barry, D., Hartigan, J.A., 1993. A bayesian analysis for change point problems. Journal of the American Statistical Association 88, 309–319.

[25] Bartholomew, D., Knott, M., Moustaki, I., 2011. Latent Variable Models and Factor Analysis: A Unified Approach. Wiley.

[26] Berkes, I., Gabrys, R., Horváth, L., Kokoszka, P., 2009. Detecting changes in the mean of functional observations. Journal of the Royal Statistical Society: Series B (Statistical Methodology) 71, 927–946.

[27] Bhattacharjee, M., Banerjee, M., Michailidis, G., 2020. Change point estimation in a dynamic stochastic block model. Journal of Machine Learning Research 21, 1–59.

[28] Bhattacharya, P., Brockwell, P., 1976. The minimum of an additive process with applications to signal estimation and storage theory. Zeitschrift für Wahrscheinlichkeitstheorie und verwandte Gebiete 37, 51–75.

[29] Bhattacharyya, G.K., Johnson, R.A., 1968. Nonparametric tests for shift at an unknown time point. The Annals of Mathematical Statistics , 1731–1743.

[30] Bleakley, K., Vert, J.P., 2011. The group fused lasso for multiple change-point detection. arXiv preprint arXiv:1106.4199 .

[31] Bollen, K., Curran, P., 2006. Latent Curve Models: A Structural Equation Perspective. Wiley.

[32] Bollen, K.A., 2014. Structural Equations with Latent Variables. John Wiley & Sons.

[33] Bonhomme, S., Manresa, E., 2015. Grouped patterns of heterogeneity in panel data. Econometrica 83, 1147–1184.

[34] Bouis, H.E., Haddad, L.J., 1990. Agricultural Commercialization, Nutrition, and the Rural Poor: A Study of Philippine Farm Households. Lynne Rienner.

[35] Bouis, H.E., Haddad, L.J., 1992. Are estimates of calorie-income fxelasticities too high?: A recalibration of the plausible range. Journal of Development Economics 39, 333–364.

[36] Bouveyron, C., Celeux, G., Murphy, T., Raftery, A., 2019. Model-based Clustering and Classification for Data Science. Cambridge University Press.

[37] Breheny, P., Huang, J., 2011. Coordinate descent algorithms for nonconvex penalized regression, with applications to biological feature selection. The Annals of Applied Statistics 5, 232–253.

[38] Breiman, L., 2017. Classification and Regression Trees. Routledge.

[39] Breiman, L., Friedman, J., Olshen, R., Stone, C., 1984. Classification and regression trees. CRC Press.

[40] Cadima, J., Calheiros, L.F., Preto, I.P., 2010. The eigenstructure of block-structured correlation matrices and its implications for principal component analysis. Journal of Applied Statistics 37, 577–589.

[41] Cai, T., Tian, L., Wong, P.H., Wei, L., 2011. Analysis of randomized comparative clinical trial data for personalized treatment selections. Biostatistics 12, 270–282.

[42] Cai, T.T., Yuan, M., 2011. Optimal estimation of the mean function based on discretely sampled functional data: Phase transition. The Annals of Statistics 39, 2330–2355.

[43] Caner, M., Hansen, B.E., 2004. Instrumental variable estimation of a threshold model. Econometric Theory 20, 813–843.

[44] Cao, H., Biao Wu, W., 2015. Changepoint estimation: another look at multiple testing problems. Biometrika 102, 974–980.

[45] Cao, H., Li, J., Fine, J., 2016. On last observation carried forward and asynchronous longitudinal regression analysis. Electronic Journal of Statistics 10, 1155–1180.

[46] Carlin, B.P., Louis, T.A., 2008. Bayesian Methods for Data Analysis. 3rd ed., CRC Press, Boca Raton, FL.

[47] Chang, J., Kolaczyk, E.D., Yao, Q., 2020. Discussion of "network cross-validation by edge sampling". Biometrika 107, 277–280.

[48] Chang, J., Kolaczyk, E.D., Yao, Q., 2022. Estimation of subgraph densities in noisy networks. Journal of the American Statistical Association 117, 361–374.

[49] Chen, E.Y., Fan, J., Zhu, X., 2023a. Community network auto-regression for high-dimensional time series. Journal of Econometrics in press.

[50] Chen, H., Ren, H., Yao, F., Zou, C., 2023b. Data-driven selection of the number of change-points via error rate control. Journal of the American Statistical Association 118, 1415–1428.

[51] Chen, J., Gupta, A.K., 2012. Parametric Statistical Change Point Analysis. Parametric Statistical Change Point Analysis.

[52] Chen, S., Tian, L., Cai, T., Yu, M., 2017. A general statistical framework for subgroup identification and comparative treatment scoring. Biometrics 73, 1199–1209.

[53] Chen, X., 1988. Testing and interval estimation in a change-point model allowing at most one change. Sci China Ser A-Math 30, 817–827.

[54] Chen, Y., Li, B., 2011. Forecasting yield curves in an adaptive framework. Central European Journal of Economic Modelling and Econometrics 3, 237–259.

[55] Chin Choy, J., Broemeling, L., 1980. Some bayesian inferences for a changing linear model. Technometrics 22, 71–78.

[56] Chiou, J.M., 2012. Dynamical functional prediction and classification, with application to traffic flow prediction. The Annals of Applied Statistics 6, 1588–1614.

[57] Chiou, J.M., Li, P.L., 2007. Functional clustering and identifying substructures of longitudinal data. Journal of the Royal Statistical Society: Series B (Statistical Methodology) 69, 679–699.

[58] Chiou, J.M., Müller, H.G., 2014. Linear manifold modelling of multivariate functional data. Journal of the Royal Statistical Society: Series B (Statistical Methodology) 76, 605–626.

[59] Cho, H., 2016. Change-point detection in panel data via double cusum statistic. Electronic Journal of Statistics 10, 2000–2038.

[60] Cho, H., Fryzlewicz, P., 2012. Multiscale and multilevel technique for consistent segmentation of nonstationary time series. Statistica Sinica , 207–229.

[61] Cho, H., Kirch, C., 2022. Bootstrap confidence intervals for multiple change points based on moving sum procedures. Computational Statistics & Data Analysis 175, 107552.

[62] Choy, J.H.l.T.C., 1977. A Bayesian analysis of a changing linear model. Oklahoma State University.

[63] Clark, G., Zborowski, D., Culbertson, J., et al., 2006. Clinical utility of epidermal growth factor receptor expression for selecting patients with advanced non-small cell lung cancer for treatment with erlotinib. Journal of Thoracic Oncology 1, 837–846.

[64] Cleynen, A., Koskas, M., Lebarbier, E., Rigaill, G., Robin, S., 2014. Segmentor3isback: an r package for the fast and exact segmentation of seq-data. Algorithms for Molecular Biology 9, 1–11.

[65] Coe, B.P., Lockwood, W.W., Girard, L., Chari, R., Macaulay, C., Lam, S., Gazdar, A.F., Minna, J.D., Lam, W.L., 2006. Differential disruption of cell cycle pathways in small cell and non-small cell lung cancer. British Journal of Cancer 94, 1927–1935.

[66] Corneli, M., Latouche, P., Rossi, F., 2018. Multiple change points detection and clustering in dynamic networks. Statistics and Computing 28, 989–1007.

[67] Crammer, K., Singer, Y., 2001. On the algorithmic implementation of multiclass kernel-based vector machines. Journal of Machine Learning Research 2, 265–292.

[68] Crane, H., et al., 2016. Dynamic random networks and their graph limits. The Annals of Applied Probability 26, 691–721.

[69] Csörg**H**o, M., 1981. Strong approximations in probability and statistics.

[70] Csörgo, M., Horváth, L., 1987. Nonparametric tests for the changepoint problem. Journal of Statistical Planning and Inference 17, 1–9.

[71] Csörgö, M., Horváth, L., 1997. Limit theorems in change-point analysis. volume 18. John Wiley & Sons Inc.

[72] Csörgo, M., Révész, P., 2014. Strong approximations in probability and statistics. Academic press.

[73] d'Aspremont, A., Bach, F., El Ghaoui, L., 2008. Optimal solutions for sparse principal component analysis. Journal of Machine Learning Research 9.

[74] Dave, S.S., Wright, G., Tan, B., Rosenwald, A., Gascoyne, R.D., Chan, W.C., Staudt, L.M., 2004. Prediction of survival in follicular lymphoma based on molecular features of tumor-infiltrating immune cells. New England Journal of Medicine 351, 2159–2169.

[75] Davis, R.A., Huang, D., Yao, Y.C., 1995. Testing for a change in the parameter values and order of an autoregressive model. The Annals of Statistics , 282–304.

[76] Davis, R.A., Lee, T.C.M., Rodriguez-Yam, G.A., 2006. Structural break estimation for non-stationary time series models. Journal of the American Statistical Association 101, 223–239.

[77] Degras, D.A., 2011. Simultaneous confidence bands for nonparametric regression with functional data. Statistica Sinica , 1735–1765.

[78] Dehning, J., Zierenberg, J., Spitzner, F.P., Wibral, M., Neto, J.P., Wilczek, M., Priesemann, V., 2020. Inferring change points in the spread of covid-19 reveals the effectiveness of interventions. Science 369, eabb9789.

[79] Diebold, F.X., Li, C., 2006. Forecasting the term structure of government bond yields. Journal of Econometrics 130, 337–364.

[80] Ding, J., Li, J., Han, Y., McKeague, I.W., Wang, X., 2023. Fitting additive risk models using auxiliary information. Statistics in Medicine 42, 894–916.

[81] Ding, J., Li, J., Wang, X., 2024a. Efficient auxiliary information synthesis for cure rate model. Journal of the Royal Statistical Society: Series C (Applied Statistics) 73, 497–521.

[82] Ding, J., Li, J., Wang, X., 2024b. Renewable risk assessment of heterogeneous streaming time-to-event cohorts. Statistics in Medicine 43, 3761–3777.

[83] Donnat, C., Holmes, S., 2018. Tracking network dynamics: A survey of distances and similarity metrics. The Annals of Applied Statistics 12, 971–1012.

[84] Doove, L.L., Dusseldorp, E., Van Deun, K., Van Mechelen, I., 2014. A comparison of five recursive partitioning methods to find person subgroups involved in meaningful treatment—subgroup interactions. Advances in Data Analysis and Classification 8(4), 403–425.

[85] Durante, D., Dunson, D.B., et al., 2016. Locally adaptive dynamic networks. The Annals of Applied Statistics 10, 2203–2232.

[86] Ellenberg, J.H., 1973. The joint distribution of the standardized least squares residuals from a general linear regression. Journal of the American Statistical Association 68, 941–943.

[87] Erdman, C., Emerson, J.W., 2008. bcp: an r package for performing a bayesian analysis of change point problems. Journal of statistical Software 23, 1–13.

[88] Everitt, B., Landau, S., Leese, M., Stahl, D., 2011. Cluster Analysis. Wiley.

[89] Fan, A., Song, R., Lu, W., 2017. Change-plane analysis for subgroup detection and sample size calculation. Journal of the American Statistical Association, 1–10.

[90] Fan, J., Gijbels, I., 1996. Local polynomial modelling and its applications: monographs on statistics and applied probability 66. volume 66. CRC Press.

[91] Fan, J., Li, R., 2001. Variable selection via nonconcave penalized likelihood and its oracle properties. Journal of the American statistical Association 96, 1348–1360.

[92] Fan, J., Wang, W., Zhu, Z., 2021. A shrinkage principle for heavy-tailed data: High-dimensional robust low-rank matrix recovery. Annals of statistics 49, 1239.

[93] Fan, J., Yao, Q., 2003. Nonlinear Time Series: Nonparametric and Parametric Methods. Springer, New York.

[94] Fan, Y., Tang, C.Y., 2013. Tuning parameter selection in high dimensional penalized likelihood. Journal of the Royal Statistical Society: Series B (Statistical Methodology) 75, 531–552.

[95] Fang, H.B., Fang, K.T., Kotz, S., 2002. The meta-elliptical distributions with given marginals. Journal of Multivariate Analysis 82, 1–16.

[96] Fearnhead, P., Grose, D., 2024. Cpop: Detecting changes in piecewise-linear signals. Journal of Statistical Software 109, 1–27.

[97] Fearnhead, P., Maidstone, R., Letchford, A., 2019. Detecting changes in slope with an l0 penalty. Journal of Computational and Graphical Statistics 28, 265–275.

[98] Foster, J., Taylor, J., Ruberg, S., 2011. Subgroup identification from randomized clinical trial data. Statistics in Medicine 30, 2867–2880.

[99] Franzolini, B., Beskos, A., De Iorio, M., Poklewski Koziell, W., Grzeszkiewicz, K., 2024. Change point detection in dynamic Gaussian graphical models: The impact of COVID-19 pandemic on the U.S. stock market. Annals of Applied Statistics 18, 555–584.

[100] Friel, N., Rastelli, R., Wyse, J., Raftery, A., 2016. Interlocking directorates in irish companies using a latent space model for bipartite networks. Proceedings of the national academy of sciences 113, 6629–6634.

[101] Fryzlewicz, P., 2007. Unbalanced haar technique for nonparametric function estimation. Journal of the American Statistical Association 102, 1318–1327.

[102] Fryzlewicz, P., 2014. Wild binary segmentation for multiple change-point detection. The Annals of Statistics 42, 2243–2281.

[103] Fryzlewicz, P., 2018. Tail-greedy bottom-up data decompositions and fast multiple change-point detection. The Annals of Statistics 46, 3390–3421.

[104] Fryzlewicz, P., 2020. Detecting possibly frequent change-points: Wild binary segmentation 2 and steepest-drop model selection. Journal of the Korean Statistical Society 49, 1027–1070.

[105] Fu, W., Song, L., Xing, E.P., 2009. Dynamic mixed membership blockmodel for evolving networks, in: Proceedings of the 26th Annual International Conference on Machine Learning, pp. 329–336.

[106] Gasser, T., Muller, H., 1984. Estimating regression functions and their derivatives by the kernel method. Scandinavian Journal of Statistics 11, 171–185.

[107] Gautier, L., Cope, L., Bolstad, B.M., Ra., I., 2004. Affy–analysis of affymetrix genechip data at the probe level. Bioinformatics. 20, 307–15.

[108] Gijbels, I., Lambert, A., Qiu, P., 2007. Jump-preserving regression and smoothing using local linear fitting: a compromise. Annals of the Institute of Statistical Mathematics 59, 235–272.

[109] Goldberg, Y., Kosorok, M.R., 2012. Q-learning with censored data. Annals of statistics 40, 529.

[110] Golub, G., Pereyra, V., 2003. Separable nonlinear least squares: the variable projection method and its applications. Inverse problems 19, R1.

[111] Gonzalo, J., Pitarakis, J.Y., 2002. Estimation and model selection based inference in single and multiple threshold models. Journal of Econometrics 110, 319–352.

[112] Górecki, T., Horváth, L., Kokoszka, P., 2018. Change point detection in heteroscedastic time series. Econometrics and Statistics 7, 63–88.

[113] Gramer, H., 1946. Mathematical Methods of Statistics. Princeton University Press, Princeton.

[114] Greene, W.H., 2003. Econometric Analysis. Pearson Education India.

[115] Grégoire, G., Hamrouni, Z., 2002. Change point estimation by local linear smoothing. Journal of Multivariate Analysis 83, 56–83.

[116] Gromenko, O., Kokoszka, P., Reimherr, M., 2017. Detection of change in the spatiotemporal mean function. Journal of the Royal Statistical Society: Series B (Statistical Methodology) 79, 29–50.

[117] Guo, C., Li, J., 2021. Homogeneity and structure identification in semiparametric factor models. Journal of Business & Economic Statistics In press.

[118] Hall, A.R., Han, S., Boldea, O., 2012. Inference regarding multiple structural changes in linear models with endogenous regressors. Journal of Econometrics 170, 281–302.

[119] Hall, A.R., Osborn, D.R., Sakkas, N., 2015. Structural break inference using information criteria in models estimated by two stage least squares. Journal of Time Series Analysis 36, 741–762.

[120] Hall, P., Heyde, C.C., 2014. Martingale limit theory and its application. Academic press.

[121] Hanneke, S., Fu, W., Xing, E.P., 2010. Discrete temporal models of social networks. Electronic Journal of Statistics 4, 585–605.

[122] Hansen, B.E., 2000. Sample splitting and threshold estimation. Econometrica 68, 575–603.

[123] Harchaoui, Z., Lévy-Leduc, C., 2010. Multiple change-point estimation with a total variation penalty. Journal of the American Statistical Association 105, 1480–1493.

[124] Holbert, D., 1982. A bayesian analysis of a switching linear model. Journal of Econometrics 19, 77–87.

[125] Horowitz, J.L., 1992. A smoothed maximum score estimator for the binary response model. Econometrica: journal of the Econometric Society , 505–531.

[126] Horowitz, J.L., 2002. Bootstrap critical values for tests based on the smoothed maximum score estimator. Journal of Econometrics 111, 141–167.

[127] Horváth, L., Kokoszka, P., 2012. Inference for functional data with applications. volume 200. Springer Science & Business Media.

[128] Hsing, T., Eubank, R., 2015. Theoretical foundations of functional data analysis, with an introduction to linear operators. volume 997. John Wiley & Sons.

[129] Huang, X., Sun, Y., Trow, P., et al., 2017. Patient subgroup identification for clinical drug development. Statistics in Medicine 36, 1414–1428.

[130] Huh, J., Park, B., 2004. Detection of a change point with local polynomial fits for the random design case. Australian & New Zealand Journal of Statistics 46, 425–441.

[131] Ing, C.K., Lai, T.L., 2011. A stepwise regression method and consistent model selection for high-dimensional sparse linear models. Statistica Sinica , 1473–1513.

[132] Jackson, B., Scargle, J.D., Barnes, D., Arabhi, S., Alt, A., Gioumousis, P., Gwin, E., Sangtrakulcharoen, P., Tan, L., Tsai, T.T., 2005. An algorithm for optimal partitioning of data on an interval. IEEE Signal Processing Letters 12, 105–108.

[133] James, N.A., Matteson, D.S., 2013. ecp: An r package for nonparametric multiple change point analysis of multivariate data. arXiv preprint arXiv:1309.3295 .

[134] Jiang, B., Li, J., Yao, Q., 2023. Autoregressive networks. Journal of Machine Learning Research 24, 1–69.

[135] Jin, B., Shi, X., Wu, Y., 2013. A novel and fast methodology for simultaneous multiple structural break estimation and variable selection for nonstationary time series models. Statistics and Computing 23, 221–231.

[136] Jin, B., Wu, Y., Shi, X., 2016. Consistent two-stage multiple change-point detection in linear models. Canadian Journal of Statistics 44, 161–179.

[137] Johnson, L.M., Strawderman, R.L., 2009. Induced smoothing for the semiparametric accelerated failure time model: asymptotics and extensions to clustered data. Biometrika 96, 577–590.

[138] Kang, X., Ganguly, A., Kolaczyk, E.D., 2020. Dynamic networks with multi-scale temporal structure. Sankhya **A** 84, 218–260.

[139] Ke, Y., Li, J., Zhang, W., 2016. Structure identification in panel data analysis. The Annals of Statistics 44, 1193–1233.

[140] Killick, R., Fearnhead, P., Eckley, I.A., 2012a. Optimal detection of changepoints with a linear computational cost. Journal of the American Statistical Association 107, 1590–1598.

[141] Killick, R., Fearnhead, P., Eckley, I.A., 2012b. Optimal detection of changepoints with a linear computational cost. Journal of the American Statistical Association 107, 1590–1598.

[142] Killick, R., Haynes, K., Eckley, I., Fearnhead, P., Lee, J., 2016. Changepoint: Methods for changepoint detection. See https://cran.r-project.org/web/packages/changepoint/changepoint.pdf .

[143] Kim, S., Koh, K., Boyd, S., Gorinevsky, D., 2009. l1 trend filtering. SIAM Review 51, 339–360.

[144] Knight, M., Nunes, M., Nason, G., 2016. Modelling, detrending and decorrelation of network time series. arXiv preprint arXiv:1603.03221 .

[145] Kolaczyk, E.D., 2017. Topics at the Frontier of Statistics and Network Analysis. Cambridge University Press.

[146] Kolmogorov, A., 1933. Sulla determinazione empirica delle leggi di probabilita. Giorn. Ist. Ital. Attuari 4.

[147] Komlós, J., Major, P., Tusnády, G., 1975. An approximation of partial sums of independent rv'-s, and the sample df. i. Zeitschrift für Wahrscheinlichkeitstheorie und verwandte Gebiete 32, 111–131.

[148] Kourtellos, A., Stengos, T., Tan, C.M., 2016. Structural threshold regression. Econometric Theory 32, 827–860.

[149] Kovács, S., Bühlmann, P., Li, H., Munk, A., 2023. Seeded binary segmentation: a general methodology for fast and optimal changepoint detection. Biometrika 110, 249–256.

[150] Krivitsky, P.N., Handcock, M.S., 2014. A separable model for dynamic networks. Journal of the Royal Statistical Society, **B** 76, 29.

[151] Kuelbs, J., Philipp, W., 1980. Almost sure invariance principles for partial sums of mixing b-valued random variables. The Annals of Probability , 1003–1036.

[152] Laber, E., YQ, Z., 2015. Tree-based methods for individualized treatment regimes. Biometrika 102-3, 501–514.

[153] Lai, T., 1973. Gaussian processes, moving averages and quick detection problem. Ann. Probab. 1, 825–837.

[154] Lai, T., 1974. Control charts based on weighted sum. Ann. Statist 2, 134–147.

[155] Lebarbier, É., 2005. Detecting multiple change-points in the mean of gaussian process by model selection. Signal processing 85, 717–736.

[156] Lee, S., Seo, M.H., Shin, Y., 2016. The lasso for high dimensional regression with a possible change point. Journal of the Royal Statistical Society: Series B (Statistical Methodology) 78, 193–210.

[157] Lee, Y., Lin, Y., Wahba, G., 2004. Multicategory support vector machines: Theory and application to the classification of microarray data and satellite radiance data. Journal of the American Statistical Association 99, 67–81.

[158] Li, D., Ling, S., 2012. On the least squares estimation of multiple-regime threshold autoregressive models. Journal of Econometrics 167, 240–253.

[159] Li, J., Fine, J., Brookhart, A., 2015. Instrumental variable additive hazards models. Biometrics 71, 122–130.

[160] Li, J., Huang, C., Hongtu, Z., Initiative, A.D.N., 2017. A functional varying-coefficient single-index model for functional response data. Journal of the American Statistical Association 112, 1169–1181.

[161] Li, J., Jin, B., 2018. Multi-threshold accelerated failure time model. The Annals of Statistics 46, 2657–2682.

[162] Li, J., Li, Y., Hsing, T., 2022. On functional processes with multiple discontinuities. Journal of the Royal Statistical Society: Series B (Statistical Methodology) 84, 933–972.

[163] Li, J., Li, Y., Jin, B., Kosorok, M., 2021. Multi-threshold change plane model: Estimation theory and applications in subgroup identification. Statistics in Medicine 40, 3440–3459.

[164] Li, J., Ma, S., 2013. Survival Analysis in Medicine and Genetics. CRC Press, Boca Raton, FL.

[165] Li, J., Wong, W.K., 2009. A semi-parametric analysis for identifying scleroderma patients responsive to an anti-fibrotic agent. Contemporary clinical trials 30, 105–113.

[166] Li, J., Xu, M., Zhong, P.S., Li, L., 2019. Change point detection in the mean of high-dimensional time series data under dependence. arXiv preprint arXiv:1903.07006 .

[167] Li, J., Zhang, W., Kong, E., 2018. Factor models for asset returns based on transformed factors. Journal of Econometrics 207, 432–448.

[168] Li, Y., Hsing, T., 2010. Uniform convergence rates for nonparametric regression and principal component analysis in functional/longitudinal data. The Annals of Statistics 38, 3321–3351.

[169] Lin, C.C., Ng, S., 2012. Estimation of panel data models with parameter heteogeneity when grouop membership is unknown. Journal of Econometric Methods 1, 42–55.

[170] Lipkovich, I., Dmitrienko, A., D'Agostino, R., 2017. Tutorial in biostatistics: data-driven subgroup identification and analysis in clinical trials. Statistics in medicine 36, 136–196.

[171] Lipkovich, I., Dmitrienko, A., Denne, J., Enas, G., 2011. Subgroup identification based on differential effect search - a recursive partitioning method for establishing response to treatment in patient subpopulations. Statistics in Medicine 30, 2601–2621.

[172] Loh, W., 2002a. Regression trees with unbiased variable selection and interaction detection. Statistica Sinica 12, 361–386.

[173] Loh, W., 2011. Classification and regression trees. Wiley Interdisciplinary Reviews: Data Mining and Knowledge Discovery 1, 14–23.

[174] Loh, W., 2014. Fifty years of classification and regression trees. International Statistical Review 82(3), 329–348.

[175] Loh, W., He, X., Man, M., 2015a. A regression tree approach to identifying subgroups with differential treatment effects. Statistics in Medicine 34, 1818–1833.

[176] Loh, W.Y., 2002b. Regression tress with unbiased variable selection and interaction detection. Statistica Sinica 12, 361–386.

[177] Loh, W.Y., He, X., Man, M., 2015b. A regression tree approach to identifying subgroups with differential treatment effects. Statistics in Medicine 34, 1818–1833.

[178] Lu, W., Zhang, H.H., Zeng, D., 2013. Variable selection for optimal treatment decision. Statistical methods in medical research 22, 493–504.

[179] Ludkin, M., Eckley, I., Neal, P., 2018. Dynamic stochastic block models: parameter estimation and detection of changes in community structure. Statistics and Computing 28, 1201–1213.

[180] Maeng, H., Fryzlewicz, P., 2024. Detecting linear trend changes in data sequences. Statistical Papers 65, 1645–1675.

[181] Maidstone, R., Hocking, T., Rigaill, G., Fearnhead, P., 2017. On optimal multiple change-point algorithms for large data. Statistics and computing 27, 519–533.

[182] Massart, P., Birgé, L., 2001. Gaussian model selection. Journal of the European Mathematical Society 3, 203–268.

[183] Matias, C., Miele, V., 2017. Statistical clustering of temporal networks through a dynamic stochastic block model. Journal of the Royal Statistical Society, **B** 79, 1119–1141.

[184] Matias, C., Rebafka, T., Villers, F., 2018. A semiparametric extension of the stochastic block model for longitudinal networks. Biometrika 105, 989–1007.

[185] Meier, A., Kirch, C., Cho, H., 2021. mosum: A package for moving sums in change point analysis. Journal of Statistical Software 97, 1–42.

[186] Messenger, R., Mandell, L., 1972. A modal search technique for predictive nominal scale multivariate analysis. Journal of the American Statistical Association 67, 768–772.

[187] Moghaddam, B., Weiss, Y., Avidan, S., 2006. Generalized spectral bounds for sparse lda, in: Proceedings of the 23rd international conference on Machine learning, pp. 641–648.

[188] Morgan, J., Sonquist, J., 1963. Problems in the analysis of survey data and a proposal. Journal of the American Statistical Association 58, 415–434.

[189] Muggeo, V.M., Adelfio, G., 2011. Efficient change point detection for genomic sequences of continuous measurements. Bioinformatics 27, 161–166.

[190] Müller, H.G., Song, K.S., 1997. Two-stage change-point estimators in smooth regression models. Statistics & Probability Letters 34, 323–335.

[191] Murphy, S.A., 2003. Optimal dynamic treatment regimes. Journal of the Royal Statistical Society: Series B (Statistical Methodology) 65, 331–355.

[192] de Oliveira, R.P., Achcar, J.A., Chen, C., Rodrigues, E.R., 2022. Non-homogeneous poisson and linear regression models as approaches to study time series with change-points. Communications in Statistics: Case Studies, Data Analysis and Applications 8, 331–353.

[193] Onnela, J.P., Chakraborti, A., Kaski, K., Kertész, J., Kanto, A., 2003. Dynamics of market correlations: Taxonomy and portfolio analysis. Physical Review E 68, 056110.

[194] Page, E., 1955. A test for a change in a parameter occurring at an unknown point. Biometrika 42, 523–527.

[195] Page, E., 1957. On problems in which a change in a parameter occurs at an unknown point. Biometrika 44, 248–252.

[196] Page, E.S., 1954. Continuous inspection schemes. Biometrika 41, 100–115.

[197] Pensky, M., 2019. Dynamic network models and graphon estimation. Annals of Statistics 47, 2378–2403.

[198] Postlethwaite, A.E., Wong, W.K., Clements, P., Chatterjee, S., Fessler, B.J., Kang, A.H., Korn, J., Mayes, M., Merkel, P.A., Molitor, J.A., et al., 2008. A multicenter, randomized, double-blind, placebo-controlled trial of oral type i collagen treatment in patients with diffuse cutaneous systemic sclerosis: I. oral type i collagen does not improve skin in all patients, but may improve skin in late-phase disease. Arthritis & Rheumatology 58, 1810–1822.

[199] Prado, R., Ferreira, M.A.R., West, M., 2021. Time Series: Modeling, Computation, and Inference, Second Edition. CRC Press, Boca Raton, FL.

[200] Privault, N., 2022. Introduction to Stochastic Finance with Market Examples. CRC Press.

[201] Priyadarshana, M., Polushina, T., Sofronov, G., 2015. Hybrid algorithms for multiple change-point detection in biological sequences. Signal and Image Analysis for Biomedical and Life Sciences, 41–61.

[202] Priyadarshana, M., Sofronov, G., 2012a. A modified cross entropy method for detecting multiple change points in dna count data, in: 2012 IEEE Congress on Evolutionary Computation, IEEE. pp. 1–8.

[203] Priyadarshana, W., Sofronov, G., 2012b. The cross-entropy method and multiple change-points detection in zero-inflated dna read count data, in: International Conference on Computational Methods (4th: 2012), Queensland University of Technology. pp. 1–8.

[204] Priyadarshana, W., Sofronov, G., 2016. breakpoint: an r package for multiple break-point detection via the cross-entropy method. See https://cran.r-project.org/web/packages/breakpoint/breakpoint.pdf .

[205] Qian, M., Murphy, S.A., 2011. Performance guarantees for individualized treatment rules. Annals of statistics 39, 1180.

[206] Qiu, P., 2005. Image processing and jump regression analysis. volume 599. John Wiley & Sons.

[207] Qiu, P., Asano, C., Li, X., 1991. Estimation of jump regression functions. Bulletin of Informatics and Cybernetics 24, 197–212.

[208] Qualls, C., Watanabe, H., 1972. Asymptotic properties of gaussian processes. The Annals of Mathematical Statistics , 580–596.

[209] Ramsay, J., Silverman, B., 2005. Functional Data Analysis. Springer Science & Business Media.

[210] Rao, D.R., Edelman, A., 2008. Sample eigenvalue based detection of high-dimensional signals in white noise using relatively few samples ieee transactions of signal processing 56, 2625–2638.

[211] Rastelli, R., Latouche, P., Friel, N., 2018. Choosing the number of groups in a latent stochastic block model for dynamic networks. Network Science 6, 469–493.

[212] Rice, J., 1984. Boundary modification for kernel regression. Communications in Statistics, Theory and Methods 13, 893–900.

[213] Rissanen, J., 1978. Modeling by shortest data description. Automatica 14, 465–471.

[214] Robins, J.M., 2004. Optimal structural nested models for optimal sequential decisions, in: Proceedings of the second seattle Symposium in Biostatistics, Springer. pp. 189–326.

[215] Rohe, K., Chatterjee, S., Yu, B., et al., 2011. Spectral clustering and the high-dimensional stochastic blockmodel. The Annals of Statistics 39, 1878–1915.

[216] Sato, H., Hirakawa, A., Hamada, C., 2016. An adaptive dose-finding method using a change-point model for molecularly targeted agents in phase i trials. Statistics in Medicine 35, 4093–4109. Epub 2016 May 11.

[217] Schmidt, M., Böhm, D., von Törne, C., Steiner, E., Puhl, A., Pilch, H., Lehr, H.A., Hengstler, J.G., Kölbl, H., Gehrmann, M., 2008. The humoral immune system has a key prognostic impact in node-negative breast cancer. Cancer Research 68, 5405–13.

[218] Schnell, P., Tang, Q., Offen, W., Carlin, B., 2016. A bayesian credible subgroups approach to identifying patient subgroups with positive treatment effects. Biometrics 72, 1026–1036.

[219] Schulte, P.J., Tsiatis, A.A., Laber, E.B., Davidian, M., 2014. Q-and a-learning methods for estimating optimal dynamic treatment regimes. Statistical science: a review journal of the Institute of Mathematical Statistics 29, 640.

[220] Schwarz, G., 1978. Estimating the dimension of a model. The annals of statistics, 461–464.

[221] Scott, A.J., Knott, M., 1974. A cluster analysis method for grouping means in the analysis of variance. Biometrics , 507–512.

[222] Seo, M.H., Linton, O., 2007. A smoothed least squares estimator for threshold regression models. Journal of Econometrics 141, 704–735.

[223] Serfling, R.J., 2009. Approximation theorems of mathematical statistics. volume 162. John Wiley & Sons.

[224] Shao, X., Zhang, X., 2010. Testing for change points in time series. Journal of the American Statistical Association 105, 1228–1240.

[225] Sheather, S.J., Jones, M.C., 1991. A reliable data-based bandwidth selection method for kernel density estimation. Journal of the Royal Statistical Society: Series B (Methodological) 53, 683–690.

[226] Shen, J., He, X., 2015. Inference for subgroup analysis with a structured logistic-normal mixture model. Journal of the American Statistical Association 110, 303–312.

[227] Shen, J.J., Zhang, N.R., 2012. Change-point model on nonhomogeneous poisson processes with application in copy number profiling by next-generation dna sequencing .

[228] Shewhart, W., 1920. The application of statistics as an aid in maintaining quality of a manufactured product. Journal of the American Statistical Association 20, 85–93.

[229] Shi, C., Fan, A., Song, R., Lu, W., 2017. High-dimensional a-learning for dynamic treatment regimes. The Annals of Statistics, (in press) .

[230] Shi, X., Beaulieu, C., Killick, R., Lund, R., 2022. Changepoint detection: An analysis of the central england temperature series. Journal of Climate 35, 6329–6342.

[231] Simon, R., 2002. Bayesian subset analysis: appliation to studying treatment-by-gender interactions. Statistics in Medicine 21, 2909–2916.

[232] Small, D.S., 2007. Sensitivity analysis for instrumental variables regression with overidentifying restrictions. Journal of the American Statistical Association 102, 1049–1058.

[233] Snijders, T.A.B., 2005. Models for longitudinal network data, in: Carrington, P., Scott, J., Wasserman, S.S. (Eds.), Models and Methods in Social Network Analysis. Cambridge University Press, New York. chapter 11.

[234] Song, R., Kosorok, M., Zeng, D., Zhao, Y., Laber, E., Yuan, M., 2015. On sparse representation for optimal individualized treatment selection with penalized outcome weighted learning. Stat 4, 59–68.

[235] Spokoiny, V.G., et al., 1998. Estimation of a function with discontinuities via local polynomial fit with an adaptive window choice. The Annals of Statistics 26, 1356–1378.

[236] Steinbach, M., Ertz, L., Kumar, V., 2003. The challenges of clustering high dimensional data in New Vistas in Statistical PhysicsApplications in Econophysics Bioinformatics and Pattern Recognition. Springer-Verlag, Germany, Berlin.

[237] Stransky, N., Vallot, C., Reyal, F., Bernard-Pierrot, I., de Medina, S.G., Segraves, R., de Rycke, Y., Elvin, P., Cassidy, A., Spraggon, C., Graham, A., Southgate, J., Asselain, B., Malfoy, B., Thiery, J.P., Chopin, D., Radvanyi, F., 2006. Regional copy number-independent deregulation of transcription in cancer. Nature Genetics 38, 1386–1396.

[238] Stute, W., 1995. The central limit theorem under random censorship. The Annals of Statistics 23, 422–439.

[239] Su, X., Tsai, C., Wang, H., Nickerson, D., Li, B., 2009a. Subgroup analysis via recursive partitioning. Journal of Machine Learning Research 10(Feb), 141–158.

[240] Su, X., Tsai, C.L., Wang, H., Nickerson, D.M., Li, B., 2009b. Subgroup analysis via recursive partitioning. Journal of Machine Learning Research 10, 141–158.

[241] Suykens, J.A., Vandewalle, J., 1999. Least squares support vector machine classifiers. Neural processing letters 9, 293–300.

[242] Takanami, T., Kitagawa, G., 1991. Estimation of the arrival times of seismic waves by multivariate time series model. Annals of the Institute of Statistical mathematics 43, 407–433.

[243] Tartakovsky, A., Nikiforov, I., Basseville, M., 2015. Sequential Analysis: Hypothesis Testing and Changepoint Detection. CRC Press.

[244] Thies, S., Molnár, P., 2018. Bayesian change point analysis of bitcoin returns. Finance Research Letters 27, 223–227.

[245] Tibshirani, R., 1996. Regression shrinkage and selection via the lasso. Journal of the Royal Statistical Society Series B: Statistical Methodology 58, 267–288.

[246] Tibshirani, R., 2014. Adaptive piecewise polynomial estimation via trend filtering. The Annals of Statistics 42, 285–323.

[247] Tong, H., 1990. Non-linear time series: a dynamical system approach. Oxford University Press.

[248] Tsiatis, A.A., Davidian, M., Zhang, M., Lu, X., 2008. Covariate adjustment for two-sample treatment comparisons in randomized clinical trials: A principled yet flexible approach. Statistics in medicine 27, 4658–4677.

[249] Vapnik, V., 1998. Statistical Learning Theory. Wiley, New York.

[250] Venkatraman, E.S., 1992. Consistency results in multiple change-point problems. Stanford University.

[251] Vert, J.P., Bleakley, K., 2010. Fast detection of multiple change-points shared by many signals using group lars. Advances in neural information processing systems 23.

[252] Wager, S., Athey, S., 2017. Estimation and inference of heterogeneous treatment effects using random forests. Journal of the American Statistical Association .

[253] Wald, A., 1940. The fitting of straight lines if both variables are subject to error. The Annals of Mathematical Statistics 11, 284–300.

[254] Wald, A., 1947. Sequential Analysis. New York: Wiley.

[255] Wald, A., Wolfowitz, J., 1948. Optimum character of the sequential probability ratio test. Ann. Math. Statist. 19, 326–339.

[256] Wang, D., Yu, Y., Rinaldo, A., 2021a. Optimal change point detection and localization in sparse dynamic networks. Annals of statistics 49, 203–232.

[257] Wang, D., Yu, Y., Rinaldo, A., 2021b. Optimal change point detection and localization in sparse dynamic networks. The Annals of Statistics 49, 203–232.

[258] Wang, J., Li, J., 2023. Multi-threshold structural equation model. Journal of Business & Economic Statistics 41, 377–387.

[259] Wang, J., Li, J., Li, Y., Wong, W.K., 2019. A model-based multithreshold method for subgroup identification. Statistics in Medicine 38, 2605–2631.

[260] Wang, T., Samworth, R.J., 2018. High dimensional change point estimation via sparse projection. Journal of the Royal Statistical Society Series B: Statistical Methodology 80, 57–83.

[261] White, H., 1982. Instrumental variables regression with independent observations. Econometrica 50, 483–499.

[262] Wilson, J.D., Stevens, N.T., Woodall, W.H., 2019. Modeling and detecting change in temporal networks via the degree corrected stochastic block model. Quality and Reliability Engineering International 35, 1363–1378.

[263] Worsley, K.J., 1979. On the likelihood ratio test for a shift in location of normal populations. Journal of the American Statistical Association 74, 365–367.

[264] Wu, J., Chu, C., 1993. Kernel-type estimators of jump points and values of a regression function. The Annals of Statistics 21, 1545–1566.

[265] Wu, W.B., Zhao, Z., 2007. Inference of trends in time series. Journal of the Royal Statistical Society: Series B (Statistical Methodology) 69, 391–410.

[266] Xia, Z., Qiu, P., 2015. Jump information criterion for statistical inference in estimating discontinuous curves. Biometrika 102, 397–408.

[267] Xu, K., 2015. Stochastic block transition models for dynamic networks, in: Artificial Intelligence and Statistics, PMLR. pp. 1079–1087.

[268] Xu, K.S., Hero, A.O., 2014. Dynamic stochastic blockmodels for time-evolving social networks. IEEE Journal of Selected Topics in Signal Processing 8, 552–562.

[269] Xu, M., Zhong, P.S., Wang, W., 2016. Detecting variance change-points for blocked time series and dependent panel data. Journal of Business & Economic Statistics 34, 213–226.

[270] Xu, Y., Yu, M., Zhao, Y.Q., Li, Q., Wang, S., Shao, J., 2015. Regularized outcome weighted subgroup identification for differential treatment effects. Biometrics 71, 645–653.

[271] Yang, T., Chi, Y., Zhu, S., Gong, Y., Jin, R., 2011. Detecting communities and their evolutions in dynamic social networks?a bayesian approach. Machine learning 82, 157–189.

[272] Yao, Q., 1993. Tests for change-points with epidemic alternatives. Biometrika 80, 179–191.

[273] Yin, Y., 1988. Detection of the number, locations and magnitudes of jumps. Communications in Statistics. Stochastic Models 4, 445–455.

[274] Yu, G., Wang, L., Han, Y., He, Q., 2012a. Clusterprofiler: an r package for comparing biological themes among gene clusters. OMICS: A Journal of Integrative Biology 16, 284–287.

[275] Yu, P., Phillips, P.C.B., 2018. Threshold regression with endogeneity. Journal of Econometrics 203, 50–68.

[276] Yu, T., Li, J., Ma, S., 2012b. Adjusting confounders in ranking biomarkers: a model-based roc approach. Briefings In Bioinformatics 13, 513–523.

[277] Yudovina, E., Banerjee, M., Michailidis, G., 2015. Changepoint inference for erdös-rényi random graphs, in: Stochastic Models, Statistics and Their Applications. Springer, pp. 197–205.

[278] Zeileis, A., Hothorn, T., Hornik, K., 2008. Model-based recursive partitioning. Journal of Computational and Graphical Statistics 17(2), 492–514.

[279] Zhang, B., Tsiatis, A.A., Laber, E.B., Davidian, M., 2012. A robust method for estimating optimal treatment regimes. Biometrics 68, 1010–1018.

[280] Zhang, B., Tsiatis, A.A., Laber, E.B., Davidian, M., 2013. Robust estimation of optimal dynamic treatment regimes for sequential treatment decisions. Biometrika 100, 681–694.

[281] Zhang, C.H., 2010. Nearly unbiased variable selection under minimax concave penalty. The Annals of statistics 38, 894–942.

[282] Zhang, N.R., Siegmund, D.O., 2007. A modified bayes information criterion with applications to the analysis of comparative genomic hybridization data. Biometrics 63, 22–32.

[283] Zhang, T., 2016. Testing for jumps in the presence of smooth changes in trends of nonstationary time series. Electronic Journal of Statistics 10, 706–735.

[284] Zhang, W., Jin, B., Bai, Z., 2021. Learning block structures in u-statistic-based matrices. Biometrika 108, 933–946.

[285] Zhang, X., Wang, J.L., 2016. From sparse to dense functional data and beyond. The Annals of Statistics 44, 2281–2321.

[286] Zhao, L., Tian, L., Cai, T., Claggett, B., Wei, L.J., 2013. Effectively selecting a target population for a future comparative study. Journal of the American Statistical Association 108, 527–539.

[287] Zhao, Y., Zeng, D., Rush, A.J., Kosorok, M.R., 2012. Estimating individualized treatment rules using outcome weighted learning. Journal of the American Statistical Association 107, 1106–1118.

[288] Zhao, Y.Q., Zeng, D., Laber, E.B., Kosorok, M.R., 2015. New statistical learning methods for estimating optimal dynamic treatment regimes. Journal of the American Statistical Association 110, 583–598.

[289] Zhao, Z., Chen, L., Lin, L., 2019. Change-point detection in dynamic networks via graphon estimation. arXiv preprint arXiv:1908.01823 .

[290] Zheng, S., Yang, L., Härdle, W.K., 2014. A smooth simultaneous confidence corridor for the mean of sparse functional data. Journal of the American Statistical Association 109, 661–673.

[291] Zhu, T., Li, P., Yu, L., Chen, K., Chen, Y., 2020a. Change point detection in dynamic networks based on community identification. IEEE Transactions on Network Science and Engineering .

[292] Zhu, X., Huang, D., Pan, R., Wang, H., 2020b. Multivariate spatial autoregressive model for large scale social networks. Journal of Econometrics 215, 591–606.

[293] Zhu, X., Pan, R., Li, G., Liu, Y., Wang, H., 2017. Network vector autoregression. The Annals of Statistics 45, 1096–1123.

[294] Zhu, X., Wang, W., Wang, H., Härdle, W.K., 2019. Network quantile autoregression. Journal of Econometrics 212, 345–358.

[295] Zou, H., 2006. The adaptive lasso and its oracle properties. Journal of the American statistical association 101, 1418–1429.

Index

FSC
www.fsc.org
MIX
Papier | Fördert gute Waldnutzung
FSC® C083411

Kolibri 360 GmbH
Ferdinand-Jühlke-Straße 7
99095 Erfurt, Deutschland
produktsicherheit@kolibri360.de